POWER ELECTRONICS: THEORY AND PRACTICALS

AUTHOR

Dr. Lalit Chandra Saikia
Professor
National Institute of Technology Silchar
Assam

REVIEWER

Dr. Asha Rani M. A.
Assistant Professor
National Institute of Technology Silchar
Assam

KHANNA BOOK
PUBLISHING CO. (P) Ltd.
4C/4344, Ansari Road, Darya Ganj,
New Delhi, India

All India Council for
Technical Education
Nelson Mandela Marg, Vasant Kunj,
New Delhi, India

Dear Readers,

To prevent the piracy, this book is secured with HIGH SECURITY HOLOGRAM on the front title cover. In case you don't find the hologram on the front cover title, please write us to at contact@khannabooks.com or whatsapp us at +91-99109 09320 and avail special gift voucher for yourself.

Specimen of Hologram on front Cover title:

Moreover, there is a SPECIAL DISCOUNT COUPON for you with EVERY HOLOGRAM.

How to avail this SPECIAL DISCOUNT:

Step 1: Scratch the hologram

Step 2: Under the scratch area, your "coupon code" is available

Step 3: Logon to www.khannabooks.com

Step 4: Use your "coupon code" in the shopping cart and get your copy at a special discount

Step 5: Enjoy your reading!

ISBN: 978-93-5538-595-6

Book Code: UG433EN

Power Electronics: Theory And Practicals

by Dr. Lalit Chandra Saikia

First Edition: 2026

Originally Published by:

All India Council for Technical Education, Ministry of Education, GOI

Exclusively Published & Distributed by:

Khanna Book Publishing Co. (P) Ltd.
4C/4344, Ansari Road, Daryaganj,
New Delhi (INDIA)

Visit us at: **www.khannabooks.com**
Write us at: **contact@khannabooks.com**

To view complete list of books,
Please scan the QR Code:

KPH

Printed in Bharat.

प्रो. टी. जी. सीताराम
अध्यक्ष
Prof. T. G. Sitharam
Chairman

सत्यमेव जयते

आज़ादी का अमृत महोत्सव

अखिल भारतीय तकनीकी शिक्षा परिषद्
(भारत सरकार का एक सांविधिक निकाय)
(शिक्षा मंत्रालय, भारत सरकार)
नेल्सन मंडेला मार्ग, वसंत कुंज, नई दिल्ली–110070
दूरभाष : 011–26131498
ई–मेल : chairman@aicte-india.org

ALL INDIA COUNCIL FOR TECHNICAL EDUCATION
(A STATUTORY BODY OF THE GOVT. OF INDIA)
(Ministry of Education, Govt. of India)
Nelson Mandela Marg, Vasant Kunj, New Delhi-110070
Phone : 011-26131498
E-mail : chairman@aicte-india.org

FOREWORD

Engineers are the backbone of any modern society. They are the ones responsible for the marvels as well as the improved quality of life across the world. Engineers have driven humanity towards greater heights in a more evolved and unprecedented manner.

The All India Council for Technical Education (AICTE), have spared no efforts towards the strengthening of the technical education in the country. AICTE is always committed towards promoting quality Technical Education to make India a modern developed nation emphasizing on the overall welfare of mankind.

An array of initiatives has been taken by AICTE in last decade which have been accelerated now by the National Education Policy (NEP) 2020. The implementation of NEP under the visionary leadership of Hon'ble Prime Minister of India envisages the provision for education in regional languages to all, thereby ensuring that every graduate becomes competent enough and is in a position to contribute towards the national growth and development through innovation & entrepreneurship.

One of the spheres where AICTE had been relentlessly working since past couple of years is providing high quality original technical contents at Under Graduate & Diploma level prepared and translated by eminent educators in various Indian languages to its aspirants. For students pursuing 2nd year of their Engineering education, AICTE has identified 87 books, which shall be translated into 12 Indian languages - Hindi, Tamil, Gujarati, Odia, Bengali, Kannada, Urdu, Punjabi, Telugu, Marathi, Assamese & Malayalam. In addition to the English medium, books in different Indian Languages are going to support the students to understand the concepts in their respective mother tongue.

On behalf of AICTE, I express sincere gratitude to all distinguished authors, reviewers and translators from the renowned institutions of high repute for their admirable contribution in a record span of time.

AICTE is confident that these outcomes based original contents shall help aspirants to master the subject with comprehension and greater ease.

T. G. Sitharam

(Prof. T. G. Sitharam)

ACKNOWLEDGMENT

The author is grateful to the authorities of AICTE, particularly Prof. T G Sitharam, Chairman; Dr. Abhay Jere, Vice-Chairman; Prof. Rajive Kumar, Member-Secretary, Dr. Sunil Luthra, Director and Reena Sharma, Hindi Officer Training and Learning Bureau, for their planning to publish the books on **Power Electronics: Theory and Practicals**. I sincerely acknowledge the valuable contributions of the reviewer of the book Dr. Asha Rani M.A, Assistant Professor, National Institute of Technology Silchar, Assam for making it students' friendly and artistically giving a better shape. I also offer my sincere thanks to my students Dr. Satish Kumar Ramoji, Dr. Sanjeev Kumar Bhagat, and Dr. Manoja Kumar Behera, for their help and support in preparing this book. At last, it is wise to thank my wife Mrs. Krishna Dutta Saikia, and my sons Abhinav and Anubhav for their all-around support during the writing of this book.

This book is an outcome of various suggestions of AICTE members, experts, and authors who shared their opinions and thoughts to further develop engineering education in our country. Acknowledgments are due to the contributors and different workers in this field whose published books, review articles, papers, photographs, footnotes, references, and other valuable information enriched us at the time of writing the book.

Dr. Lalit Chandra Saikia

PREFACE

The book titled "Power Electronics: Theory and Practicals" is an outcome of the rich experience of my teaching of basic courses in Electrical Engineering. The initiation of writing this book is to expose the fundamentals of Power Electronics to Electrical Engineering students. The basics of power electronics including semiconductor physics, power semiconductor devices, turn-on and turn-off methods of SCRs, phase controlled rectifies, and industrial applications also presented.

Keeping in mind the purpose of wide coverage as well as to provide essential supplementary information, I have included the topics recommended by AICTE, in a very systematic and orderly manner throughout the book. Efforts have been made to explain the fundamental concepts of the subject in the simplest possible way.

During the process of preparation of the manuscript, I have considered the various standard textbooks and accordingly, I have developed sections like critical questions, solved and supplementary problems, etc. While preparing the different sections emphasis has also been laid on definitions and principles of operation and also on comprehensive synopsis of formulae for a quick revision of the basic principles. The book covers all types of medium and advanced level problems and these have been presented in a very logical and systematic manner. The gradations of those problems have been tested over many years of teaching to a wide variety of students.

Apart from illustrations and examples as required, I have enriched the book with numerous solved problems in every unit for proper understanding of the related topics. Laboratory experiments on the basics of power electronics as recommended by AICTE are included in the units.

In addition, besides some essential information for the users under the heading "Know More" I have clarified some essential basic information in the appendix section.

This book is concerned with the basics of power electronics covering construction, principle of operations, various characteristics of power transistors and thyristors, SCR turned-on and turned-off methods, and industrial control circuits using power electronics. All are grouped in separate units. The subject matters are presented in a constructive manner so that an electrical engineering degree prepares students to work in different sectors at the very forefront of technology.

I sincerely hope that the book will inspire electrical engineering students to learn and discuss the ideas behind the basics of power electronics and will surely contribute to the development of a solid foundation for the subject. I would be thankful for all beneficial comments and suggestions that will contribute to the improvement of future editions of the book. It gives me immense pleasure to place this book in the hands of the teachers and students. It was indeed a great pleasure to work on different aspects covered in the book.

Dr. Lalit Chandra Saikia

OUTCOME BASED EDUCATION

For the implementation of an outcome based education the first requirement is to develop an outcome based curriculum and incorporate an outcome based assessment in the education system. By going through outcome based assessments evaluators will be able to evaluate whether the students have achieved the outlined standard, specific and measurable outcomes. With the proper incorporation of outcome based education there will be a definite commitment to achieve a minimum standard for all learners without giving up at any level. At the end of the programme running with the aid of outcome based education, a student will be able to arrive at the following outcomes:

PO1. Engineering knowledge: Apply the knowledge of mathematics, science, engineering fundamentals, and an engineering specialization to the solution of complex engineering problems.

PO2. Problem analysis: Identify, formulate, review research literature, and analyze complex engineering problems reaching substantiated conclusions using first principles of mathematics, natural sciences, and engineering sciences.

PO3. Design / development of solutions: Design solutions for complex engineering problems and design system components or processes that meet the specified needs with appropriate consideration for the public health and safety, and the cultural, societal, and environmental considerations.

PO4. Conduct investigations of complex problems: Use research-based knowledge and research methods including design of experiments, analysis and interpretation of data, and synthesis of the information to provide valid conclusions.

PO5. Modern tool usage: Create, select, and apply appropriate techniques, resources, and modern engineering and IT tools including prediction and modeling to complex engineering activities with an understanding of the limitations.

PO6. The engineer and society: Apply reasoning informed by the contextual knowledge to assess societal, health, safety, legal and cultural issues and the consequent responsibilities relevant to the professional engineering practice.

PO7. Environment and sustainability: Understand the impact of the professional engineering solutions in societal and environmental contexts, and demonstrate the knowledge of, and need for sustainable development.

PO8. Ethics: Apply ethical principles and commit to professional ethics and responsibilities and norms of the engineering practice.

PO9. Individual and team work: Function effectively as an individual, and as a member or leader in diverse teams, and in multidisciplinary settings.

PO10. Communication: Communicate effectively on complex engineering activities with the engineering community and with society at large, such as, being able to comprehend and write effective reports and design documentation, make effective presentations, and give and receive clear instructions.

PO11. Project management and finance: Demonstrate knowledge and understanding of the engineering and management principles and apply these to one's own work, as a member and leader in a team, to manage projects and in multidisciplinary environments.

PO12. Life-long learning: Recognize the need for, and have the preparation and ability to engage in independent and life-long learning in the broadest context of technological change.

COURSE OUTCOMES

After completion of the course, the students will be able to:

CO-1: Select Power Electronic Devices for specific applications.

CO-2: Explain the construction of various power semiconductor devices and power electronics circuits.

CO-3: Understand the operation of various semiconductor devices and their circuits.

CO4: Analyse various circuits for turn-on and turn-off of thyristors and phase-controlled rectifiers.

CO-5: Design appropriate power electronic circuits for industrial control.

Mapping of Course Outcomes with Programme Outcomes to be done according to the matrix given below:

Course Outcomes	Expected Mapping with Programme Outcomes *(1- Weak Correlation; 2- Medium correlation; 3- Strong Correlation)*											
	PO-1	PO-2	PO-3	PO-4	PO-5	PO-6	PO-7	PO-8	PO-9	PO-10	PO-11	PO-12
CO-1	3	2	2	2	1	-	-	-	-	-	-	-
CO-2	3	1	2	1	-	-	-	-	-	-	-	-
CO-3	3	3	2	2	2	-	-	-	-	-	-	-
CO-4	3	3	3	2	1	-	-	-	-	-	-	-
CO-5	3	3	3	1	2	-	-	-	-	-	-	-

GUIDELINES FOR TEACHERS

To implement Outcome Based Education (OBE) knowledge level and skill set of the students should be enhanced. Teachers should take a major responsibility for the proper implementation of OBE. Some of the responsibilities (not limited to) for the teachers in OBE system may be as follows:

- Within reasonable constraint, they should manoeuvre time to the best advantage of all students.
- They should assess the students only upon certain defined criterion without considering any other potential ineligibility to discriminate them.
- They should try to grow the learning abilities of the students to a certain level before they leave the institute.
- They should try to ensure that all the students are equipped with the quality knowledge as well as competence after they finish their education.
- They should always encourage the students to develop their ultimate performance capabilities.
- They should facilitate and encourage group work and team work to consolidate newer approach.
- They should follow Blooms taxonomy in every part of the assessment.

Bloom's Taxonomy

Level	Teacher should Check	Student should be able to	Possible Mode of Assessment
Create	Students ability to create	Design or Create	Mini project
Evaluate	Students ability to justify	Argue or Defend	Assignment
Analyse	Students ability to distinguish	Differentiate or Distinguish	Project/Lab Methodology
Apply	Students ability to use information	Operate or Demonstrate	Technical Presentation/ Demonstration
Understand	Students ability to explain the ideas	Explain or Classify	Presentation/Seminar
Remember	Students ability to recall (or remember)	Define or Recall	Quiz

GUIDELINES FOR STUDENTS

Students should take equal responsibility for implementing the OBE. Some of the responsibilities (not limited to) for the students in OBE system are as follows:

- Students should be well aware of each UO before the start of a unit in each and every course.
- Students should be well aware of each CO before the start of the course.
- Students should be well aware of each PO before the start of the programme.
- Students should think critically and reasonably with proper reflection and action.
- Learning of the students should be connected and integrated with practical and real life consequences.
- Students should be well aware of their competency at every level of OBE.

ABBREVIATIONS AND SYMBOLS

List of Abbreviations

General Terms			
Abbreviations	**Full form**	**Abbreviations**	**Full form**
AC	Alternating current	NPT	Non-punch-through
AT	Ambient temperature	ODF	Overdrive factor
BEJ	Base-emitter junction	PIV	Peak inverse voltage
DIAC	Bidirectional diode	PCC	Phase control converter
TRIACs	Bidirectional triode thyristors	PCRs	Phase-controlled rectifiers
BJT	Bipolar junction transistor	PEs	Power electronics
CBJ	Collector-base junction	PF	Power factor
CB	Common base	PSDs	Power semiconductor devices
CC	Common collector	PUT	Programable unijunction transistor
CE	Common emitter	PO	Programme outcome
CSCR	Complementary SCR	POs	Programme outcomes
COMFET	Conductively – modulated field effect transistor	PTR	Pulse transformer
COs	Course outcomes	PT	Punch-through
CF	Creast factor	QD	Quantum dot
DC	Direct current	RBM	Reverse blocking mode
DPF	Displacement power factor	RCTs	Reverse conducting thyristors
DF	Displacement power factor	RCM	Reverse conduction mode
EP	Electrical power	RRC	Reverse recovery current
ETOs	Emitter turn-off thyristors	RBSOA	Reverse-biased SOA
FET-CTHs	FET-controlled thyristors	RF	Ripple factor
FET	Field effect transistor	SOA	Safe operating area
FF	Form factor	SCSs	Silicon control switches
FBM	Forward blocking mode	SCR	Silicon-controlled rectifier
FCM	Forward conduction mode	SET	Single electron transistor
FBSOA	Forward-biased SOA	SET	Single electron transistor
GEMFET	Gain-modulated field effect transistor	SACCB	Static AC circuit breakers
GTOs	Gate turn off thyristors	SDCCB	Static DC circuit breakers
HF	Harmonic factor	SITHs	Static induction thyristors
IRED	Infrared emitter diode	SITs	Static induction transistors
IGBT	Insulated gate bipolar transistor	SMPS	Switched mode power supply
IC	Integrated circuit	SMPS	Switched mode power supply
IGCTs	Integrated gate-commutated thyristors	TR	Thermal resistance
JFET	Junction field effect transistor	THD	Total harmonic distortion
kV.	Kilo volt	TUF	Transformer utilization factor

General Terms			
Abbreviations	**Full form**	**Abbreviations**	**Full form**
LASCRs	Light activated silicon-controlled rectifiers	UJT	Unijunction transistor
MT	Main terminal	UPS	Uninterrupted power supplies
MOSIGT	Metal oxide insulated gate transistor	UPS	Uninterrupted power supply
MOSFETs	Metal oxide semiconductor field effect transistors	VST	Voltage synchronizing transformer
MTOs	MOS turn off thyristors	V-I	Voltage-current
MCTs	MOS-controlled thyristors	w.r.t	With respect to
NASA	National Aeronautics and Space Administration	R_{DS}	Drain-to-source resistance

List of Symbols

Symbols	Description	Symbols	Description
n^+	Heavily doped silicon	β_F	Forced current gain
p^-	Lightly doped silicon	D	Drain
n	Emission coefficient	S	Source
k	Boltzmann's constant	G	Gate
q	Electron charge	K	Cathode
T	Absolute temperature	J	Junction
E	Emitter	t	Time
B	Base	D	Duty cycle
C	Collector	η	Intrinsic standoff ratio
V	Volt	f	Frequency
A	Ampere or anode	ω	Angular frequency
α	Forward current gain	T	Time period
β	Current gain	V_{gtri}	Gate triggering voltage
ΔI_C	Change in the collector's current	v_d	Voltage across diode
ΔV_{BE}	Change in base-to-emitter voltage	V_{BD}	Breakdown voltage
g_m	Transconductance	V_{BO}	Break-over voltage
t_d	Delay time	V_f	Voltage safety factor
t_r	Rise time	P_{av}	Average rate of heat produced
t_{on}	Turn-on time	P_{gav}	Gate power dissipation
I_e	Excess current	I_{gm}	Maximum limit of current
t_{off}	Turned-off time	V_{gm}	Maximum limit of voltage
t_f	Fall time	t_{rr}	Reverse recovery time
t_o	Off period	t_{gr}	Gate recovery time
t_n	Conduction period	I_{CBO}	Collector base leakage current
t_s	Storage time	I_A	Anode current
P_T	Power dissipation	R_P	Emitter shorting resistance
t_q	Turn-off time	V_{GTH}	Gate threshold voltage
I_G	Gate current	V_{TH}	Threshold voltage
R_{drift}	Drift region resistance	I_G	Gate current
P_{off}	Power loss during switching off	R_{SA}	sink to ambient thermal resistance
I_{DS}	Drain to source current	T_C	Case temperature

Symbols	Description	Symbols	Description
I_D	Drain current	T_S,	Sink temperature
R_{JC}	Junction to case thermal resistance	SiO_2	Silicon oxide
R_{CS}	Case to sink thermal resistance	V_{DS}	Drain to source voltage
T_J	Junction temperature	V_{diode}	Diode voltage
R_{DS}	Drain-to-source resistance	I_{diode}	Diode current
I_{DS}	Drain to source current	I_{leak}	Leakage current
V_{GS}	Gate to source voltage	I_B	Base current
V_{RRV}	Repetitive reverse voltage	I_E	Emitter current
V_{TD}	Thermal voltage	V_{CESAT}.	Saturation voltage
I_C	Collector current	V_{BESAT}	Saturated base-emitter voltage
V_{BE}	Base-emitter voltage	I_{BSAT}	Saturated base current
h_{FE}	Current gain h-parameter	V_{CESAT}	Saturated collector emitter voltage
E_m	Peak input AC voltage	I_{CEO}	Collector-emitter leakage current
E_{dc}	Average load volatge	E_{dm}	Maximum value of output voltage
E_{rms}	RMS load voltage	I_{rms}	RMS load current
P_{rms}	Power delivered to the load	V_{CE}	Collector emitter voltage

LIST OF FIGURES

Unit 1 Power Electronic Devices

Unit 2 Thyristor Family Devices

Unit 3 Turn-on and Turn-off Methods of Thyristors

Unit 4 Phase Controlled Rectifiers

Unit 5 Industrial Control Circuits

CONTENTS

Unit 5: Industrial Control Circuits **264-289**

1 Power Electronic Devices

UNIT SPECIFICS

This unit covers the following aspects:

- *Introduction to power electronics and its history.*
- *A brief introduction to semiconductor physics along with p-n junction diode.*
- *Power diodes and their characteristics.*
- *Power transistors and its classification.*
- *Construction, types, working principle, V-I characteristics, and uses of bipolar junction transistors (BJTs).*
- *Construction, types, working principles, V-I characteristics, and uses of power metal oxide semiconductor field effect transistors (MOSFETs).*
- *Construction, types, working principles, V-I characteristics, and uses of insulated gate bipolar transistor (IGBT).*
- *Introduction to single electron transistor (SET), quantum dot, and coulomb blockade.*
- *Working principle, and V-I characteristics of SET.*
- *Practical experiments on power BJT and IGBT.*

The various power semiconductor devices, their construction, V-I characteristics, working principle, use, and practical applications of the topics are discussed here to generate further curiosity and creativity as well as improving problem-solving capacity along with some numerical problems.

Besides giving a large number of multiple-choice questions as well as questions of short and long answer types marked in two categories following the lower and higher order of Bloom's taxonomy, assignments through several numerical problems, a list of references, and suggested readings are given in the unit so that one can go through them for practice.

Some practical experiments related to the courses covered in Unit-1 are also appended at the end of this unit to make the students aware of the hands-on on these topics.

After the related practical experiments on the topic, based on the content, there is a "Know More" section appended. This section has been designed to supplement additional information and higher learning skills on the topic.

RATIONALE

This fundamental unit on power electronics devices helps students to get a primary idea about power electronics, power diodes, power transistors, and types of power transistors. The construction, V-I characteristics, working principles, and uses of various power transistors like bipolar junction transistors (BJTs), metal oxide semiconductor field effect transistors (MOSFETs), insulated gate bipolar transistors (IGBTs) are discussed and analysed. The physics behind the construction of single electron transistor (SET), quantum dot, and coulomb blockade are discussed in short to develop the basic idea about the same.

Some related problems are pointed out after each section with their solutions which can help further for getting a clear idea of the concerned topics on power transistors used in power electronics. The V - I characteristics, the working principles etc will certainly help students with numerical problem-solving.

As a student in the field of electrical engineering, this unit on power electronics devices helps students to grasp the basic knowledge of power semiconductor devices.

PRE-REQUISITES

ESC101: Basic Electrical Engineering

UNIT OUTCOMES

After completion of Unit-1 students will be able to:

U1-O1: *Describe the semiconductor physics behind power electronics devices.*
U1-O2: *Explain the construction of power diodes, and power transistors.*
U1-O3: *Explain the working principle of various types of power transistors like BJTs, MOSFETs, and IGBTs.*
U1-O4: *Analyse the V-I characteristics of various types of power transistors like BJTs, MOSFETs, and IGBTs.*
U1-O5: *Identify the uses of power transistors and analyse the performance of them.*

Unit-1 Outcomes	EXPECTED MAPPING WITH COURSE OUTCOMES (1- Weak Correlation; 2- Medium correlation; 3- Strong Correlation)				
	CO-1	CO-2	CO-3	CO-4	CO-5
U1-O1	1	3	3	1	1
U1-O2	1	3	2	2	1
U1-O3	1	1	2	1	1
U1-O4	1	-	3	3	2
U1-O5	3	-	-	-	2

1.1 INTRODUCTION TO POWER ELECTRONICS

The power system is divided into three subsystems. They are generation, transmission, and distribution. Electrical energy is generated via a generation system, transmitted using a transmission system, and distributed to consumers using a distribution system. The word "Power" in "Power Electronics (PEs)" is associated with the power appliances for the subsystems of the power system. The "Electronics" in "Power Electronics" is associated with the solid-state devices and circuits used for controlling as well as the conversion of electrical power (EP). Thus, "Power Electronics" can be defined as the application of solid-state electronics for the control and conversion of EP. **Figure 1.1 (a)** shows the relationships among power, electronics, and control.

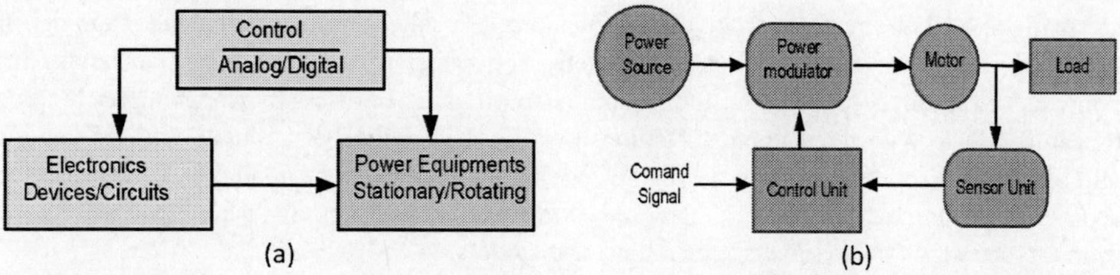

Figure1.1 (a) Relationship among power, electronics, and control, (b) Power electronics system

The block diagram of a simple PEs system is shown in **Figure.1.1 (b).** The power source shown in the figure may be AC or DC. The motor together with the power modulator and control unit is the electrical drive that is used to drive the mechanical load. The motor may be a DC motor like separately excited, shunt, series, or compound; an AC motor like an induction motor, synchronous motor; stepper motor, reluctance motor, etc. The power modulator has many functions such as the conversion of electrical energy received from the power source as required by the motor drive, selection of mode of operation such as motor or braking, modulation of power from the power source to the motor, starting, braking and reversal of speed, etc. Power modulators are controlled by the control unit. The sensor unit measures the load parameters and sends the parameter to the control unit.

Solid-state devices are nothing but semiconductor devices. When these semiconductor devices are used for power modulation and control, such devices are called power semiconductor devices. Power electronics mainly work based on the switching of power semiconductor devices. The power handling capacity and the switching speed of power devices increase with the advancement of power semiconductor technology. The control and design of control strategies for power semiconductor devices are improved with the introduction of microelectronics. The construction, working principles, V-I characteristics, and application of various power transistors are discussed in Unit-I.

1.2 HISTORY OF POWER ELECTRONICS

Various power electronic devices have been developed in the past. Power electronics started with the mercury arc rectifiers (1900). Afterwards, the rectifiers like metal tank types and vacuum tubes were introduced. These rectifiers as well as other devices like thyratron were in use until 1950s. The transistor was invented by Bardeen, Brattain, and Shockley in 1948. This is the starting point of the electronic revolution. The thyristor was invented in 1957. A thyristor is also called a silicon-controlled rectifier (SCR). These are invented in Bell telephone laboratories. The second revolution of power electronics started in 1958 when the commercial thyristor was introduced by General Electric Company. Many power semiconductor devices of different types and several control techniques have been developed since 1958. Conventional thyristor was developed in 1958 and it was used in Industry till

1970. After 1970, many other semiconductor devices were developed. Nowadays power electronics incorporate power semiconductor devices and microelectronics.

1.3 SEMICONDUCTOR PHYSICS

Power semiconductor devices are made from semiconductor material. High-purity, single-crystal silicon is used for power semiconductor devices. Many processes are adopted to convert single-crystal material into semiconductor devices. The elements are divided into various groups such as Group - I, Group - II, Group - III, Group - IV, Group - V, etc in the periodic table. Elements such as Boron, Gallium, or Indium belong to Group - III. Elements such as Silicon, Germanium, etc under Group - IV. The Elements such as Phosphorus, Arsenic, Antimony, etc. belong to Group - V. Group - IV elements are semiconductors. Pure silicon or pure germanium are called intrinsic semiconductors that have a resistivity in between insulators and conductors. The resistivity and the charge carriers of intrinsic semiconductors can be changed by adding some impurities. The process by which impurities are added to semiconductors is called doping. Doping involves the addition of a single atom of impurity per millions of semiconductor atoms. Using various impurities, levels of doping and shapes, insulation, appropriate technology for photolithography, cutting, etc., power semiconductor devices are developed from *n*-type and *p*-type extrinsic semiconductor materials.

1.3.1 *n*-type extrinsic semiconductor material

When pure silicon is doped with a small amount of Group - V element, *n*-type material or *n*-type extrinsic semiconductor material is produced. The elements of Group -V have five valance electrons. Four of five electrons of the dopant (Group-V element) atom form covalent bonds with its nearby silicon atoms. In this process, one electron is produced from one dopant (Group-V element) atom. The excess electrons increase the conductivity of the material. Thus, free electrons are available in an *n*-type material. In *n*-type material, the free electrons are the majority carriers and holes are the minority carriers. Thus, the number of electrons is more than the number of holes in *n*-type materials. When the silicon is lightly doped with Group - V element (called impurity) such as phosphorous, the doping is called *n* doping, and the material is called an *n*-type semiconductor. When the silicon is heavily doped with Group - V element, the doping is called n^+ type doping, and the resultant material is called an n^+ type semiconductor. The n-type semiconductor is also called a doped n-type semiconductor.

1.3.2 *p*-type extrinsic semiconductor material

When pure silicon is doped with a small amount of elements under Group - II, *p-type* material or *p*-type extrinsic semiconductor material is produced. Group-III elements have three valance electrons. When it is added to silicon, three covalent bonds are produced with nearby silicon atoms. In this doping process, a vacant location is produced from each dopant atom. The vacant location is called a hole. Like electrons, holes are also considered mobile carriers because they can be filled by adjacent electrons. The holes increase the conductivity of the material. Thus, free holes are found in *p*-type material. In *p*-type material, the free holes and electrons are the majority carriers and electrons are the minority carriers. Thus, the number of holes is more than the number of electrons in *p-type* materials. When silicon is lightly doped with the elements under Group - III like boron, the doping is called *p*-doping and the resultant material is called a *p*-type semiconductor. When the silicon is heavily doped with Group - III (impurity), the doping is called p^+ doping and the material is called a p^+ type semiconductor. The p-type semiconductor is also called a doped p-type semiconductor. The structure of *n*-type as well as *p*-type materials are shown in **Figure.1.2.**

In both n-type and p-type materials, the carriers are incessantly generated by thermal agitations. They combine and recombine as per their lifetime. They achieve an equilibrium carrier density in numbers/cm^3 from 10^{10} to 10^{13} in a temperature range of about 0^0C to 1000^0C. If an electric field is applied current flows, p-type and n-type materials.

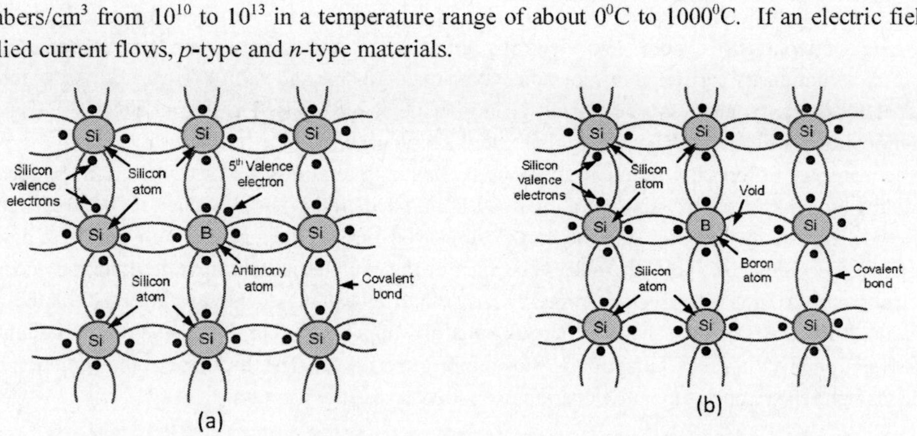

Figure.1.2 Structure of atom of the semiconductor (a) n-type, and (b) p-type

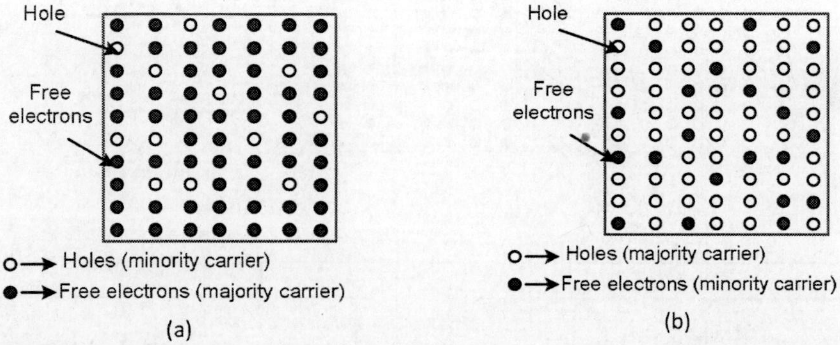

Figure.1.3 Majority and minority carriers in extrinsic semiconductor (a) n-type, (b) p-type

1.3.3 Drift and diffusion

The flow of current in a semiconductor is the sum of the net flow of holes and the net flow of electrons. The flow of holes and electrons is in opposite directions. The movement of the free carrier is through two mechanisms called drift and diffusion. When an electric field is applied across the semiconductor, holes move in parallel with the electric field and free electrons move antiparallel to the electric field. The velocity of the free carriers is called drift velocity. This velocity is proportional to the strength of the field. This mechanism is called drift. On the other hand, the free carriers move due to variations in the spatial density of free carriers. In this case, carriers are moving from higher density to lower density. This mechanism is called diffusion. The diffusion is because of the random thermal velocity that each carrier has. The spatial density is created by using many methods. Variation in doping density is also one such method.

1.3.4 *p-n* junction

Some covalent bonds break because of thermal energy and create some free electrons and immobile positive ions called holes in p-type and n-type semiconductors. In an n-type extrinsic semiconductor, the number of free electrons is more than the holes because of the extra free electrons by adding an impurity from the Group-V element. Similarly, in *p*-type extrinsic semiconductors, there are more holes than free electrons because of the addition of an impurity (element) from Group - III. Thus, in *n*-type majority carriers are electrons and minority carriers are holes. Similarly, in *p*-type extrinsic semiconductors, holes and electrons are majority and minority carriers respectively. The majority and minority carriers in *n*-type and *p*-type materials are shown in **Figure.1.3.** A *p-n* junction is created using *p*-type and *n*-type extrinsic semiconductor materials. The *p-n* junction is the basis for any bipolar device. The *p-n* junction is also called a bipolar diode. The technique of joining *p*-type and *n*-type materials to form a *p-n* junction may be by heating them in wafer (a very thin layer) form and gradually by cooling or diffusion of the two. After the formation of the *p-n* junction, the free electrons from the *n*-type semiconductor start moving to the *p*-type through the junction. Thus, near the junction, the *n*-side becomes positively charged. The holes of the *p*-type material near the junction accept the electrons and become negatively charged. Thus, a barrier called a depletion layer is created at the junction. This depletion layer has no charge carrier and opposes the current flow through the junction. The depletion layer of the *p-n* junction is shown in **Figure.1.4 (a).** The thickness of the depletion layers depends upon the temperature and the voltage applied across the junction.

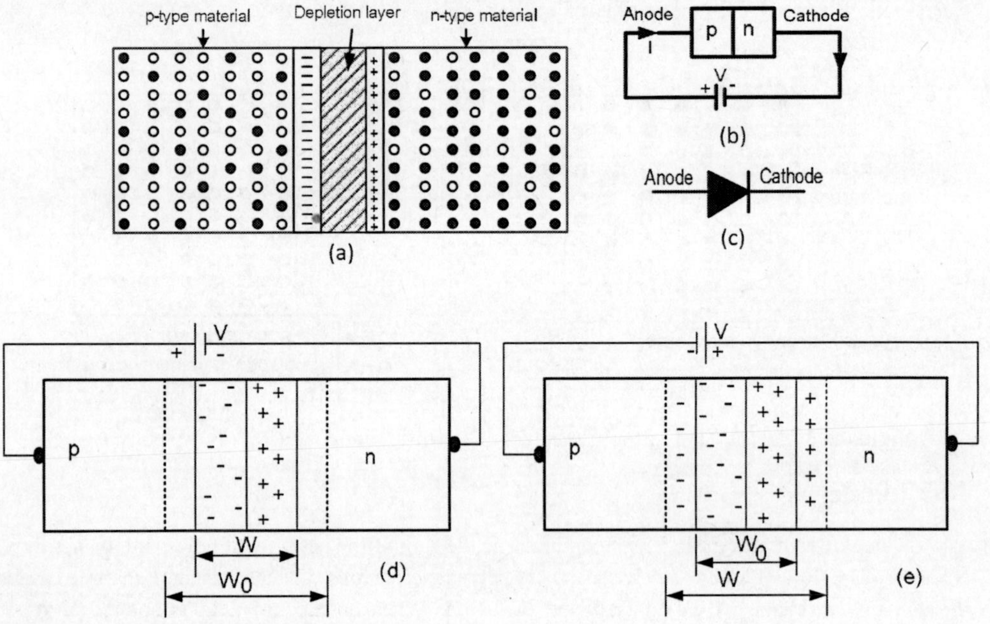

Figure.1.4 (a) Depletion layer in a *p-n* junction, (b) *p-n* junction with a voltage source, (c) diode symbol, (d) diode in forward bias, and (e) diode in reverse bias

1.3.5 Forward and reverse bias

When an external voltage is applied across the *p-n* junction, it appears entirely across the space charge region. It is because of the large resistance offered by the depletion layer. The positive terminal of the applied voltage is connected to the *p*-side and the negative to the *n*-side of the *p-n* junction, the applied voltage opposes the contact potential of the barrier. The height of the deletion layer reduces. The junction is said to be forward-biased.

When the negative terminal of the applied voltage is connected to the *p*-side and the positive terminal to the *n*-side, the barrier height increases. The depletion layer must grow or shrink as the barrier height grows or shrink. The *p-n* junction is said to be reverse bias. The forward bias and reverse are shown in **Figure.1.4 (d) and (e)** respectively. In these figures, W_0 is the width of the depletion layer at zero bias, and W is width of the depletion layer after the application of bias voltage.

1.4 INTRODUCTION TO POWER ELECTRONICS DEVICES

Power semiconductor devices are made of silicon and silicon carbide. Silicon is widely used. There are three types of silicon-based devices. They are diodes, transistors, and thyristors. In power application, all these three are further classified into (a) power diodes, (b) thyristors, (c) power bipolar transistors, (d) power metal oxide semiconductor field effect transistors, and (e) insulated gate bipolar transistors and static induction transistors.

1.4.1 Introduction to power diode

A diode also called as a *p-n*-junction diode is a two-terminal semiconductor device. The terminals are called anode and cathode. It has two layers of extrinsic semiconductor material, *p*-type and *n*-type. A power diode is an uncontrolled semiconductor device. The turn-on and turn-off characteristics are not under control. The *p-n*-junction and the symbol of the diode are shown in **Figure.1.4 (b) and (c).**

Power semiconductor devices have more complicated structures than low-power semiconductor devices. It is because of the modification of low-power devices to make them suitable for high-power applications. The same basic modifications have been done in low-power semiconductor devices to make them suitable for high-power applications. A diode or *p-n* junction is the basic semiconductor device and it is the simplest of all semiconductor devices. Hence, if we understand the medication in the diode, the modification in other semiconductor devices will be understood easily.

A diode acts as a switch in applications like rectifiers, freewheeling in regulators, trapped energy recovery, capacitor charge reversal, energy transfer between components, voltage isolation, etc. Power diodes are similar to *p-n* junction diodes but the power diodes have larger handling capabilities of voltage, current, and power than ordinary diodes. The switching speed of the power diode is low compared to the ordinary signal diode.

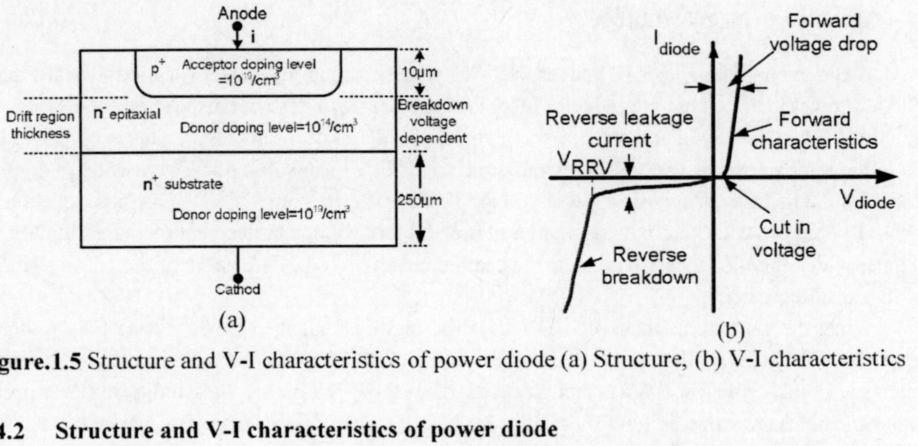

Figure.1.5 Structure and V-I characteristics of power diode (a) Structure, (b) V-I characteristics

1.4.2 Structure and V-I characteristics of power diode

The basic structure of the power diode is shown in **Figure.1.5(a).** It is made up of a heavily doped *n*-type substrate on the top of which is grown a lightly doped n^- epitaxial layer of specified thickness. A *p-n* junction is formed by diffusing a heavily doped *p* region on the n^- epitaxial layer. The thickness and doping levels for each layer for a typical *p-n* junction diode are shown in **Figure.1.5(a).** The drift region n^- is not found in low-power diodes. The function of this drift region is to absorb the depletion layer of the reverse bias p^+n^- junction. This drift region is quite wide with a large reverse voltage and establishes the required breakdown voltage. The doping level of the drift region is light.

The V-I characteristic of the diode is shown in **Figure.1.5 (b).** In this plot, the voltage across the diode (diode) is the abscissa and the diode current (I_{diode}) is the ordinate. The difference between a low-power diode and a high-power diode is that the current grows linearly in a power diode. The large current in the power diode results in an ohmic voltage drop. The voltage drops in the lightly doped drift region account for this ohmic drop. The V-I characteristics as shown in **Figure.1.5** (b) is represented by equation (1.1). This equation is called the *Schockley* diode equation under steady-state operation,

$$I_{diode} = I_{leak}(e^{V_{diode}/nV_{TD}} - 1) \tag{1.1}$$

where, I_{leak}= leakage current or reverse saturation current (for the diode), n= emission coefficient or identity factor, V_{TD} = Thermal voltage given by equation (1.2), where k = Boltzmann's constant (=1.3806 x 10^{-23} Joule/Degree kelvin), T= absolute temperature in kelvin, q = electron charge (= 1.6022x10^{-19} coulomb)

$$V_{TD} = \frac{kT}{q} \tag{1.2}$$

The V-I characteristics of the diode shown in Figure.1.5 (b) have three regions. They are the forward-biased region, reverse-biased region, and breakdown region. In the forward biased region, the diode voltage. The diode current is very small when the diode voltage V_{diode} is less than the cut-in voltage (of the order of 0.7 for a silicon diode). At cut-in voltage, the diode fully conducting. In the reverse-biased region, the $V_{diode} < 0$, which shows that diode current is negative and approximately equal to the negative of the reverse leakage current (I_{leak}) and hence flows in the reverse direction. In the breakdown region, The reverse voltage is very high, generally greater than 1000V. If the reverse voltage exceeds the V_{RRV}, the diode current increases rapidly.

1.4.3 Working Principle of Diode

When the anode is positive concerning (w.r.t) the cathode, the diode is called forward bias. If this anode-to-cathode voltage is gradually increased the depletion layer vanishes and the device starts conducting. When the diode conducts, a small forward voltage drops across it. This drop depends on the junction temperature as well as the manufacturing process. The value of forward voltage drop is in the range of 0.8 to 1.0V. When the applied voltage V_s is increased from 0 to cut-in voltage (or threshold or turn-on voltage) the current is very small and beyond this voltage the current increases rapidly. The voltage at which the diode starts conducting is called cut-in voltage. The value of the cut-in voltage is 0.7V for the silicon diode.

When the cathode is positive w.r.t anode the diode is called to be in reverse bias. When the reverse bias voltage is applied across the diode terminals, the depletion layer widens. The reason for the increase in the depletion layer is that electrons from the *n*-region are attracted toward the positive terminal and holes are attracted to the negative terminal of the applied voltage. The increase in the width of the depletion layer opposes the current flow through the diode. A small current called reverse leakage current or leakage current (in the micro or milliampere range) flows due to minority carriers. Increasing the reverse voltage up to a value at which reverse breakdown of the device occurs. The voltage at which the breakdown occurs is called the breakdown voltage. This breakdown is called avalanche breakdown.

With the application of reverse voltage, there will be a slow increase in leakage current until the avalanche or breakdown voltage is reached. At this voltage, the diode is turned on in the reverse direction. The current at this voltage is very high and is limited by some resistance otherwise this may destroy the diode. The avalanche breakdown is avoided by operating the diode below the repetitive reverse voltage (V_{RRV}).

1.4.4 Types of power diodes

There are three types of power diodes. They are general purpose, high speed or fast recovery, and Schottky.

(a) **General purpose diodes:** This diode has a relatively high recovery time of the order of 25μs. The current rating of general-purpose diodes ranges from 1A to several thousand amperes. The voltage rating ranges from 50V to about 5kV. This is used in battery chargers, uninterrupted power supplies (UPS), welding, traction, electroplating, etc.

(b) **High-speed recovery diodes:** This type of diode has a relatively low reverse recovery time of the order of 5μs or less. The current rating of general-purpose diodes ranges from 1A to several thousand amperes. The voltage rating ranges from 50 V to about 3 kV. These are used in power converters like choppers, communication circuits, switched-mode power supplies, induction heating, etc.

(c) **Schottky diodes:** These diodes have a very fast recovery time and low forward voltage drop. The voltage rating of Schottky diodes ranges up to 100 V and the current rating is from 1A to 300 A. These are used in high-frequency instrumentation, and switching power supplies.

1.5 POWER TRANSISTORS

1.5.1 Introduction

Unlike diodes, power transistors are controlled devices. The turn-on and turn-off characteristics of transistors are under control. The transistor is a three-terminal semiconductor device. The transistor remains in on-state only when the control signal is present. When the control signal is removed, the transistor is turned off. The switching speed of modern power transistors is much higher than that of thyristors.

1.5.2 Classification of Power Transistors

Power transistors are classified into the following categories.

1. Bipolar junction transistors (BJTs)
2. Metal oxide semiconductor field effect transistors (MOSFETs)
3. Static induction transistors (SITs)
4. Insulated gate bipolar transistors (IGBTs)
5. COOLMOS

All the above transistors are assumed as ideal switches in power electronics converters. The bipolar junction transistors (BJTs) and metal oxide semiconductor field effect transistors (MOSFETs)s can be chosen for the power converter and one can be a replacement for another provided the voltage and current ratings of the same meet the output requirements of the particular converter. The construction, working, V-I characteristics, etc. of a few power transistors such BJTs, MOSFETs, and IGBTs are discussed in this Unit.

1.6 BIPOLAR JUNCTION TRANSISTORS

1.6.1 Introduction

A bipolar junction transistor (BJT) is a three-layer, two-junction semiconductor device. It has three terminals. The terminals are named emitter, base, and collector. The emitter terminal is indicated by an arrow. It has alternate *p*-type and *n*-type layers diffused to form *npn* and *pnp* semiconductor devices. Accordingly, transistors are classified into two. They are *npn* and *pnp* transistors. The middle layer is the base (B) of the transistor and the other two regions are the emitter (E) and collector (C). The junction between the emitter and base is called the base-emitter junction (BEJ), and the junction between the collector and base is called the collector-base junction (CBJ). In power applications in general, the base is the input terminal and collector is the output terminal and the emitter is the common terminal and this configuration is called the common emitter (CE) configuration. The symbol of *npn* and *pnp* transistors are shown in **Figure.1.6.** It is formed by using either one *p*-type and two *n*-type semiconductor material regions or one *n*-type and two *p*-type material regions. When there is one *p*-region sandwiched by two *n*-region to form two junctions, the transistor is called the *npn* transistor. On the other with one *n*-region sandwiched by two *p*-region, a *pnp* transistor is formed.

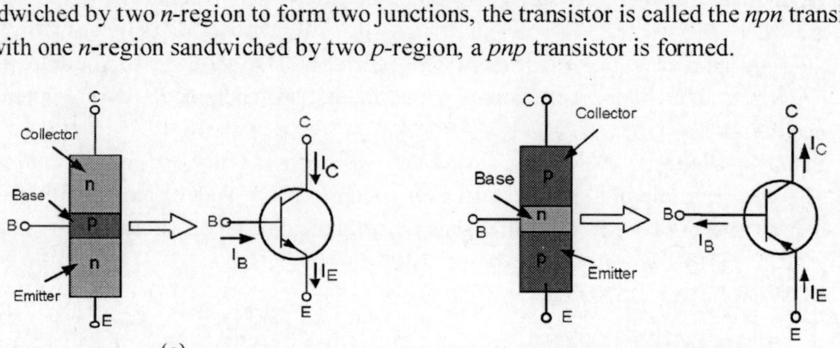

(a) (b)

Figure.1.6 Symbols of transistors (a) *npn*- transistor, (b) *pnp*-transistor

1.6.2 Construction of BJTs

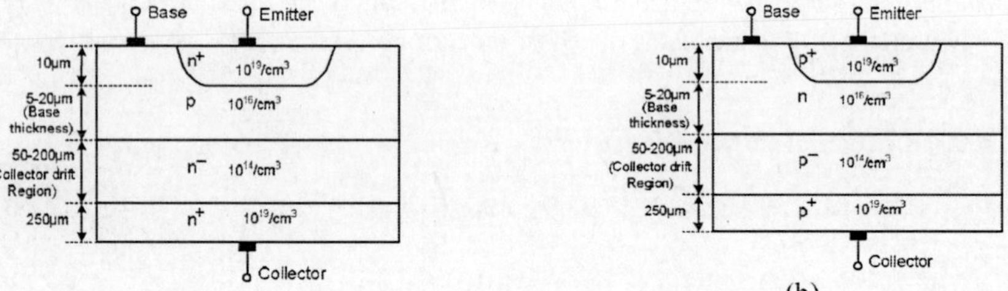

(a) (b)

Figure.1.7 Construction of transistor (a) *npn* transistor structure, (b) *pnp* transistor structure

The vertical cross-section of *npn* and *pnp* transistors is shown in Figure.1.7(a) and Figure.1.7(b) respectively. The vertical structure maximizes the cross-sectional are and thus minimizes the on-state resistance. This facilitates the reduction of power loss. A large cross-sectional area also reduces the thermal resistance of the BJT. The thickness and doping level of each region are shown in the figure for a typical power transistor. There is a drift region in the collector region of n^- doping in the case of the *npn* transistor and p^- in the case of *pnp* type transistor. This drift region is lightly dopped and its thickness is less than the collector n^+ (in case of npn) or p^+ (in case of pnp) region. The base region is moderately dopped and the thickness is small. The emitter is heavily doped.

1.6.3 BJT configuration

The transistor has three possible configurations. They are common emitter (CE), common collector (CC), and common base (CB) configurations. The configurations for an *npn* and *pnp* – transistors are shown in **Figure.1.8** and **Figure.1.9** respectively. The common emitter configuration is commonly used in switching operations. In common base, common emitter, and common collector, the common terminal is base, emitter, and collector respectively. In the common base configuration, the emitter is the input terminal and the collector is the output terminal. In the case of common collector configuration base is the input and the collector is the output terminal.

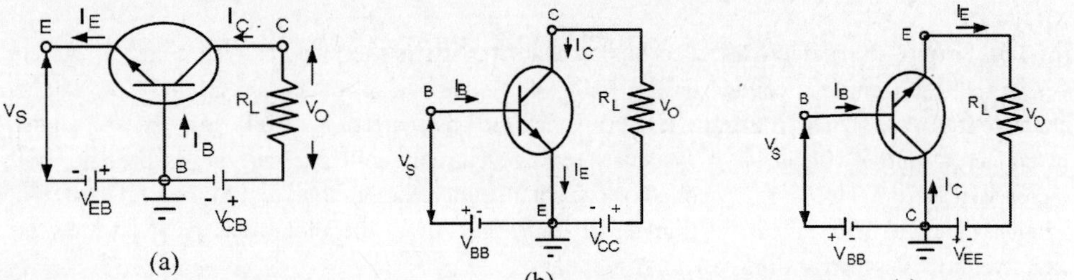

Figure.1.8 Various configurations of BJT using *npn* transistor (a) Common base, (b) Common emitter, and (c) Common collector

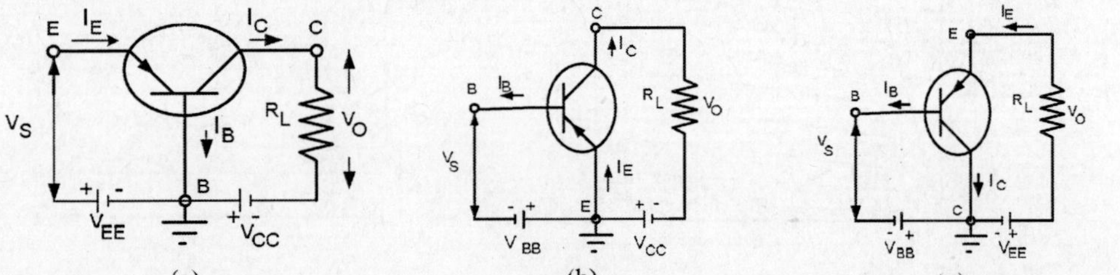

Figure.1.9 Various configurations of BJT using *pnp* transistor (a) Common base, (b) Common emitter, and (c) Common collector

1.6.4 V-I characteristics of BJTs

(a) Steady-state characteristics:

The common emitter (CE) configuration is taken here for the analysis of V-I characteristics. The circuit diagram of a *npn* transistor in CE configuration is shown in **Figure.1.10 (a).**

There are mainly two characteristics. They are input and output characteristics.

(b) The input characteristics: The plot between base current (I_B) versus base-emitter voltage (V_{BE}) is the input characteristic. Since the base-emitter junction is like a *p-n* junction diode, hence the input characteristics are just like a diode. The input characteristic is shown in **Figure.1.10 (b)**

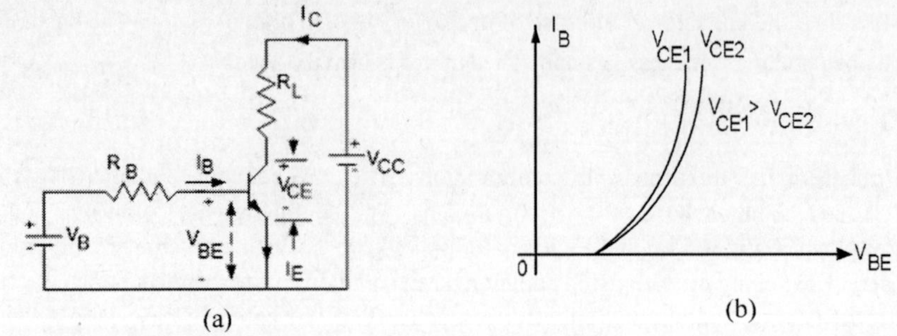

Figure.1.10 Circuit diagram and input characteristics of CE configuration of BJT (*npn*), (a) Circuit diagram (b) Input characteristics.

(c) The output characteristics: The output characteristic is the plot between collector current *(I_C)* versus collector-emitter voltage *(V_{CE})* for a particular value of base current *(I_B)*. The output characteristics are shown in **Figure.1.11(a)**. When the base current is zero i.e. $I_B = 0$, with an increase in collector-emitter voltage *(V_{CE})*, a small leakage current (collector current) is present. With the increase in I_B from 0 to I_{B1}, I_{B2}, ... collector current also increases as shown in **Figure.1.11 (a)**. There is another characteristic called the transfer characteristic. It is the plot between V_{CE} versus I_B. The characteristic is shown in **Figure.1.11 (b).**

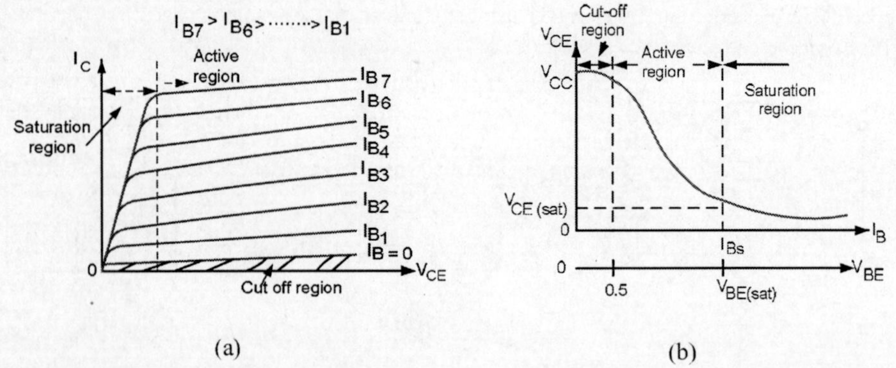

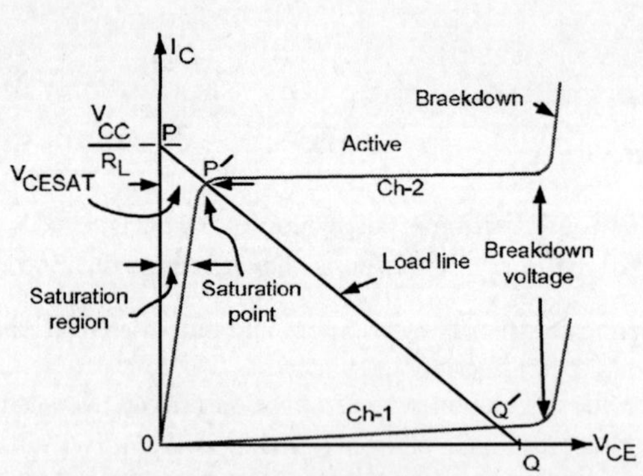

(c)

Figure.1.11 Output and transfer characteristics of BJT (*npn*) (a) output characteristics, (b) transfer characteristics, (c) Output characteristic only for two characteristics

1.6.5 The Working Principle of BJT

There are three operating regions of a BJT. They are the cut-off region, the saturation region, and the active region. In the cut-off region, the transistor is in off condition and the base current is not sufficient enough to turn the transistor to on condition. Both CBJ and BEJ are reversed-biased. In the active region, as shown in **Figure.1.11 (a)**, the BJT acts as an amplifier in which the base current is amplified by a gain. The V_{CE} decreases with the base current. The collector base junction is reversed biased and the base-emitter junction is forward-biased. The BJT in the saturation region works as a switch. In this region, the collector current is sufficiently high so that V_{CE} is low. Both CBJ and BEJ are forward-biased.

The working of the transistor can be analyzed by taking two output characteristics. One is at $I_{B0} = 0$ (Ch-1), and the other is at $I_{Bn} \neq 0$ (Ch-2) as shown in **Figure.1.11 (c)**. In the early part of the curve ch-2, the V_{CE} is very low. This part is called the saturation region. The BJT in the saturation region works as a switch. The flat part of Ch-2 represents a nearly constant collector current, I_C with an increase in V_{CE}. This region is called the active region. In this region, the BJT works as an amplifier. With further increase in V_{CE} at $I_{Bn} \neq 0$ (Ch-2), there is a vertical rise in the curve and this represents the breakdown of the transistor. The breakdown of the transistor is avoided at any cost. A line connecting points P and Q is also shown in **Figure.1.11 (c)**. This line is called the load line. The load line is the locus of all operating points. The line is derived from equation (1.3), where R_L is the load resistance and I_C collector current.

$$I_C = \frac{V_{CC} - V_{CE}}{R_L} \tag{1.3}$$

When the transistor is on, the collector-emitter voltage *(V_{CE})* is zero, and the collector current is given by, $I_C = \frac{V_{CC}}{R_L}$ which is shown by point P. When the transistor is in off condition, the collector current is zero and the collector-emitter voltage is equal to V_{CC}. This situation is represented by point Q. Most of the electrons given out by the emitter reach the collector. Though the collector current is somewhat less than the emitter current, they are almost equal. This relation between the emitter and collector currents (I_E and I_C) is related by the term forward current gain and denoted by α. The α is given by equation (1.4). The value of α varies from 0.95 to 0.99.

$$\alpha = \frac{I_C}{I_E} \tag{1.4}$$

In BJT, the base current is effectively the input current and the collector current is the output current. The relation between collector and base current is related to the term "current gain", denoted by the symbol β, and is defined as the ratio between collector and base current. It is given by equation (1.5).

$$\beta = \frac{I_C}{I_B} \tag{1.5}$$

The value of the base current is very much smaller than the collector current, and the value of β is greater than one. Generally, it varies from 50 to 300. In some cases, β is called the h parameter, denoted h_{FE}. Thus, in such cases, equation (1.5) becomes equation (1.6).

$$\beta = h_{FE} = \frac{I_C}{I_B} \tag{1.6}$$

Using Kirchhoff's current law, in the CE circuit of the *npn* transistor shown in Figure.1.10 (a), we get (1.7).

$$I_E = I_B + I_C \tag{1.7}$$

Dividing both sides of equation (1.7), we get

$$\frac{I_E}{I_C} = 1 + \frac{I_B}{I_C}$$

$$\text{Or } \frac{1}{\alpha} = 1 + \frac{1}{\beta}$$

$$\text{Or } \beta = \frac{\alpha}{1-\alpha} \quad (1.8)$$

$$\text{And } \alpha = \frac{\beta}{1+\beta} \quad (1.9)$$

1.6.5.1 Working of transistor as a switch

It is learned from the previous discussion that when the BJT operates in the cut-off region, the same is in off condition and when BJT operates in the saturation region, the collector current is sufficiently high so that V_{CE} is low. Thus, the operation of BJT as a switch means that it may operate either in the cut-off region or in the saturation region. When operates at point *P*, it works as an ideal close switch since the collector-emitter voltage is zero. When operates at point *Q*, it works as an open switch since the collector current is zero at this point. In practice, because of the large base current, the BJT works in saturation region at point *P'* instead of *P* with a small saturation voltage V_{CESAT}. The voltage V_{CESAT} represents the on-state voltage drop of the BJT. When the control signal i.e., the base current is set to zero, the BJT is turned off. The operating point is now shifted to *Q'* from *Q*. This is because of the leakage current that flows in the collector circuit when BJT is in off condition.
Application of Kirchhoff's voltage law for the portion of the circuit comprising of V_B, R_B, and emitter as shown in **Figure.1.10 (a)**.

$$V_B - R_B I_B - V_{BE} = 0$$

$$\text{Or } I_B = \frac{V_B - V_{BE}}{R_B}$$

$$\text{Again, } V_{CC} = V_{CE} + I_C R_L$$

$$\text{Or } V_{CE} = V_{CC} - \beta I_B R_L$$

$$\text{Or } V_{CE} = V_{CC} - \frac{\beta R_L}{R_B}(V_B - V_{BE}) \quad (1.10)$$

Now, $V_{CE} = V_{CB} + V_{BE}$ from which $V_{CB} = V_{CE} - V_{BE}$

(1.11)

Now, the current produced due to saturation voltage, V_{CESAT} is given by (1.12)

$$I_{CSAT} = \frac{V_{CC} - V_{CESAT}}{R_L} \quad (1.12)$$

Thus, the base current which produces saturation is given by (1.13).

$$I_{BSAT} = \frac{I_{CSAT}}{\beta} \quad (1.13)$$

The BJT operates in the active region when the base current is less than I_{BSAT}. On the other hand, if the base current is greater than I_{BSAT}, V_{CESAT} is nearly zero, and in this case, $I_{CSAT} = \frac{V_{CC}}{R_L}$. Thus, the collector current is almost constant even if base current is increased. The BJT is in a hard drive situation when the base current is more than I_{BSAT}. The V_{CESAT} is very small and hence the on-state losses are reduced. Another factor called overdrive factor (ODF) is defined in hard drive condition and is given by (1.14).

$$ODF = \frac{I_B}{I_{BSAT}} \tag{1.14}$$

Another term called forced current gain, β_F is also defined by equation (1.15). The value is β_F is less than β or h_{FE}.

$$\beta_F = \frac{I_{CSAT}}{I_B} \tag{1.15}$$

The power loss in the junctions of BJT is given by (1.16).

$$P_{Total} = V_{BE} I_B + V_{CE} I_C \tag{1.16}$$

Under saturated region, the base-emitter voltage (V_{BESAT}) is more than the collector-emitter voltage (V_{CESAT}), it is seen the hat collector-base voltage is negative. It implies that in saturation region junctions of the BJT are forward biased.

1.6.5.2 Switching performance of BJT

There are two parallel capacitances in a forward-biased *p-n* junction. They are depletion layer capacitance and diffusion capacitance. There is only depletion layer capacitance in reversed biased *p-n* junction. The values of these capacitances depend on the junction voltage and physical construction. Under normal steady-state conditions, these capacitances do not play any role. Under transient conditions, these capacitances play a vital role in turn-on and turn-off characteristics. The transconductance (g_m) also plays an important role in the transient performance of BJT. The g_m is defined as the ratio between the change in collector current (ΔI_C) and the change in base-to-emitter voltage (ΔV_{BE}), and is given by **(1.17).**

$$g_m = \frac{\Delta I_C}{\Delta V_{BE}} \tag{1.17}$$

The circuit models under transient conditions are shown in **Figure.1.12**(a) and (b), in which C_{CB} and C_{BE} are the effective capacitances of CBJ and BEJ respectively, R_{BE} and R_{CB} are the resistances of BEJ and CBJ respectively. Due to the internal capacitances, the BJT does not turn on instantaneously. The *npn* transistor circuit at CE mode shown in **Figure.1.10 (a)** is considered to study the switching performance. A resistive load (R_L) is connected between the collector and given by the emitter. The waveforms under switching are shown **in Figure.1.12 (c).** When the base-to-emitter voltage (V_{BE}) is applied, the base current rises to I_{BS} and the collector current is equal to 0 or collector-emitter leakage current (I_{CEO}). After some time, called delay time (t_d), the collector current start to increase. This delay is due to the time required to charge the internal capacitance to V_{BES}. After this delay, the collector current rises to a steady state value equal to I_{CSAT} and takes time t_r. The time (t_r) is called rise time and it depends on the internal capacitance of BEJ. During rise time collector-emitter voltage (V_{CE}) falls to V_{CESAT}. The sum of t_d and t_r is called the turn-on time (t_{on}) (i.e $t_{on} = t_d + t_r$) of the transistor. The transistor remains in on condition if the forward base current is maintained.

In general, the base current is more than that required to bring the transistor to saturation. As a result, excess minority carriers are stored in the base region. If the overdrive factor is more, the greater is the charge storage. This extra charge is called saturating charge. The saturating charge is proportional to the excess base drive and corresponding current, I_e. The excess current I_e is given by equation (1.18).

$$I_e = I_B - \frac{I_{CSAT}}{\beta} = ODF.I_{BS} - I_{BS} = I_{BS}(ODF - 1) \tag{1.18}$$

The saturating charge, Q_S is given by equation (1.19), where τ_s is the storage time constant of the BJT.

$$Q_S = \tau_S I_e = \tau_s I_{BS}(ODF - 1) \tag{1.19}$$

When V_{BE} is removed at time t_1, the collector current does not change for time t_s. This time is called storage time. During this time the saturating charge is removed from the base and after this time period, the collector current begins to fall and the collector voltage begins to build up till t_f. This time is called fall time. After t_f, the collector current falls to I_{CEO} or almost zero and the collector voltage is equal to V_{CC}. The sum of t_s and t_f gives the turned-off time *(t_{off})* of the BJT (i.e $t_{off} = t_s + t_f$). The conduction period *(t_n)* and off period *(t_o)*, and time period *(T)* is also shown in **Figure.12 (c)**.

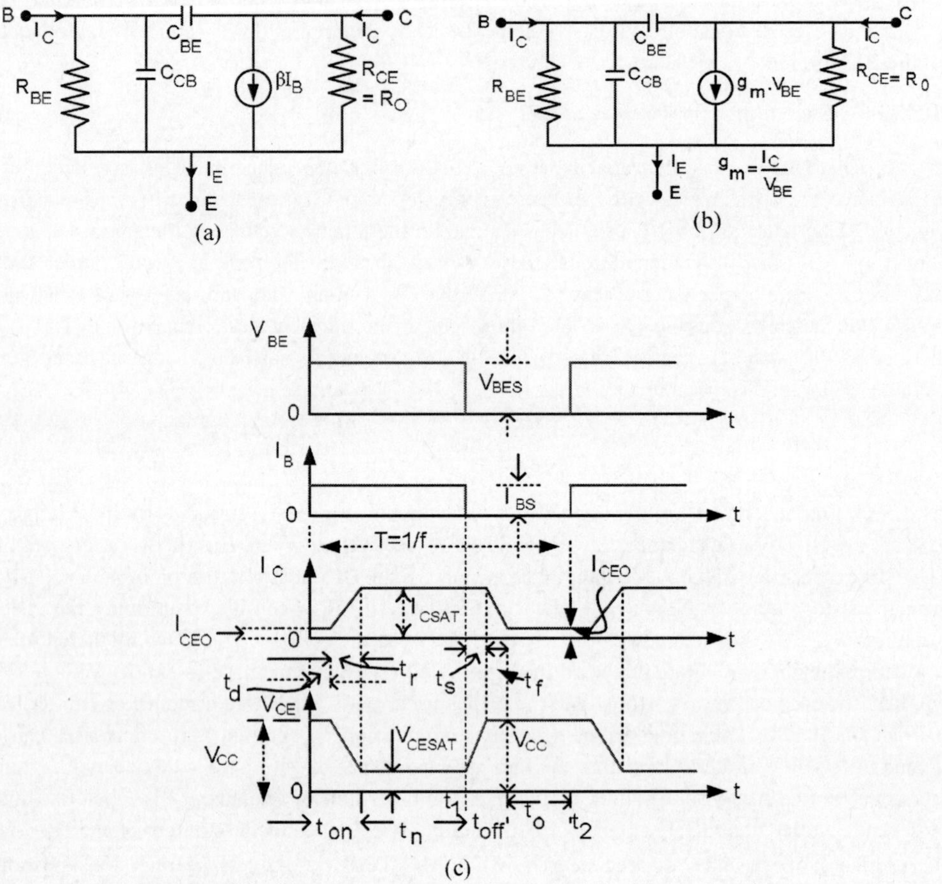

Figure.1.12 (a) Circuit model with current gain, (b) circuit model with transconductance, and (c) *npn* transistor switching waveform

1.6.5.3 Switching limits and safe operating area

Secondary breakdown: The transistor may undergo a second breakdown (SB). The SB is a destructive phenomenon. It results from the current flow to a small portion of the base and produces local hot spots. The energy of these hot spots is sufficient for local heat production. This heat may damage the device. The resulting high current concentration may lead to a secondary breakdown. The high current concentration may be caused by defects in the n transistor. SB involved some combination of voltage, current, and time. Since time is one of the factors for SB, it is energy dependent.

There are operating limits of collector current and collector-emitter voltage for reliable operation of BJT. These limits are specified by the safe operating area (SOA). Most manufacturers of BJT provide two SOA. They are forward-biased SOA (FBSOA) and reverse-biased SOA (RBSOA).

FBSOA: FBSOA is related to the operation of BJT when BEJ is forward-biased to turn the transistor on. The power handling capability of the power transistor is limited by junction temperature and second breakdown. The manufacturer of the BJT provides the FBSOA curve for some particular test conditions. FBSOA indicates the collector current – collector-emitter voltage limits of the BJT. A typical FBSOA is shown in **Figure.1.13(a).** The scale for collector current and collector-emitter voltage is in logarithm.

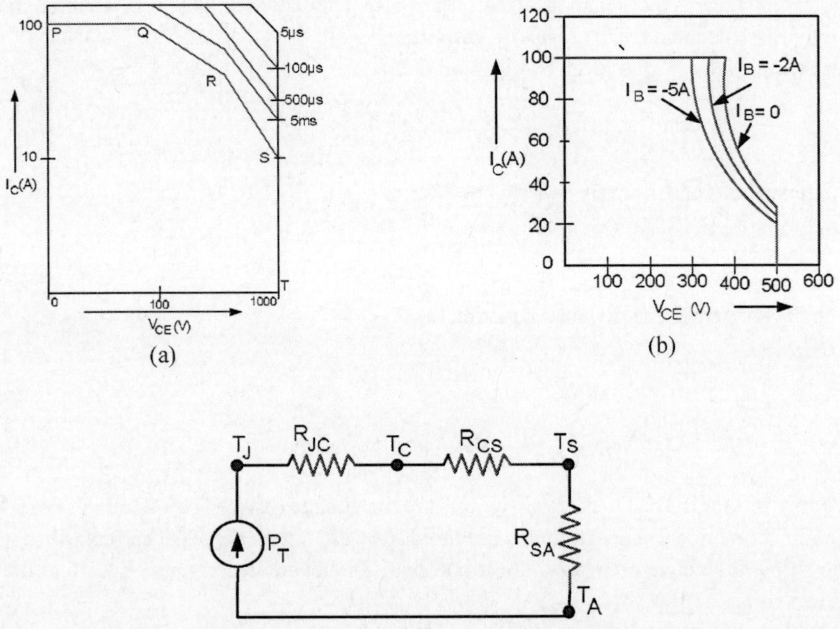

(c)

Figure.1.13 (a) Typical FBSOA of BJT, (b) Typical RBSOA of BJT, (c) Thermal equivalent circuit BJT

In **Figure.1.13(a),** PQ is the boundary (maximum limit) for DC and continuous current for collector-emitter voltage V_{CE} less than 80V. If V_{CE} is more than 80V, the collector current need to be reduced to the limit shown by QR so that the junction temperature is limited to a safe value. Further, higher value of V_{CE}, the collector current is reduced to boundary RS. The boundary ST represents the maximum voltage capability of the transistor. In **Figure.1.13(a),** 5ms, 500μs, 100μs, etc. are the pulse wides for pulsed operation. The manufacturer of BJT specifies the limit for temperature 25^0 and for DC and single operation. For actual working temperature and the repetitive nature of pulses, these cures are modified the using thermal impedance of the BJT.

RBSOA: During turn-off, the BJT is subjected to high current and high voltage with BEJ reversed biased. The collector-emitter voltage (V_{CE}) must be at a safe value or below a specified value of collector current during the turn – off a reversed biased condition. The manufacturer provides an RBSOA at specified conditions. **Figure.1.13(b)** shows a typical RBSOA. SOA during reverse biased is specified as reverse blocking SOA (RBSOA). It is a plot between I_C and V_{CE}. RBSOA specifies the limits for BJT operation during turn–off. During turn-off $I_B = 0$ or when BEJ is reverse biased. It is seen that with an increase in reverse bias, SOA decreases in size.

Derating of power: While deciding the rating of the device, it is necessary to consider ambient temperature (AT) and thermal resistance (TR). The power dissipation P_T is normally specified at a particular temperature say 25^0 C. If the ambient temperature increases to say 150^0 C, the BJT can dissipate zero power. On the other hand, if the ambient temperature is reduced to 0^0 C, the power dissipation is maximum which is not practical. This is the reason that AT and TR are considered to decide the rating of BJT. The thermal equivalent circuit of a transistor is shown in **Figure.1.13 (c),** where R_{JC} = TR from junction to case, R_{CS}= TR from case to sink, and R_{SA} = TR from sink to ambient. All the thermal resistances are at ^{0}C/W. The case temperature, sink temperature, and junction temperatures are denoted by T_C, T_S, and T_J respectively.
The case temperature (T_C) is given by equation (1.20).

$$T_C = T_J - P_T R_{JC} \tag{1.20}$$

The sink temperature is given by equation (1.21).

$$T_S = T_C - P_T R_{CS} \tag{1.21}$$

The ambient temperature is given by equation (1.22).

$$T_A = T_S - P_T R_{SA} \tag{1.22}$$

Thus,

$$T_J - T_A = P_T (R_{JC} + R_{CS} + R_{SA}) \tag{1.23}$$

Breakdown voltages: The breakdown voltages are the voltage between two terminals of BJT while the other terminals are open, short-circuited, or biased (forward or reverse). The current rises rapidly at the breakdown but the voltage remains almost constant. The manufacturers have specified the following breakdown voltages.

(a) Maximum voltage between emitter and base while collector open (V_{EBO})
(b) Maximum voltage between collector and emitter at specified negative voltage applied between base and emitter (V_{CEV} or V_{CEX})
(c) Maximum sustaining voltage between collector and emitter with a base open circuited ($V_{CEO(SUS)}$)

1.6.5.4 Uses of BJTs

Power BJT is extensively used in Choppers (DC to DC converters), and inverters (DC to AC converters). It is mainly preferred in low and medium-power applications.

1.7 POWER MOSFETs

1.7.1 Introduction

The metal oxide semiconductor field effect transistor (MOSFET) is a member of the transistor family. It was invented in the late 1970s and available in various power electronics applications since the 1980s. A MOSFET is a voltage-controlled semiconductor device. It is a unipolar device. It is a three-terminal semiconductor device. The terminals are named drain *(D)*, source *(S)*, and gate *(G)*.

These names *i.e.*, drain, source, and gate are similar to the collector, emitter, and base of BJT. The source and drain of MOSFETs are identified by relative magnitudes of the bias voltage but cannot be differentiated by structural properties.

. It requires a small input current (gate current). The switching speed of MOSFET is very high. The switching time is in the nanoseconds range. It has no secondary breakdown problem. But MOSFET suffers from electrostatic discharge. It is also difficult to protect it from short circuits. Hence, some special care must be taken while handling the same. The types of MOSFETs are shown in **Figure.1.14.**

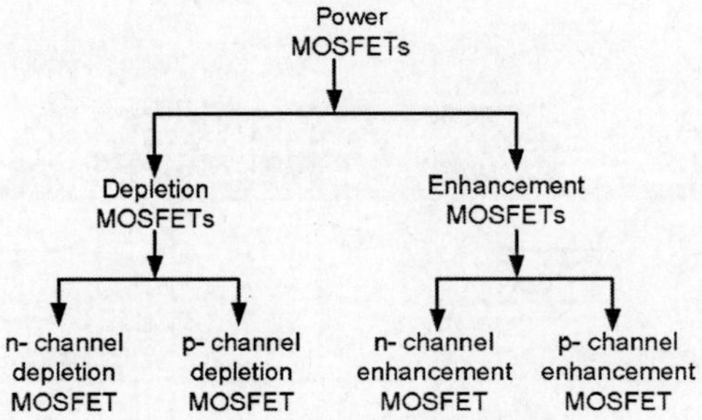

Figure.1.14 Types of MOSFETs

There are mainly two types of MOSFETs. They are depletion MOSFETs and enhancement MOSFETs. Both have two types. They are further subdivided into two types. They are *n*-channel and *p*-channel MOSFETs. The *n*-channel MOSFET is commonly used because of the higher mobility of electrons.

1.7.2 Construction of MOSFET

1.7.2.1 Basic Construction MOSFETs

The basic construction of an *n*-channel and *p*-channel MOSFETs are shown in **Figure.1.15(a)** and **(b)** respectively. The MOSFET has four vertically oriented layers. The layers are alternate *p* and *n* type semiconductors. The doping layers are n^+, p, n^-, and n^+ in n-channel type MOSFETs and p^+, n, p^-, and p^+ in p-channel type MOSFETs. The two ends of doping of each type are labelled as drain and source. The source and drain layers of *n*-channel, and *p*-channel are heavily doped and hence denoted by n^+ in *n*-channel and p^+ in the p-channel type of MOSFETS. The doping level of the source and drain region is nearly same of the order of $10^{19}/cm^3$. The *p* middle layer in *n*-channel and *n* middle layer in p-channel type is called the body of the MOSFETs. The body of the MOSFETs is the region in which the channel is formed between source and drain. The typical doping level of the body is $10^{16}/cm^3$. The n^- (in case of n-channel) and p^- (in case of p-channel) region is the drift region. This region is an epitaxially grown layer. The density of this drift layer is less than the normal doping level of *p*-body (in case of *n*-channel) and *n*-body (in case of *p*-channel) and hence denoted by n^- and p^-. The breakdown voltage of the MOSFETs is decided by the drift region. The typical doping level of the drift region is in the range of 10^{14} - $10^{16}/cm^3$.

There are two *pn* junctions. One is between body and the drain and other is body and the source. These junctions are reverse bias with either positive or negative polarity voltage and hence no way current can flow from source to drain. The gate is insulated from the body with a very thin (of the order of $1000\,\overset{0}{A}$ (angstroms) layer of silicon oxide. On application of a voltage biases positive polarity to the gate with respect to source, the silicon surface beneath the gate oxide is converted to *n*-type layer channel in *n*-channel MOSFET. It allows current to flow from source to drain. The value of source-to-drain current at a particular gate to source voltage is dependent on gate oxide thickness, wide of gate, and parallelly connected gate/ source regions.

In case of n channel MOSFET, the n^+ substrate on the drain side onto which the n^- drift region is grown epitaxially. The p type body region is diffused into the wafer from the source side of the wafer after which n^+ source diffusion. The portion of the wafer is protected by using silicon dioxide so that dopant cannot reach the wafer. Finally, the metallization of gate and source.

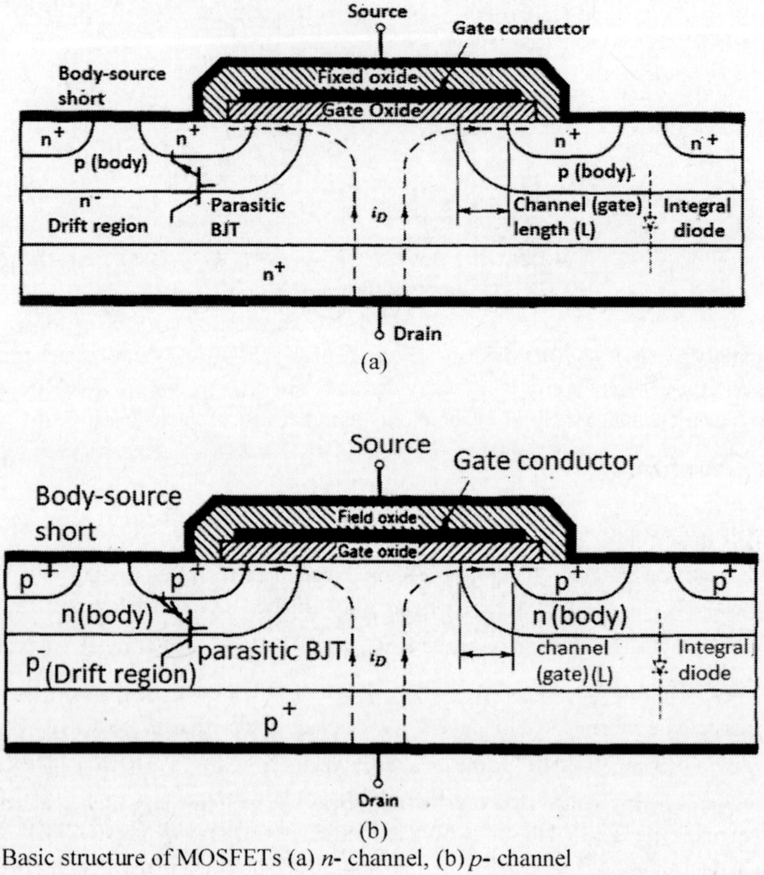

Figure.1.15 Basic structure of MOSFETs (a) *n*- channel, (b) *p*- channel

The MOSFET has a parasitic BJT in its structure. The body serves as base, source as emitter, and drain is the collector of this BJT. The forward current gain of this BJT is much greater than 1. The reason for high forward current gain is due to the length of the body region. The channel of the MOSFET is formed. The length of the body is kept short as possible to minimize the on-state resistance. To ensure the BJT cut off all the times, the potential of the base is kept equal to source potential and to do this the body-source short is provided as shown in Figure.1.15. As a result of providing a short between body and the source a parasitic diode is connected between drain and source. This integral diode is used in half-bridge and full-bridge converters.

1.7.2.2 Construction of depletion-type MOSFETs

The structure of *n*-channel depletion MOSFET and its symbol is shown in **Figure.1.16(a)** and **(b)** respectively and the same for *p*-channel depletion MOSFET is shown in **Figure.1.16 (c)** and **(d)** respectively. The *n*- channel depletion type MOSFET is constructed on a *p-type* substrate with two n^+ type substrates. The *p* - channel depletion type MOSFET is constructed on an *n-type* substrate with two p^+ type substrates. A thin layer of oxide is used to isolate the gate (G). The oxide layer is in general silicon oxide (SiO_2). In general, the substrate is connected to the source. MOSFET is heavily doped in the drain side creating an n^+ buffer below the *n*- drift layer. In the case of *p* - channel depletion type MOSFET, polarities of V_{GS}, V_{DS}, and I_{DS} are made reversed as shown in **Figure.1.16(c)** and **(d).**

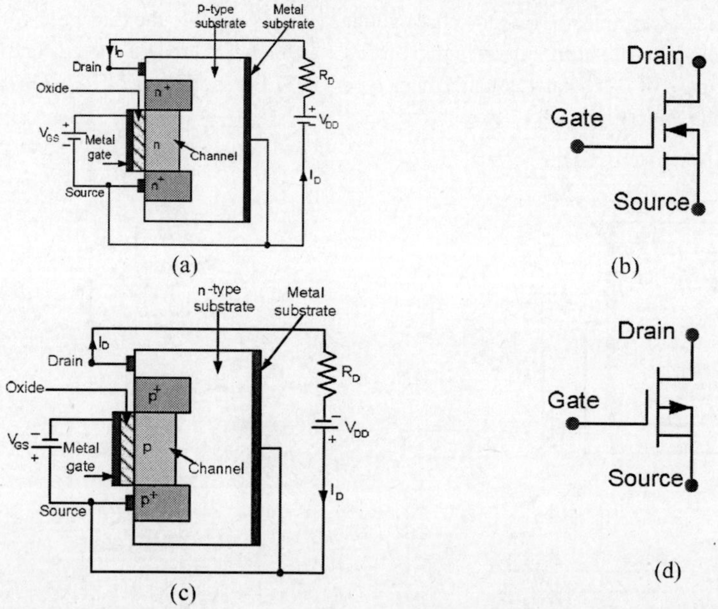

Figure.1.16 Structure and symbol of depletion MOSFET (a) *n*-channel depletion MOSFET, (b) symbol of *n*- channel depletion MOSFET, (c) *p*-channel depletion MOSFET, (d) symbol of *p*- channel depletion MOSFET

1.7.2.3 Construction of enhancement-type MOSFETs

In the enhancement type MOSFETs have no physical channel. The structure and symbol of *n*-channel enhancement MOSFET are shown in **Figure.1.17(a)** and **(b)** respectively. Similarly, the structure and symbol of *p*-channel enhancement MOSFET are shown in **Figure.1.17 (c)** and **(d)** respectively. The *n*- channel enhancement type MOSFET is constructed on a *p*-type substrate with two n^+ type substrate. The *p* - channel enhancement type MOSFET has constructed an *n*-type substrate with two p^+ type substrates. A thin layer of oxide is used to isolate the gate from the substrate. The oxide layer is in general silicon oxide. This oxide layer is in the between gate and the n+ and p junction. In general, the substrate is connected to the source. MOSFET is heavily doped on the drain side creating an n^+ buffer below the n- drift layer. This layer does not allow the depletion layer to reach the metal. Also, it reduces the forward voltage during conduction. In the case of *p* - channel depletion type MOSFET, polarities of V_{GS}, V_{DS}, and I_{DS} are made reversed as shown in **Figure.1.17(c)** and **(d).**

1.7.3 Working principle of MOSFETs

In case of *n*-channel depletion type MOSFET, the gate to source voltage (denoted by V_{GS}) is either positive or negative. When the V_{GS} is negative, some electrons from the *n*-channel are repelled. As a result, a depletion layer is formed just below the oxide layer and the drain-to-source resistance (R_{DS}) increases and a narrow channel is formed. If V_{GS} is made more negative, R_{DS} is very high and the channel is depleted, and there is no drain to source current i.e $I_{DS} = 0$. The value of V_{GS} at which the I_{DS} is zero is called pinch-off voltage (V_{POFF}). With V_{GS} being positive, the channel becomes wider and R_{DS} becomes very low, and hence I_{DS} increases.

In case of *n*-channel enhancement type MOSFET, if the gate to source voltage (denoted by V_{GS}) is positive, induced voltage attracts the electron from the *p*-type substrate and accumulates the electrons just beneath the oxide layer. In case, gate to source voltage (V_{GS}) is equal to the threshold voltage (V_{TH}), good numbers of electrons are accumulated and a virtual *n* - channel is created and current flows from drain to source. In case of *p*- channel enhancement type MOSFET, polarities of V_{GS}, V_{DS}, and I_{DS} are made reversed as shown in **Figure.1.17(c)** and **(d).**

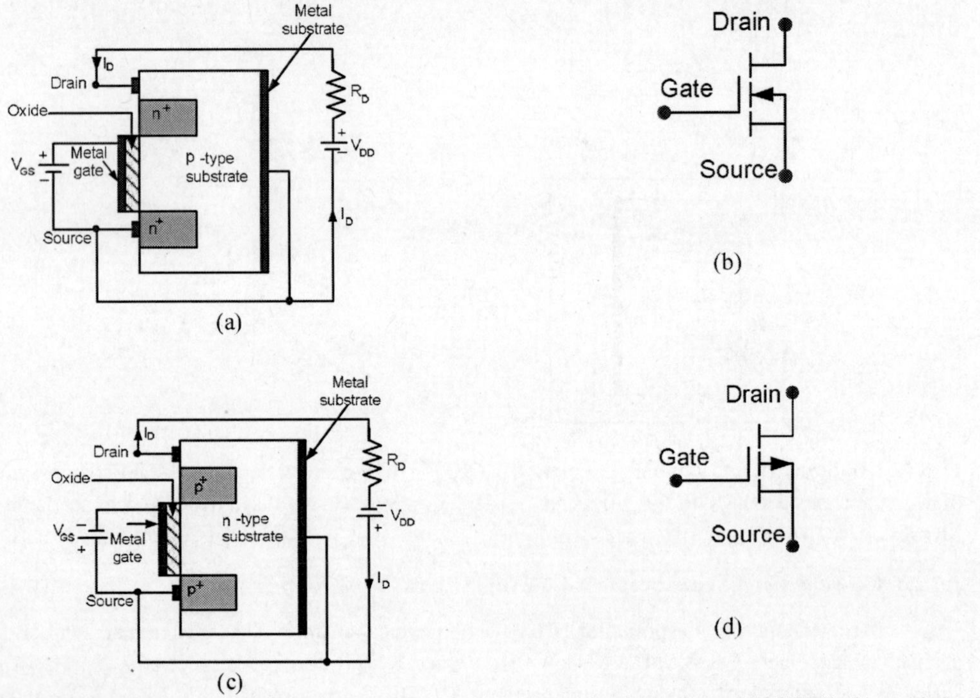

Figure.1.17 Structure of enhancement MOSFET (a) *n*-channel enhancement MOSFET, (b) symbol of *n*-channel enhancement MOSFET, (c) Structure *p*- channel enhancement MOSFET, (d) symbol of *p*-channel enhancement MOSFET

A depletion type MOSFETs remain on at zero gate voltage. On the other hand, enhancement MOSFETs, remain off at zero gate voltage. Hence, enhancement-type MOSFETs are normally used for switching devices in power electronics.

When gate is more positive with respect to source, the electrons are pulled from n^- layer into *p*-layer. This action opens a channel closest to the gate and thus allows current to flow from drain to source. The oxide layer is in general silicon oxide. This oxide layer is in between gate and the n^+ and *p* junction.

1.7.4 V-I characteristics of MOSFET

The static characteristics of power MOSFET is discussed here. In this connection, the circuit (shown in **Figure.1.18(a)**) of an *n*-channel enhancement MOSFET is considered. The various static characteristics are

(a) Transfer characteristics
(b) Output characteristics
(c) Switching characteristics

(a) Transfer characteristics: MOSFET is a voltage-controlled semiconductor device. The input impedance is very high. The gate draws a very small leakage current. The order of this leakage current is in the nano ampere range. The transconductance of MOSFET is the ratio of drain current (I_D) and gate current (I_G). The transconductance of MOSFET is very high in the order of 10^9. The transfer characteristics is the plot between drain to source current (I_{DS}) and the drain to source voltage (V_{DS}). The transconductance defines the transfer characteristics of MOSFET. Typical transfer characteristics of *n*-channel power MOSFET is shown in **Figure.1.18**(b). It is clear from the figure that the device is off below the threshold voltage (V_{TH}).

Similarly, the transfer characteristics of *p*-channel enhancement type MOSFET can be obtained (shown in **Figure 1.18**(c). The transfer characteristics of depletion type *n*-channel and *p*-channel MOSFETs are shown in **Figure 1.18** (d) and (e) respectively.

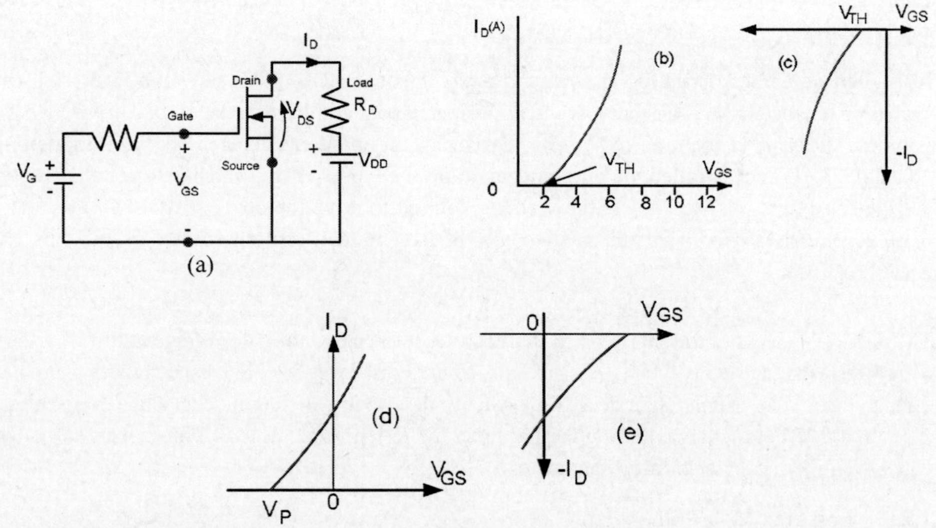

Figure.1.18 (a) Circuit diagram of an n-channel enhancement type MOSFET (b) transfer characteristics n- channel enhancement type MOSFET, (c) transfer characteristics of *p*- channel enhancement type MOSFET, (d) transfer characteristics *n*- channel depletion type MOSFET, (e) transfer characteristics *p* - channel depletion type MOSFET

(b) Output characteristics: The output characteristics of power MOSFET is the plot between drain current, I_D and drain to source voltage, V_{DS}. The plot is shown in **Figure.1.19(a).** The variation of I_D is the function of V_{GS}. It is seen from the figure that at a low value of V_{DS}, I_D increases, and the graph is linear. Thus, the value of drain to source resistance R_{DS} is constant and given by equation **(1.24).**

$$R_{DS} = \frac{V_{DS}}{I_D} \quad \textbf{(1.24)}$$

With an increase in V_{DS}, for a particular value of gate to source voltage V_{GS}, the plot is almost flat. Hence, the I_D is almost constant. A load line is also shown in **Figure.1.19(a)**. The load line intersects the output characteristics corresponding to gate to source voltage, V_{GN} at point *P* and it cut the V_{DS} axis at *Q*. The point *P* corresponds to the fully on state and *Q* corresponds to the fully off state. Like BJT, MOSFET also operates as a switch when it is either at *P* or *Q*.

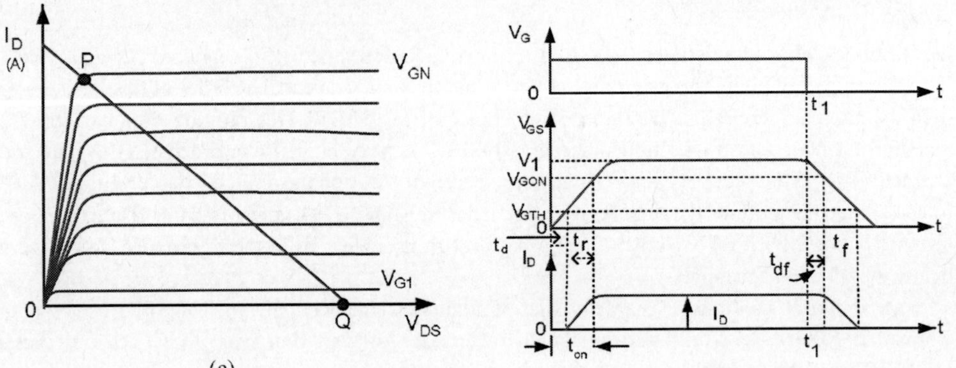

(a) (b)

Figure.1.19 Output and switching characteristics of a *n*-channel enhancement type MOSFET (a) Output characteristics, (b) Switching characteristics

(c) Switching Characteristics: The switching waveform of MOSFET is shown in Figure.1.18(b). This switching characteristic is affected by internal capacitances and impedance of gate drive circuit. There is a time delay, t_d at the time of starting. During t_d, the input capacitance charges to gate threshold voltage, V_{GTH}. The t_d is called the turn on delay time. The time during which gate voltage rises to V_{GON} is called rise time, t_r. The V_{GON} is the voltage required to drive the device (MOSFET) on state. In t_r, drain current rises from 0 to full on state current I_D. The total turn-on voltage (t_{on}) is thus given by equation (1.25).

$$t_{on} = t_d + t_r \tag{1.25}$$

The turn-off process of the MOSFET is started after the removal of gate voltage at time t_1 as shown in **Figure.1.18(b).** The delay time (t_{df}) is the time taken to discharge the input capacitances from overdrive voltage V_1 to V_{GON}. After that, a delay time called fall time in which input capacitances discharge from V_{GON} to a threshold voltage. During the fall time, the I_D falls from I_D to 0. Thus, when V_{GS} is less than equal to V_{GTH}, the device turn-off completely.

1.7.5 Losses in MOSFET

Following are the losses taking place in a MOSFET

(a) Turn-on switching loss
(b) On state loss or conduction loss
(c) Turn off switching loss
(d) Off state loss

(a) Turn-on switching loss: This loss comprises losses that occur during the delay time (t_d) and rise time (t_r). While calculating this loss separately the loss during the delay time and rise time and finally added to get the total turn-on loss. Mathematically after derivation, the average power loss during turn-on is given by equation **(1.26),** where fs is the switching frequency.

$$P_{on} = \frac{V_{DS(\max)} \cdot I_D \, t_r}{6} \cdot f_s \quad (1.26)$$

(b) On state loss or Conduction loss: This loss takes place during the conduction period (t_n). The average power loss during this period is given by equation **(1.27),** where T is the time period.

$$P_n = I_D^2 . R_{DS(on)} \cdot \frac{t_n}{T} = I_D^2 . R_{DS(on)} \cdot t_n \cdot f_s \quad (1.27)$$

(c) Turn off switching loss: This loss includes the loss during the discharge time required for the capacitors (t_{df}) and fall time (t_f). Mathematically after derivation, the power loss during switching off (P_{off}) is given by equation **(1.28).**

$$P_{off} = \frac{V_{DS(\max)} \cdot I_D \cdot t_f \cdot f_s}{6} \quad (1.28)$$

(d) Off state loss: The off-state loss of MOSFET occurred during an off period. It is given by equation **(1.29).**

$$P_o = V_{DS(\max)} \cdot I_{DSS} \cdot t_o \, f_s \quad (1.29)$$

The total loss is given by equation **(1.30)**

$$P_{TOTAL} = P_{on} + P_n + P_{off} + P_0 \quad (1.30)$$

At a low value of switching frequency, the power loss in MOSFET is higher than that of BJT. With an increase in switching frequency, the power loss in MOSFET is less than that of BJT. It is a reason why MOSFET is preferred over BJT at a higher switching frequency.

1.7.6 Equivalent circuit of MOSFET

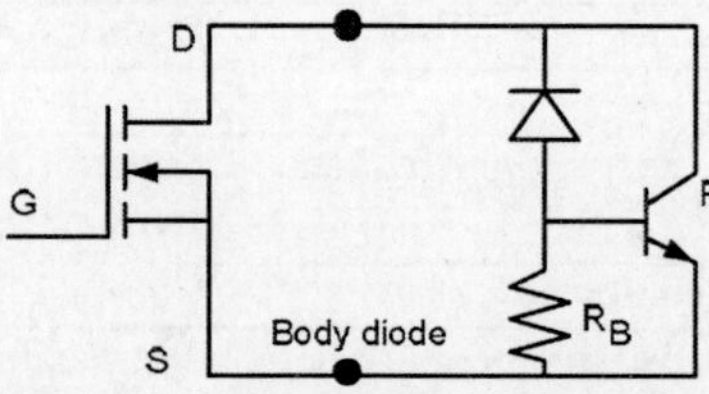

Figure.1.20 MOSFET equivalent circuit

The equivalent circuit of power MOSFET is shown in **Figure.1.20.** There is a pn^-n^+ diode simply called the p-i-n diode which is between p region and drain region is also called body diode. This diode allows reverse current to flow when the MOSFET is in off state. The p-region is in contact with the source metal. The integral diode is also shown. The MOSFET has a parasitic BJT (n^-pn^+) in its structure. The body is serves as base, source as the emitter, and the drain is the collector of this BJT. The body (base of the BJT) is shorted to (source n^+) emitter to facilitate the parasitic BJT cut-off all the time. The resistance of the p-region is denoted by R_B in Figure.1.20. With the application of the drain voltage nearer to avalanche voltage, current flows into p-region of BJT in addition to normal MOSFET current through the n-channel. There is a voltage drop in R_B which is appearing as forward biasing voltage across the base to emitter junction. When this drop is beyond 0.7V, parasitic BJT is no longer capable of supporting p-base/n^- collector breakdown voltage. During turn-off transient reverse, reverse peak current flows through the body diode and R_B. The voltage drop across R_B makes the B_E junction of the parasitic BJT to operate in the cut-off state. This results in further increase in reverse peak current. Thus, reverse recovery time also increases.

1.7.7 Safe operating area of MOSFET

The safe operating area (SOA) of a MOSFET is decided by the maximum drain current ($I_{D(max)}$), maximum power dissipation (P_{max}), and forward blocking voltage (V_{DSS}). The $I_{D(max)}$ is limited by the power dissipation limit at low value of drain voltage. On the other hand, the V_{DSS} is limited by the avalanche breakdown phenomenon at low value of drain current. The P_{max} is limited by the maximum permissible temperature (T_{jmax}). **Figure.1.21** shows the SOA of MOSFET.

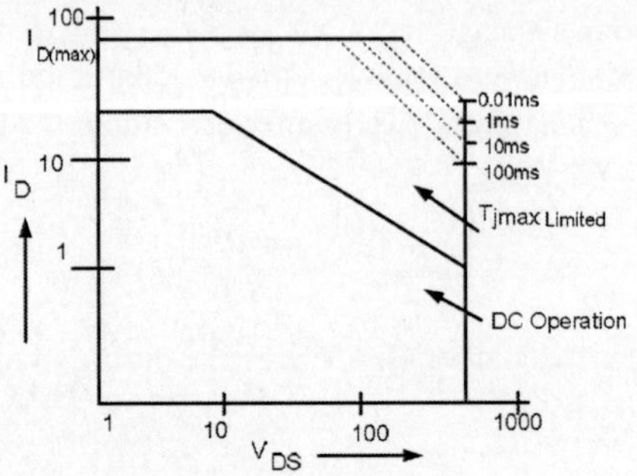

Figure.1.21 SOA of MOSFET

1.7.8 Comparison of Power MOSFET and Power BJT

The comparison of MOSFET and BJT are shown in **Table.1.1**

Table.1.1 Comparison of MOSFET and BJT

Sl No	Power MOSFET	Power BJT
1	Lower switching loss	Higher switching loss
2	High on state loss	Low on-state loss
3	First choice for high-frequency application	Superior for lower operating frequency (10 to 30 kHz) application
4	Voltage controlled device	Current control device
5	It has a positive temperature co-efficient of resistance	Temperature co-efficient of resistance
6	Secondary breakdown not occurs.	Secondary breakdown occurs
7	Very high input impedance	Low Input impedance
8	It is the majority carrier device	It is majority as well as minority carrier device.
9	Peak current capability more than BJT	Peak current capability less than MOSFET
10	More sensitive to a voltage spike	Less sensitive to voltage spike
11	On state voltage is higher than BJT	On state voltage is lower than MOSFET
12	Lower on state loss	Higher on state loss
13	Conduction loss higher	Conduction loss lower

1.7.9 Uses of MOSFET

The IGBTs are used in various applications. Some of them are listed below.

(a) Computers

(b) Uninterrupted power supplies (UPSs)
(c) Switched mode power supply (SMPS)
(d) High frequency-based Inverters
(e) Motor control

1.8 INSULATED GATE BIPOLAR TRANSISTOR

1.8.1 Introduction

The insulated gate bipolar transistor (IGBT) is a newly developed semiconductor device. It is also known as a metal oxide insulated gate transistor (MOSIGT), conductively – modulated field effect transistor (COMFET) or gain-modulated field effect transistor (GEMFET). IGBT combines the advantages of both MOSFET and BJT. It has three terminals called gate, emitter, and collector. It has high input resistance like MOSFET and low on-state power loss like BJT. Also, the second breakdown problem is not present in IGBT. The cost of IGBT is somewhat higher than BJTs but it is popular because of its lower switching losses as well as smaller snubber circuit requirement.

1.8.2 Construction of IGBT

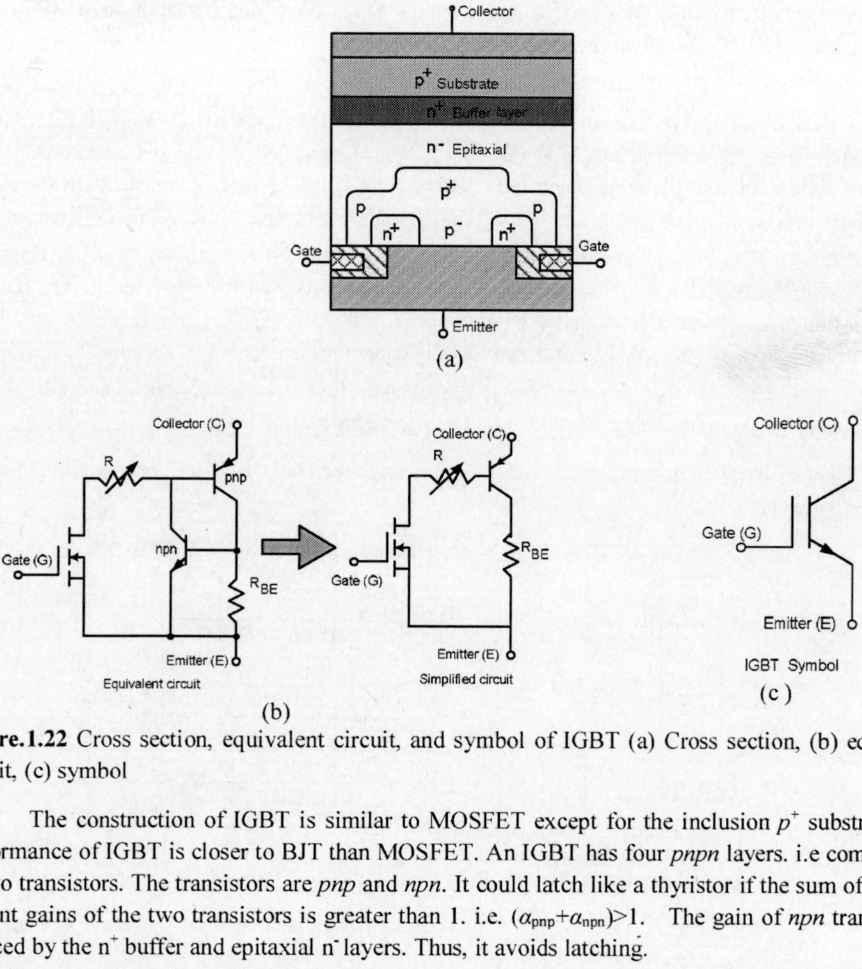

Figure.1.22 Cross section, equivalent circuit, and symbol of IGBT (a) Cross section, (b) equivalent circuit, (c) symbol

The construction of IGBT is similar to MOSFET except for the inclusion p^+ substrate. The performance of IGBT is closer to BJT than MOSFET. An IGBT has four *pnpn* layers. i.e combination of two transistors. The transistors are *pnp* and *npn*. It could latch like a thyristor if the sum of forward current gains of the two transistors is greater than 1. i.e. $(\alpha_{pnp}+\alpha_{npn})>1$. The gain of *npn* transistor is reduced by the n^+ buffer and epitaxial n^- layers. Thus, it avoids latching.

There are two structures of IGBT. They are punch-through denoted by PT and non-punch-through, denoted by NPT. In PT, the switching time is reduced by using n^+ layer in the drift region nearer to the collector. The carrier life of NPT is kept more than PT type IGBT. With this provision, conductivity modulation is established in the drift region. The structure of an IGBT is shown in **Figure.1.22 (a).** The equivalent circuits and the symbol of IGBT are shown in **Figure.1.22 (b)** and **(c)** respectively.

1.8.3 Working principle of IGBT

An IGBT is a voltage control device like MOSFET. When the gate is provided with positive voltage with respect to the emitter and gate-emitter voltage is more than the threshold voltage, the *n*-channel is created in *p*-region. The n^+ and n^- regions are short circuited by this channel. The movement of electrons through this n-channel results in an injection of hole from p^+ substrate to the n^- epitaxial layer. Thus, a forward current is established. A *pnp* transistor is formed with p^+, n^-, p. In this transistor p^+ is the emitter, n^- is the base, and p is the collector region. A *npn* transistor is formed with n^-, p, n^+ layers. The p serves as a collector for the *pnp* and base for the *npn* transistor.

1.8.4 Symmetric and asymmetric IGBTs

Based on n^+ a buffer, IGBTs are classified as (a) Symmetric IGBT or non-punch through (NPT) IGBTs, (b) Asymmetric IGBTs or punch-through (PT).

(a) Symmetric IGBT or non-punch-through (NPT) IGBTs

The Symmetric IGBT or non-punch through (NPT) IGBTs are shown in **Figure.1.23.** There are three junctions J_1, J_2, J_3. The blocking capacity of the device depends on the conditions of these junctions. In the forward blocking operation (**Figure.1.23(a)**), the collector is positive with respect to the emitter, and junctions J_1 and J_3 are forward biased and J_2 is reverse biased. This J_2 junction decides the blocking capacity. The J_2 junction is in between p region (moderate doping) and n^- drift region (light doping). The thickness of decides the blocking capability. On the other on the application of reverse voltage i.e collector is negative with respect to the emitter, junctions J_1 and J_3 are reversed biased and J_2 is forward biased. The reverse blocking capability is decided by reversed biased J_1 and J_3 junctions. Since the emitter n^+ layer is heavily doped and p-layer moderately doped and thin, J_1 do not have capacity to block. Because of the existence of n^- drift region J_3 can block the reverse voltages. The n^- layer is designed to block large voltages and hence the NPT structure is compatible to forward blocking voltage.

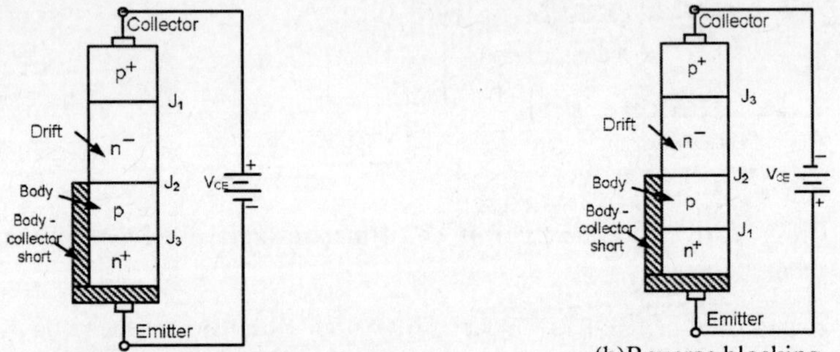

(a) Forward blocking (b)Reverse blocking

Figure.1.23 Symmetric or non-punch through (NPT) type IGBT

(b) Asymmetric IGBT or punch-through (PT) IGBTs

In this type of IGBT there is n^+ buffer layer in between the collector and n^- drift layers. The structure of PT type IGBT is shown in **Figure.1.23.** In forward biased collector is positive with respect to emitter and in case of reverse biased collector is negative with respect to emitter. When forward biased, the junctions J_1 and J_3 are forward biased and J_2 is reverse biased and J_2 decides the blocking capacity. This J_2 is in between *p*- layer and n^- drift region. The thickness of the drift region decides forward blocking capacity. In reversed conditions, junctions J_1 and J_3 are reverse biased and J_2 is forward biased. Junction J_1 cannot block the reverse voltage because the emitter n^+ layer is heavily doped and *p*-layer is moderately doped and thin, J_1 cannot block. Because of the presence of n^+ buffer layer J_3 also do not have the capacity to block large reverse voltage. Thus, PT structure have asymmetric blocking capacity.

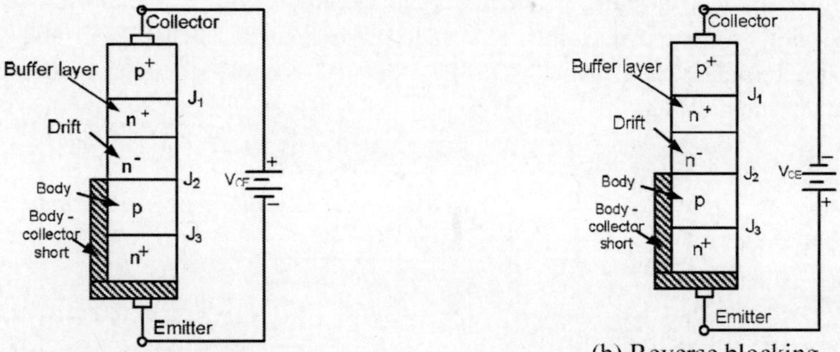

(a) Forward blocking (b) Reverse blocking

Figure.1.24 Asymmetric or punch through (PT) IGBT

1.8.5 Equivalent circuit of IGBT

An equivalent circuit of an IGBT is shown in **Figure.1.25.**

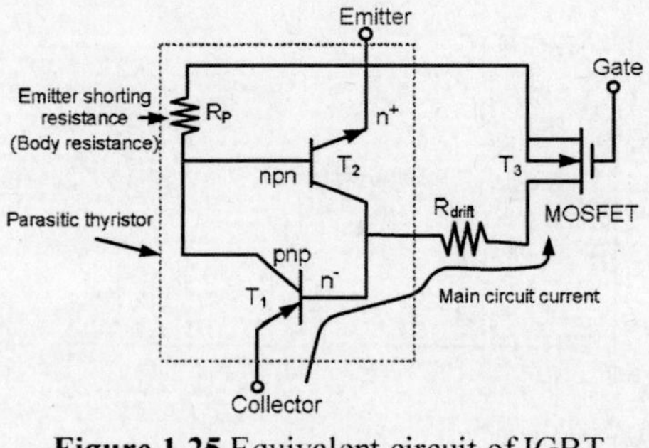

Figure.1.25 Equivalent circuit of IGBT

It consists of a transistor a coupled *pnp* and *npn* transistor. The transistor represents a four-layer parasitic thyristor structure with a MOSFET shunting the upper *npn* transistor. The structure may give rise to the latching up of the device either high current mode or low current mode. The emitter of transistor T_1 is the collector of the IGBT. The emitter of T_2 is the emitter of IGBT which connected through the emitter shorting resistance R_P to the collector of T_1. The MOSFET M_3 is shown because input side of the IGBT is like MOSFET. The resistances R_P and R_{drift} represent the body spreading and drift region resistance. R_{drift} is considerably large because the drift region is lightly doped and have low conductivity. When R_P is significant, it turned on the transistor T_2. This will turn-on the parasitic thyristor structure and bring to the on state. In this way latch-up takes place in IGBT. After the parasitic thyristor is latched on, it cannot be turned off since the gate do not have any control. The IGBT can be turned off by forced commutation similar to silicon controlled rectifier. To prevent turning on the parasitic thyristor, R_P should be kept minimum.

1.8.6 V-I characteristics of IGBT

The circuit diagram of an IGBT is shown in **Figure.1.26 (a).** The considered IGBT is of *n*-channel type.

1.8.6.1 Static characteristics of IGBT

The static V-I or output characteristics of an IGBT is the plot of collector current (I_C) versus collector-to-emitter voltage (V_{CE}) corresponding to a particular gate-to-emitter voltage. The plot is shown in **Figure.1.26 (b).** In the forward direction, the shape of the plot is similar to the bipolar junction transistor. However, the controlling parameter is gate to emitter voltage. The transfer characteristics of IGBT is the plot between collector current and gate-to-emitter voltage (V_{GE}) and it is shown in **Figure.1.26 (c).** It is identical to that of MOSFET. An IGBT is in off-state if the gate-to-emitter voltage is less than a threshold voltage.

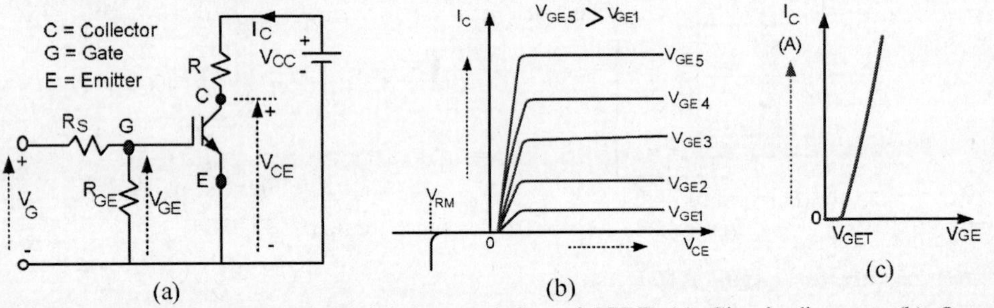

Figure.1.26 Circuit diagram and Static characteristics of IGBT (a) Circuit diagram, (b) Output characteristics, and (c) transfer characteristics

1.8.6.2 Switching characteristics of IGBT

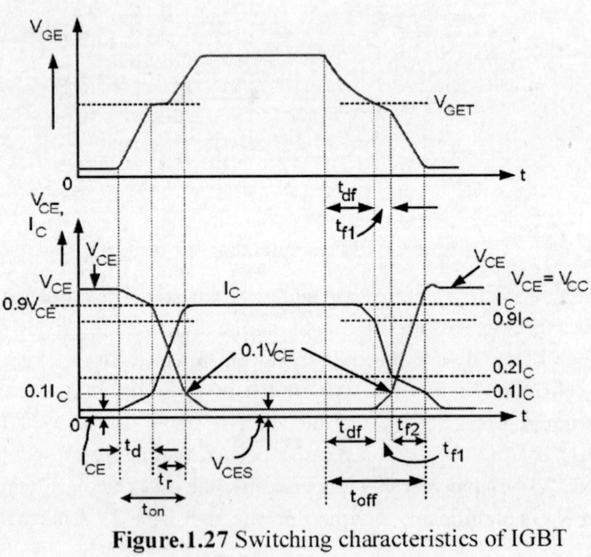

Figure.1.27 Switching characteristics of IGBT

The switching characteristics of IGBT are drawn in Figure.1.27. The turn-on time (ton) is the time between the instants of forward blocking to forward on state and it is the sum of delay time (td) and the rise time (tr). The delay time is the time taken for the initial collector- emitter voltage (VCE) falls from VCE to 0.9VCE or the time taken by the collector current to rise from its initial leakage current (ICE) to 0.1IC where IC is the final value of collector current. The rise time (tr) is the time in which VCE falls from 0.9VCE to 0.1VCE or the time taken by the collector current to rise from 0.1IC to its final value (IC). After the turn-on time collector current reached to the final value and the collector-emitter voltage falls to VCES, a small value called saturated value. The turn-off time consists of three parts. They are delay time (tdf), initial fall time (tf1), and final fall time (tf2). During the delay time gate voltage falls from VGE to the threshold voltage, VGET. The collector current also falls from IC to 0.9IC during delay time tdf. At the end of tdf, the collector-emitter voltage rises. The tf1 is the time at which collector current falls from 90% to 20% of the initial value, IC. The time tf2 is the time taken by the collector current to fall from 20% to 10% of the initial value, IC.

1.8.7 Safe operating area (SOA) of IGBT

The forward bias safe operating area (FBSOA) and reverse bias safe operating area (RBSOA) of IGBT are robust and shown in **Figure.1.28 (a) and (b)** respectively. FBSOA is identical to that of MOSFET. The FBSOA is a square shape in case of short switching time. For longer switching time the shape is not thermally limited. The RBSOA is different from that of FBSOA. The upper right-hand corner is progressively cut out and RBSOA becomes smaller as $\frac{dV_{CE}}{dt}$ becomes larger. The RBSOA is a function of $\frac{dV_{CE}}{dt}$ avoiding latch up. A large value of $\frac{dV_{CE}}{dt}$ during turn-off will cause latching up the IGBT like thyristor. With proper selection of V_{CE} and gate drive resistance, the user can easily control the reapplied $\frac{dV_{CE}}{dt}$.

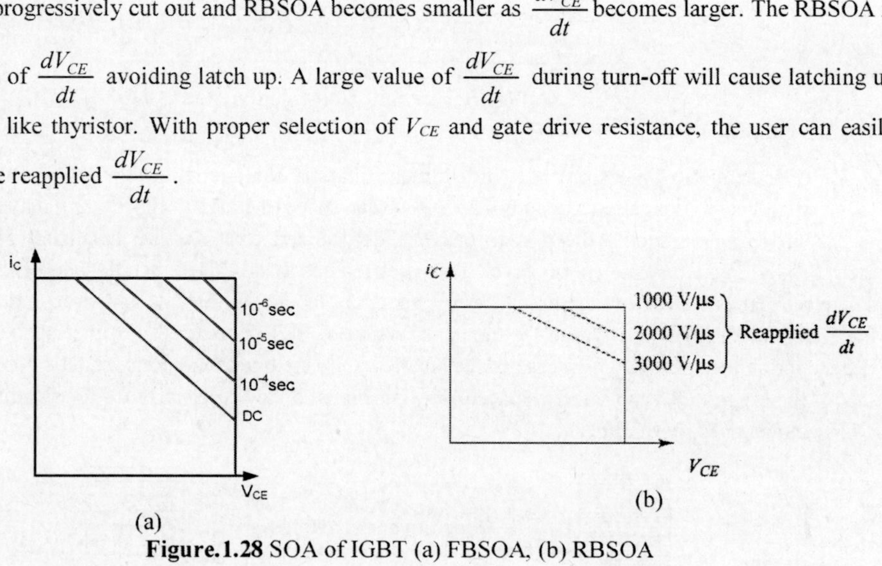

Figure.1.28 SOA of IGBT (a) FBSOA, (b) RBSOA

1.8.8 Uses of IGBT

The IGBTs are used in medium power applications. Examples of such applications are given below.

(a) DC and AC motor drive.
(b) UPS systems
(c) Power supplies
(d) Drives for solenoids, relays, and contactors.
(e) Converters

The IGBT is more expensive than BJT. But because of lower switching requirements, lower switching losses, and smaller snubber circuit requirements it very much more popular than BJT. The converters with IGBT are more efficient, smaller in size, and low in cost than the converters with BJTS.

1.9 CONCEPT OF SINGLE ELECTRON TRANSISTOR

1.9.1 Introduction

Single electron transistor (SET) is a nanodevice that utilizes the quantum phenomena. It can perform the role of a switch or as an amplifier like field effect transistor (FET). It has three terminals and they are drain, source, and gate. The SET consists of a metallic island placed between two junctions called tunnelling junctions. The tunnelling junction is nothing but a thin oxide layer between the island and the electrodes. The quantum dots are also used as an island. The construction of SET is shown in Figure.1.29

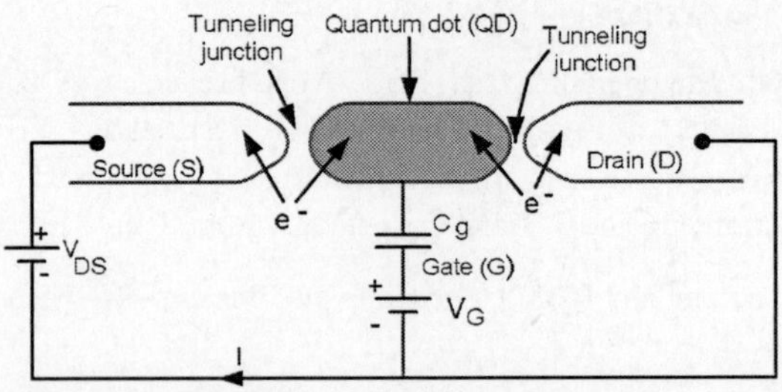

Figure.1.29 SET Construction

1.9.2 Quantum Dot

When all the dimensions of a three-dimensional material are reduced to the nano meter range, the resultant material is called a quantum dot (QD). The shape of QD may cubical or it may be spherical. In the process of reducing the dimensions, the delocalized electrons are localized. The localized electrons fall into a discrete energy level. This is called quantized. Thus, the QD behaves like an atom. The electrical as well as optical properties of this atom depend on the size and structural geometry of the QD. The metallic QD is used for the construction of SET. The density of the localized electrons dependent on the size of QD. In case of the smallest QD, the quantum energy state is compressed and that of the biggest QD, the same is spaced. The density of localized electronics for smaller and bigger QD is shown in **Figure.1.30**

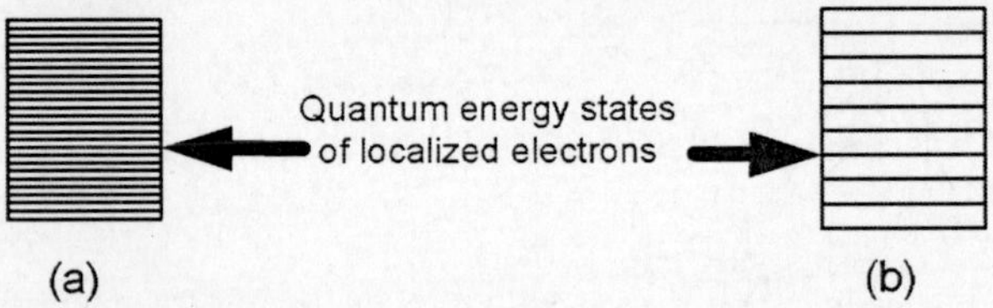

Figure.1.30 Quantum energy state (a) Smallest QD, (b) Biggest QD

1.9.3 The Coulomb Blockade and Working of SET

The SET has a metallic island. Instead of this island the quantum dot is also used. The island (or QD) is placed in between two tunnelling junctions. One junction connects the island (or QD) and the source (S) and the other connects the drain (D) and the island (or QD). The tunnelling junctions consist of a layer of oxide. The schematic diagram of the core of a SET shows the junctions, island (or

QD), the electrodes called drain (D) and source (S), and the gate (G) in **Figure.1.31(a).** The circuit diagram of the SET is shown in **Figure.1.31 (b).** Though the island of the SET is very small in the order of nanoscale, it contains large numbers of electrons ($\approx 10^9$). Through tunnelling, one can add or remove electrons from this island. When the electrons are added to the island are called excess electrons. When the electrons are removed from the island leaving a positive charge (may be called an excess hole). Let the charge of the island is Q_{QD}, the total capacitance of the island is C_T, the capacitance between the source and island is C_S, capacitance between drain and island, C_D, the charge of an electron is e, n is the number of excess electrons, then the charging energy of the SET is given by (1.31) in which C_T is given by (1.32).

$$E_{CH} = \frac{1}{2}.\frac{Q_{QD}^{\ 2}}{C_T} = \frac{1}{2}.\frac{n^2 e^2}{C_T} \tag{1.31}$$

$$C_T = C_G + C_S + C_D \tag{1.32}$$

The energy scale (E_C) of SET is defined by equation (1.33). This energy is not dependent on the Q_{QD} but on the induced charges by the gate.

$$E_C = \frac{e^2}{2C_T} \tag{1.33}$$

If the V_G is the gate voltage, the gate charge is given by equation (1.34).

$$Q_G = V_G C_G \tag{1.34}$$

If the number elementary gate charge is n_g, the electrostatic energy (E_{els}) of the system is given by (1.35).

$$E_{els} = E_C(\mathrm{n} - \mathrm{n}_g)^2 \tag{1.35}$$

From (1.33) and (1.35), the electrostatic energy (E_{els}) is given by (1.36).

$$E_{els} = \frac{1}{2}.\frac{Q^2}{C_T} = \frac{1}{2}.\frac{(ne - V_G C_G)^2}{C_T} = \frac{1}{2}.\frac{(\mathrm{ne} - Q_G)^2}{C_T} \tag{1.36}$$

The energy as defined in (1.36), decides whether the tunnelling junctions are forbidden or allowed. When an electron is added, the excess electrons increased the energy of the system, the tunnelling is energetically forbidden. In this case, the coulomb charging energy will act as a blockade. This is called a coulomb blockade.

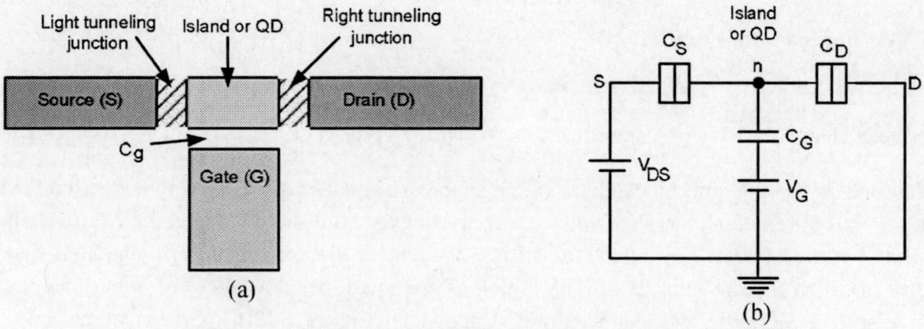

Figure.1.31 (a) Core of SET, (b) Circuit schematic of SET

Two cases are possible. Firstly, if there are n excess electrons on the QD and tunnelling of one electron causes an increase in energy as per equation (1.36). Thus, the system has n+1 electrons and QD is energetically forbidden. Hence, there is no tunnelling across the junctions. This is nothing but a coulomb blockade and, it is said as active. Secondly, say tunning of extra electrons on the QD, result in lowering the energy, and in this case, there is no coulomb blockade. Thus, tunnelling will be happened by adding an extra electron to the QD. Here, the charging of the QD negative. The same principle can be applied if one wishes to subtract electrons from QD and the charging the QD is positive. The energy on the QD before junction is determined by the drain to source voltage (V_{DS}) (shown in **Figure.1.31 (b)**). When this energy is higher than the coulomb blockade, tunnelling occurs. The gate charge and the excess electrons on the QD decide the height of the coulomb blockade.

1.9.4 The working of SET

The electrical behaviour of SET is governed by the elementary charges and the gate voltage. Based on these, the current may be flow or not flow between the source and drain. When energy on the QD overcomes the coulomb charging energy i.e. the coulomb blockade, one electron will tunnel from the source to the QD. Thus, one electron is added. In the same way the tunnelling process will occur from QD to the drain.

The tunnelling junctions are nothing but a thin insulating barrier between two conducting electrodes. In this case, the electrodes are source (S) and drain (D). These insulating barriers do not allow current to flow from S and D. However, as per quantum mechanics there is a probability (it greater than unity) for an electron located at one side (say source side) of the barrier to other side. This transfer of electrons through the barrier yield in the charging of neighbouring QD. This results in an increase in electrostatic energy given by (1.33) and (1.36). This electrostatic energy is nothing but the coulomb blockade energy. This energy is the repelling energy of the previously present energy in the QD which prevent next electron moving towards QD. The capacitance C_T, given by equation (2) is very small. As a result, as per equation (1.33), the coulomb charging energy is very high and hence electrons cannot move simultaneously but pass the barrier one by one. The suppression of electron transfer is termed as coulomb blockade tunnelling. The suppression of electrons can be eliminated by two possible ways. They are

(a) thermal excitation to overcome coulomb charging energy.
(b) externally applied voltage to overcome coulomb charging energy.

There is a voltage called threshold voltage (V_t) at which energy of electrons just above the coulomb charging energy, and current can start flowing through the tunnelling junction. When the applied voltage exceeds V_t, electrons tunnel through the junction to the QD. The applied voltage is the applied voltage between source to drain.

1.9.5 V-I characteristics of SET

The V-I characteristic of a SET is shown in **Figure.1.32**. This characteristic is the plot of current through the SET and the voltage applied across. When the voltage, $|V|$ is less than $\frac{e}{C_T}$, the current is zero. This state is the coulomb blockade at which suppresses the tunnelling of single electron for low bias condition. If the externally applied voltage increases beyond threshold voltage, there is no coulomb blockade and current flows. The behaviour of the junctions is like resistance. The junctions may be symmetric in which (i) capacitances of both junctions are equal, (ii) resistances of both junctions are equal. For such a SET, the V-I characteristic is shown in **Figure.1.32(a)**. On the other hand, for asymmetric junction, the V-I characteristic is like a staircase as shown in **Figure.1.32(b)**.

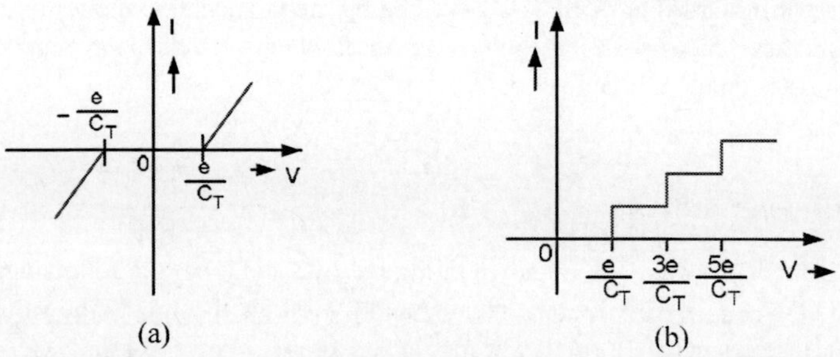

Figure.1.32 V-I characteristic of SET (a) for symmetric junction, (b) for asymmetric junction.

1.9.6 Application of SET

SET can be used in the following applications

(a) As a charge sensor.
(b) For detection of Infrared radiations.
(c) Ultrasonic microwave detector.
(d) Supersensitive electrometer
(e) Single electron spectroscopy
(f) Programmable single electron transistor logic

Unit Summary

This unit explores the structure, operating principle, V-I characteristics and switching characteristics of some selected power electronic devices such power diodes, power bipolar transistor (BJT), metal oxide semiconductor field effect transistor (MOSFET), Insulated gate bipolar transistor (IGBT) and single electron transistor (SET). The important summary of this chapter is given below.

1. The ratings of the power semiconductor devices such as a diode, BJT, etc. are much higher than that are used in an electronics circuit.
2. The power BJT is a three terminal and three-layer power semiconductor device. The terminal are emitter, base and collector.
3. There are two types of Power BJT. They are *npn* type and *pnp* type. There are various circuit configurations of each BJT. They are common base, common emitter, and common collector. The common emitter configuration is normally used.
4. Power BJT is normally off device. It can be turned on by the application of sufficient base current.
5. Power BJT ha low current gain and a larger breakdown voltage rating.
6. SOA of BJT is limited by second breakdown. The RBSOA is normally the limiting factor.
7. MOSFET have a vertically oriented structure. The terminals are drain, source and gate.
8. The MOSFET is normally off device, it is turned on by the application of a sufficiently large gate to source voltage.
9. MOSFET turns on and turns off very rapidly because it is a majority carrier device.
10. On-state losses of MOSFET rises faster with blocking voltage rating than BJT.
11. MOSFET has positive temperature coefficient of resistance.
12. Performance of IGBT is midway between MOSFET and BJT.
13. The turn-on speed of IGBT can be controlled by the rate of change of gate to source voltage.
14. Single electron transistor (SET) is a nano device that utilizes the quantum phenomena.

15. The electrical behaviour of SET is governed by the elementary charges and the gate voltage. Based on these, the current may be flow or not flow between the source and drain.
16. SET operates on the principle of coulomb blockade.

Exercises

Example.1.1

The circuit diagram of a power BJT is shown in **Figure.1.33** and it has the following parameter: V_{CC}= 200 V, R_C= 25 Ω, $V_{CE\,(sat)}$=1.2 V, $V_{BE\,(sat)}$=1.4 V and β = 10, then compute the value of (a) Collector current (Ic), (b) Base current (IB), (c) Power loss in collector (d) Power loss in base.

Solution:

(a) $I_C = \dfrac{V_{CC} - V_{CESAT}}{R_C} = \dfrac{200 - 1.2}{25} = 7.952 \text{ A}$

(b) $I_B = \dfrac{I_C}{\beta} = \dfrac{7.952}{10} = 0.7952 \text{ A}$

(c) $P_C = V_{CESAT} \times I_C = 1.2 \times 7.952 = 9.542 \text{ W}$

(d) $P_B = V_{BESAT} \times I_B = 1.4 \times 0.7952 = 1.113 \text{ W}$

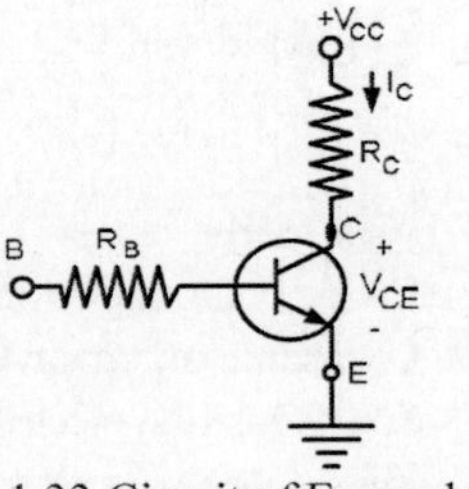

Figure.1.33 Circuit of Example 1.1

Example.1.2

The switching characteristics of a power transistor are shown below and it has the following data: V_{CC} = 200V, V_{CESAT}= 2 V, I_{CSAT}= 100A, I_{CEO}= 3 mA, t_d= 0.3 µS, T_r= 2 µS, switching frequency (f) = 10 kHz, Calculate the average power loss due to collector current during turn on.

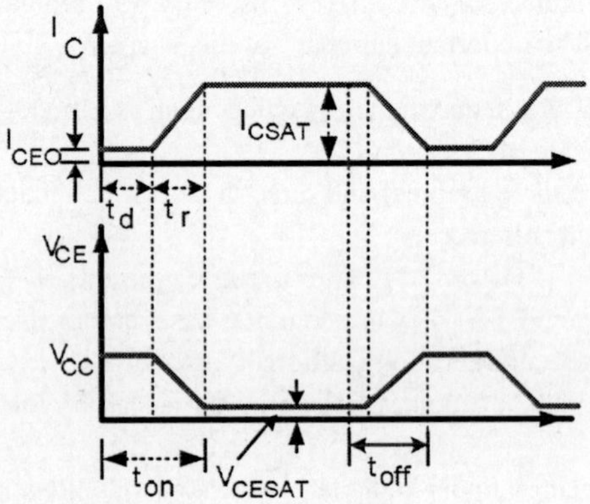

Figure.1.34 *Waveforms related to example1.2*

Solution:

Average power loss during turned on = Average power loss during delay time t_d + Average power loss during rise time t_r. Thus, for $0 \le t \le td$, the average power loss during delay time is given by

$$P_d = \frac{1}{T}\int_0^{t_d} i_C(t)\times v_{CE}(t)dt$$

$$= \frac{1}{T}\int_0^{t_d} I_{CEO}.V_{CC}dt = f\times I_{CEO}\times V_{CC}.td$$

$$= 10 \times 10^3 \times 3 \times 10^{-3} \times 200 \times 0.3 \times 10^{-6}$$
$$= 1.8 \text{ mW}$$

For $0 \leq t \leq t_r$

$$i_C(t) = \frac{I_{CSAT}}{t_r} \times t$$

$$v_{CE}(t) = \left[V_{CC} - \frac{V_{CC} - V_{CESAT}}{t_r} \times t \right]$$

The average power loss during rise time is given by

$$P_r = \frac{1}{T} \int_0^{t_r} \frac{I_{CSAT}}{t_r} \cdot t \left[V_{CC} - \frac{V_{CC} - V_{CESAT}}{t_r} \cdot t \right] dt$$

$$= f \cdot I_{CSAT} \cdot t_r \left[\frac{V_{CC}}{2} - \frac{V_{CC} - V_{CESAT}}{3} \right]$$

$$= 10 \times 10^3 \times 100 \times 2 \times 10^{-6} \left[\frac{200}{2} - \frac{200 - 3}{3} \right]$$

$$= 68.6666 \text{ W}$$

∴ Total average power loss during turns on of the transistor $= P_d + P_r$

$$= 0.0018 + 68.6666$$
$$= 68.6684 \text{ W}$$

Example.1.3

A MOSFET switch is used to control the power of a resistive load (R_L) = 15 Ω, having a supply voltage V_S = 160 V, $R_{DS(ON)}$ = 0.2 Ω, t_r = 1.6 μS, duty cycle = 0.7 and the switching frequency is 20 kHz. (a) Compute the power loss in the on state, (b) Compute the power loss during the turn-on interval.

Solution:

(a)

$$I_D = \frac{V_S}{(R_L + R_{DS(ON)})}$$

$$= \frac{160}{(15 + 0.2)}$$

$$= 10.526 \text{ A}$$

Switching period (T) is given by,

$$T = \frac{1}{f}$$

$$= \frac{1}{20 \times 10^3}$$

$$= 50 \mu S$$

On-time(t_{ON}) = duty cycle (d) × Switching period (T)

$$= 0.7 \times 50 \ \mu S = 35 \ \mu S$$

Energy loss during on time (W_{ON}) = $I_D^2 \times R_{DS(ON)}) \times t_{ON}$

$$= (10.526)^2 \times 0.2 \times 35 \times 10^{-6}$$
$$= 775.623 \ \mu J$$

Power loss during on time (P_{ON}) = $W_{ON} \times f$

$$= 775.623 \times 10^{-6} \times 20 \times 10^3$$
$$= 15.512 \text{ W}$$

(b) Energy loss during turn-on (W_{ON})

$$\frac{V_{DS(max)}I_D t_r}{6}$$

$$= = \frac{160 \times 10.526 \times 1.6}{6}$$

$$= 449.10\ \mu J$$

Power loss during turn-on (P_{ON}) = $W_{ON} \times f$

$$= 449.10 \times 10^{-6} \times 20 \times 10^3$$

$$= 8.98\ W$$

Example.1.4

An IGBT switching circuit is shown in Figure.1.35 The supply voltage is 200 V and the load resistance is 10 Ω. The switching frequency of IGBT is 2 KHz. Find the time required for the pulse if the power required for the load is 5 kW.

Solution:

$$\text{Time period (T)} = \frac{1}{f} = \frac{1}{2 \times 10^3} = 0.5\ mS$$

$$V_{L(avg)} = \frac{V_S t_{ON}}{T}$$

$$P_L = \frac{(V_{L(avg)})^2}{R_L}$$

By using above two equations:

$$t_{ON} = \frac{P_L R_L T}{V_S^2} = \frac{5 \times 10^3 \times 10 \times 0.5 \times 10^{-3}}{200 \times 200} = 0.625mS$$

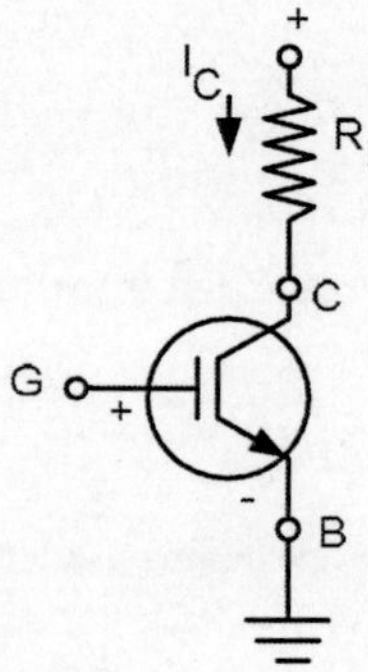

Figure.1.35 Circuit of example 1.4

Example.1.5

The IGBT switching circuit shown in Figure.1.33 has the following data: V_S = 200 V, R_L= 20 Ω, the duty cycle is = 70 %, switching frequency is 5 kHz, t_{ON}= 3 μS, t_{OFF} = 1.5 μS and $V_{CE(sat)}$ = 3.0 V. Calculate (a) Average load current, (b) Conduction power loss, (c) Power loss in watt during turn-on, (d) Power loss in watt during turn-off

Solution:

(a) Maximum value of load current, $I_{C(max)} = \frac{V_S - V_{CE(sat)}}{R_L} = \frac{200 - 3}{20} = 9.85A$

Average load current, $I_{C(avg)}$ = duty cycle × $I_{C(max)}$

$$= 0.7 \times 9.85$$

$$= 6.895\ A$$

(b) Conduction power loss = $V_{CE(sat)} \times I_{C(avg)}$

$$= 3 \times 6.895$$

$$= 20.685\ W$$

(c) Switching power loss during turn-on = $\frac{V_{CE(max)} \times I_{C(max)} \times t_{ON}}{6} \times \text{switching frequency}$

$$= \frac{200 \times 9.85 \times 3 \times 10^{-6}}{6} \times 5 \times 10^3$$

$$= 4.925\ W$$

(d) Switching power loss during turn-off = $\frac{V_{CE(max)} \times I_{C(max)} \times t_{OFF}}{6} \times \text{switching frequency}$

$$= \frac{200 \times 9.85 \times 1.5 \times 10^{-6}}{6} \times 5 \times 10^{3}$$

$$=2.462 \text{ W}$$

Example.1.6

The maximum and minimum values of current gain (β) of a BJT (shown in Figure.1.36) are 40 and 8 respectively. The value of load resistance connected in the collector circuit is 11Ω. The DC supply voltage in the collector emitter (V_{CC}) is 200V. The base voltage (V_B) is 10V. The saturated value of collector emitter voltage (V_{CESAT}) and that of base-emitter voltage (V_{BESAT}) are given as 1.0V and 1.5V respectively. Determine (i) resistance R_B, due to which saturation occurs with overdrive factor (ODF) equal to 5, (ii) forced current gain (β_F), (iii) BJT power loss.

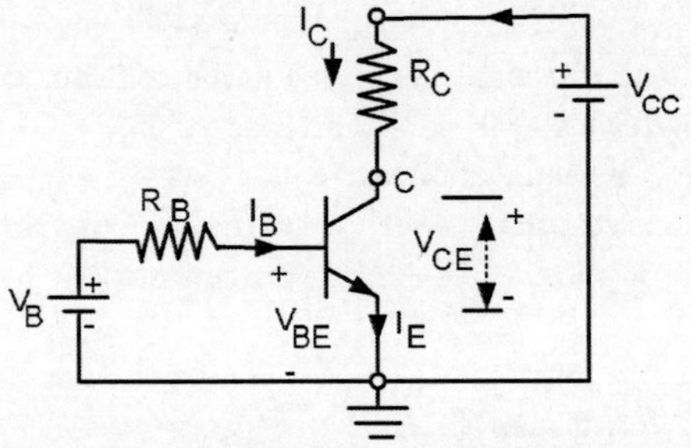

Figure.1.36 Circuit of example 1.6

Solution:

The saturated value of collector current, (I_{CSAT}) is given by

$$I_{CSAT} = \frac{V_{CC} - V_{CESAT}}{R_C} = \frac{200-1}{11} = 18.1 \text{ A}$$

The saturated value of base current (I_{BSAT}) is given by

$$I_{BSAT} = \frac{I_{CSAT}}{\beta_F} = \frac{18.1}{8} = 2.2625 \text{ A}$$

The base current is obtained by using overdrive factor (ODF),

$$I_B = ODF \text{ x } I_{BSAT} = 5 \text{ x } 2.2625 = 11.3125 \text{ A}$$

The value of R_B can be calculated from

$$R_B = \frac{V_B - V_{BESAT}}{I_B} = \frac{10-1.5}{11.3125} = 0.7514\Omega$$

The forced current gain can be obtained as follows.

$$\beta_F = \frac{I_{CSAT}}{I_B} = \frac{18.1}{11.3125} = 1.6$$

The total BJT power loss can be calculated from

$$P_{BJT} = V_{BE} I_B + V_{CE} I_C$$

$$\therefore P_{BJT} = 1.5 \text{ x } 11.3125 + 1.0 \text{ x } 18.1 = 35.07 \text{ W}$$

Example.1.7

The junction and ambient temperature of a BJT are 160°C and 25°C respectively. Thermal resistance between junction and case, case and sink, and sink and ambient are 0.4°C/W, 0.1°C/W, and 0.5°c/W respectively. Calculate (a) case temperature, (b) Maximum power dissipation.

Solution:

The given data are R_{JC}=0.4°C/W, R_{CS} = 0.1°C/W, R_{SA}= 0.5°c/W, T_J=160°C, T_A =25°C. It is required to find T_C and P_T

Now, $T_J - T_A = P_T(R_{JC} + R_{CS} + R_{SA}) = P_T R_{JA}$

Here, $R_{JA} = 0.4 + 0.1 + 0.5 = 1.0$, and $T_J - T_A = (160 - 25)^0 C = 135^0 C$

Hence, $P_T = 135 / 1.0 = 135W$

Again, $T_C = T_J - P_T R_{JC} = 160 - 135 \text{ x } 0.4 = 106^0 C$

Example.1.8

The maximum and minimum values of current gain (β) of a BJT (shown in **Figure.1.34**) are 75 and 12 respectively. The value of load resistance connected in the collector circuit is 1.5Ω. The DC supply voltage in the collector emitter (V_{CC}) is 40V. The base voltage (V_B) is 6V. The saturated value of collector emitter voltage (V_{CESAT}) and that of base-emitter voltage (V_{BESAT}) are given as 1.2V and 1.6V respectively. The value of resistance R_B is 0.7Ω. Determine (i) overdrive factor (ODF), (ii) forced current gain (β_F), (iii) BJT power loss.

Solution:

The saturated value of collector current, (I_{CSAT}) is given by

$$I_{CSAT} = \frac{V_{CC} - V_{CESAT}}{R_C} = \frac{40 - 1.2}{1.5} = 25.867 \text{ A}$$

The saturated value of base current (I_{BSAT}) is given by

$$I_{BSAT} = \frac{I_{CSAT}}{\beta_F} = \frac{25.867}{12} = 2.156 \text{ A}$$

The value of I_B can be calculated from

$$R_B = \frac{V_B - V_{BESAT}}{I_B}$$

$$0.7 = \frac{6 - 1.6}{I_B} = \frac{4.4}{I_B}$$

$$\therefore I_B = \frac{4.4}{0.7} = 6.286A$$

The Overdrive factor (ODF), can be calculated as follows

$$ODF = \frac{I_B}{I_{BSAT}} = \frac{6.286}{2.156} = 2.92 \approx 3$$

The forced current gain can be obtained as follows.

$$\beta_F = \frac{I_{CSAT}}{I_B} = \frac{25.867}{6.286} = 4.12$$

The total BJT power loss can be calculated from

$$P_{BJT} = V_{BE} I_B + V_{CE} I_C$$

$$\therefore P_{BJT} = 1.6 \text{ x } 6.286 + 1.2 \text{ x } 25.867 = 41.098 \text{ W}$$

Example.1.9

The switching waveforms of BIJ connected in CE configuration is shown in **Figure.1.37**. The various data are as follows. V_{CC}=200V, V_{BESAT} =3V, I_B = 8A; V_{CSAT} =2V, I_{CSAT} = 100A, t_d=0.5μs, t_r =5μs, t_s =

5μs, t_f = 3μs, switching frequency, f_s = 10kHz, duty cycle k =0.5, collector to emitter leakage current, I_{CEO} = 3mA, Find the power loss due to collector current during (i) t_{on}, (ii) t_n, (iii) t_{off}, (iv) t_o, (v) total average power loss.

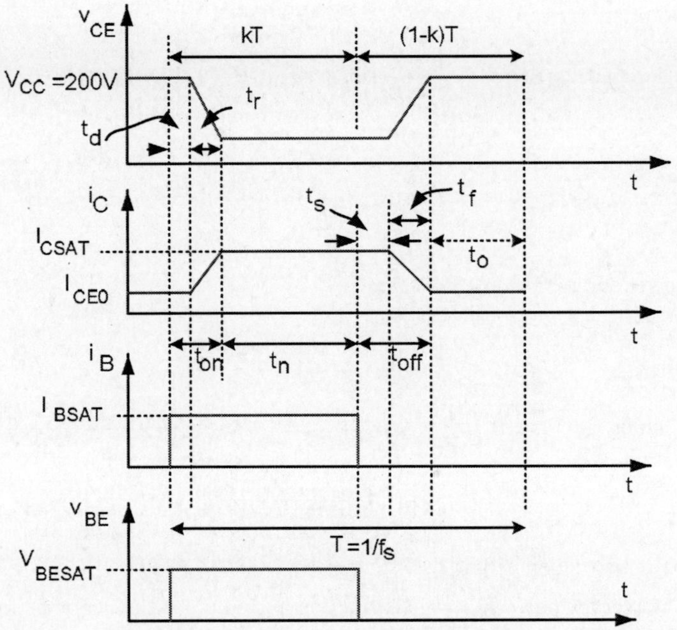

Figure.1.37 *Waveform related to example 1.9*

Solution:

Given, V_{CC}=200V, V_{BESAT}=3V, I_B = 8A; V_{CSAT}=2V, I_{CSAT} = 100A, t_d=0.5μs, t_r=1μs, t_s = 5μs, t_f = 3μs, switching frequency, f_s = 10kHz, duty cycle k =0.5

Now,

Time period, T = 1/f_s= 100μs;

$$kT = 0.5 \text{ x} 100 = 50\mu s;\ (1 - k)T = (1 - 0.5)\text{x}100 = 50\mu s$$

$$t_d + t_r + t_n = kT = 50\mu s$$

$$\therefore t_n = 50 - 0.5 - 1.0 = 48.5\mu s$$

$$t_s + t_f + t_o = (1 - k)T = 50\mu s$$

$$\therefore t_o = 50 - 5 - 3 = 42\mu s$$

The various time periods in the given switching waveforms are (1) Delay period, $0 \le t \le t_d$; (2) rise time period, $0 \le t \le t_r$, (3) conduction period, $0 \le t \le t_n$, (4) storage period, $0 \le t \le t_s$; (5) fall time period, $0 \le t \le t_f$, (6) off-period, $0 \le t \le t_0$. The sum of t_d and t_r will give us t_{on}, a sum of t_s and t_f will give us t_{off}. The total average power loss will be the sum of the average power loss during t_{on}, t_n, t_{off}, and t_o. Now we will find the power loss in each period and at last add all to get the total power loss.

A. Calculation of average power during on period, t_{on}

Delay time period, $0 \le t \le t_d$

$i_c(t) = I_{CEO} = 3x10^{-3} = 0.003,\ v_{CE}(t) = V_{CC} = 200$

Instantaneous power, $p_d(t) = i_c(t) x v_{CE}(t) = 0.003x200 = 0.6W$

$$\text{Average power during delay period} = \frac{1}{T}\int_0^{t_d} p_d(t)dt = f_s t_d x0.6 = 10x10^3 x0.5x10^{-6} x0.6$$

$$= 0.003W$$

Rise time period, $0 \le t \le t_r$

Here, $i_c(t) = \frac{I_{CSAT}}{t_r}t$; $v_{CE}(t) = V_{CC} + (V_{CESAT} - V_{CC})\frac{t}{t_r}$

Instantaneous power during rise time *(t_r=1 μs)*

$$p_r(t) = i_C(t)v_{CE}(t) = I_{CSAT}\frac{t}{t_r}\left[V_{CC} + (V_{CESAT} - V_{CC})\frac{t}{t_r}\right]$$

$$\text{Average power during rise time, } P_r = \frac{1}{T}\int_0^{t_r} i_c(t)v_{CE}(t)\,dt = f_s t_r I_{CSAT}\left[\frac{V_{CC}}{2} + \frac{(V_{CESAT} - V_{CC})}{3}\right]$$

$$= 10x10^3 x100x1x10^{-6} x\left[\frac{200}{2} - \frac{2-200}{3}\right] = 34W$$

Total average power during on period, $P_{on} = P_d + P_r = 34 + 0.003 = 34.003W$

B. Calculation of average power during conduction period (t_n), Conduction period, $0 \le t \le t_n$

Here, instantaneous collector current, $i_C(t) = I_{CSAT} = 100A$

Instantaneous value of collector -emitter voltage, $v_{CE}(t) = V_{CESAT} = 2V$

∴ Instantaneous power, $p_n(t) = v_{CE}(t) i_C(t) = 200W$

$$\text{Average power during conduction, } P_n = \frac{1}{T}\int_0^{t_r} i_c(t)v_{CE}(t)\,dt = f_s t_n V_{CESAT} I_{CSAT}$$

$$= 2x10^3 x48.5x10^{-6} x2x100 = 97W$$

C. Calculation of average power during turn-off storage time, Turn-off period, $0 \le t \le t_{off}$

During storage time, $0 \le t \le t_s$

Here, instantaneous collector current, $i_C(t) = I_{CSAT} = 100A$

Instantaneous value of collector -emitter voltage, $v_{CE}(t) = V_{CESAT} = 2V$

∴ Instantaneous power, $p_n(t) = v_{CE}(t) i_C(t) = 200W$

$$\text{Average power during conduction, } P_s = \frac{1}{T}\int_0^{t_s} i_c(t)v_{CE}(t)\,dt = f_s t_s V_{CESAT} I_{CSAT}$$

$$= 2x10^3 x5x10^{-6} x2x100 = 10W$$

During fall time, $0 \le t \le t_f$

Here, instantaneous collector current, $i_C(t) = I_{CSAT}(1 - \frac{t}{t_f})$, Neglecting I_{CEO}

Instantaneous value of collector -emitter voltage, $v_{CE}(t) = \frac{V_{CC}}{t_f}t$, Neglecting, I_{CEO}

$\therefore$ Instantaneous power, $p_f(t) = v_{CE}(t) i_C(t) = V_{CC}\, I_{CSAT} \left[\left(1 - \frac{t}{t_f} \right) \frac{t}{t_f} \right]$

Average power during conduction, $P_f = \frac{1}{T} \int_0^{t_f} i_c(t) v_{CE}(t) dt = \frac{V_{CC} I_{CSAT} t_f f_s}{6}$

$$= \frac{200x100x3x10^{-6}x10x10^3}{6} = 100W$$

Total average power during turn off period, $P_{off} = P_s + P_f = 10 + 100 = 110W$

D. Calculation of average power during off-period, $0 \le t \le t_o$

Here, instantaneous collector current, $i_C(t) = I_{CEO}$

Instantaneous value of collector -emitter voltage, $v_{CE}(t) = V_{CC}$

$\therefore$ Instantaneous power, $p_O(t) = v_{CE}(t) i_C(t) = V_{CC}\, I_{CEO} = 200x3x10^{-3} = 0.6W$

Average power during off period, $P_0 = \frac{1}{T} \int_0^{t_O} i_c(t) v_{CE}(t) dt = I_{CEO} V_{CC} t_O f_s$

$$= 0.6x42x10^{-6}x10x10^3 = 0.252W$$

E. Calculation of total average power, P_T

Total average power, $P_T = P_{on} + P_n + P_{off} + P_0 = 34.003 + 97 + 110 + 0.252 = 241.26W$

Example.1.10

The parameters of a MOSFET are as follows. V_{DS}=100V, Load resistance in drain to source circuit, RL =12Ω, f_s=30kHz, drain to source resistance, RDS(on) = 0.2Ω, duty cycle, k = 70%, t_r= 2μs. Find the power (a) loss in on state, and (b) power loss during turn-on.

Solution:

The circuit diagram is shown below.

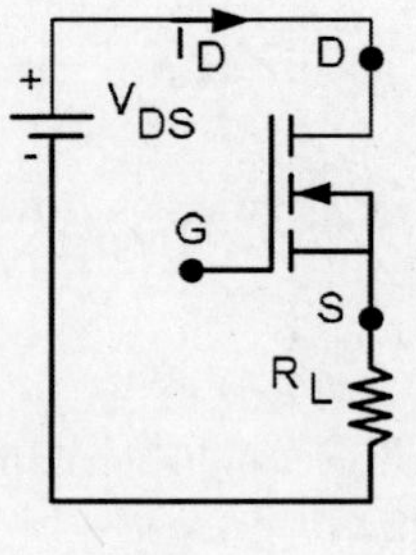

Drain current is given by $I_D = \frac{V_{DS}}{R_L + R_{DS(on)}} = \frac{100}{12 + 0.2} = 8.2A$

Time period, T=$\frac{1}{f_s} = \frac{1}{30x10^3} = 33.33\mu s$

The on time period, $t_{on} = DT = 0.7x33.33\mu s = 23.33\mu s$

(a) Power loss during conduction period = $I_D^{\,2}.R_{DS(on)}.t_{on}.f_s = 8.2^2 x0.2x23.33x10^{-6}x30x10^3 = 9.41W$

(b) The power loss during turn on, $P_{on} = \frac{V_{DS(max)} I_D t_r}{6} . f_s = \frac{100x8.2x2x10^{-6}x30x10^3}{6} = 8.2W$

Example.1.11

The parameters of a MOSFET are as follows. V_{DS} =120V, Drain current, I_D = 4A, I_{DSS} = 2mA, fs = 45kHz, drain to source resistance, $R_{DS(on)}$ =0.2Ω, duty cycle, k = 0.5, t_r = 80ns. Find the power (a) loss in on state, and (b) power loss during turn-on.

Solution:

Time period, $T = \frac{1}{f_s} = \frac{1}{45x10^3} = 22.22\mu s$

The conduction period = off period Since D = 0.5

$\therefore$ The conduction period, $t_n = t_{off} = \frac{22.22}{2} = 11.11\mu s$

The power loss during on state, $P_n = \frac{4^2 x0.2x11.11x10^{-6}}{22.22x10^{-6}} = 1.6W$

The power loss during off state, $P_o = \frac{120x2x10^{-3}x11.11x10^{-6}}{22.22x10^{-6}} = 0.12W$

The power loss during turn on, $P_{on} = \frac{120x4x80x10^{-9}x45x10^3}{6} = 0.288W$

The power loss during turn-off, $P_{off} = \frac{120x4x120x10^{-9}x45x10^3}{6} = 0.432W$

Total power loss, $P_T = P_{on} + P_n + P_{off} + P_0 = 0.288 + 1.6 + 0.432 + 0.12 = 2.44W$

Multiple Choice Questions

1. The majority charge carries in an *npn* transistors are
   (a) Electron
   (b) Holes in
   (c) Trivalent atom
   (d) Pentavalent atom
2. *npn* transistors are preferred over *pnp* transistors because they have
   (a) high mobility of holes
   (b) equal to the mobility of holes
   (c) low mobility of holes
   (d) higher mobility of electrons than the mobility of holes in *pnp* transistors
3. *pnp* transistor has the following arrangement
   (a) *p*- type base, *n*- type emitter, *p*-type collector
   (b) *p*-type emitter, *n*-type base, *p*-type collector
   (c) *p*-type collector, *n*-type base, *n*-type emitter
   (d) *p*-type emitter, *n*- type collector, *p*-type base
4. In an *npn* transistor operating in saturated mode, the output voltage V_{CE} is
   (a) Greater than $2V_{BE}$
   (b) Between $2V_{EB}$ and V_{BE}
   (c) Less than V_{BE}
   (d) Equal to V_{BE}

5. Which of the following are the charge carriers available in BJT?
   (a) Holes
   (b) Electrons
   (c) Neutrons
   (d) Both a and b
6. The amount of voltage required for forward biasing p-n junction is called
   (a) Avalanche voltage
   (b) Breakdown voltage
   (c) Breakover voltage
   (d) Cut in voltage
7. Which of the following BJT terminal controls the current flow?
   (a) Base
   (b) Collector
   (c) Emitter
   (d) Drain
8. In which of the following region do BJT operate in forward bias?
   (a) Active
   (b) Cut-off
   (c) Saturation
   (d) Both a and c
9. Which of the following BJT region does amplification?
   (a) Active
   (b) Cut-off
   (c) Saturation
   (d) Both a and c
10. In which of the following region do BJT operate in reverse bias?
   (a) Active
   (b) Cut-off
   (c) Saturation
   (d) Both a and c
11. Which of the following BJT region perform switching?
   (a) Active
   (b) Cut-off
   (c) Saturation
   (d) Both b and c
12. A power transistor is a
   (a) Three-layer, three junction device
   (b) Two-layer, three junction device
   (c) Three-layer, two junction device
   (d) Four-layer, two junction device
13. The maximum reverse voltage in a diode is called
   (a) Barrier potential
   (b) Breakover voltage
   (c) Peak inverse voltage
   (d) Average voltage
14. dv_{CE}/dt of an IGBT during turn-off should be controlled to prevent
   (a) Short circuit
   (b) Thermal run away

(c) EMI problem
(d) Latch up

15. The IGBT used in switching has T_{ON} = 3.0 μs, T_{OFF} = 1.20 μs, Duty cycle (D) = 0.70, V_{CE} (sat) = 2.0 V and f_s = 1 kHz. What are the switching power losses during turn-on and turn-off?
(a) 2.2 W and 1.7 W
(b) 1.98 W and 0.792 W
(c) 2.2 W and 0.792 W
(d) 1.98 W and 1.7 W

16. Which of the following two devices best describes the characteristics of an IGBT,
(a) SCR and TRIAC
(b) MOSFET and SCR
(c) SCR and BJT
(d) MOSFET and BJT

17. A charger can charge a laptop's battery with 100 W at 20 V. A switching frequency of 200 kHz is used by the power components in the converter inside the charger. Which electrical appliance is most appropriate for this use?
(a) IGBT
(b) MOSFET
(c) SCR
(d) BJT

18. The reverse recovery current depends on
(a) Temperature
(b) Storage charge
(c) Peak inverse voltage
(d) Forward current

19. Power MOSFET is
(a) Voltage controlled
(b) Current controlled
(c) Field control
(d) Both (a) and (c)

20. The main difference between MOSFET and BJT is
(a) MOSFET is made of silicon
(b) BJT current controlled and a MOSFET is voltage controlled
(c) MOSFET current is controlled and a BJT is voltage controlled
(d) None of the above

21. Power MOSFET is
(a) Bipolar
(b) Unipolar
(c) Voltage controlled
(d) Both (b) and (c)

22. Secondary breakdown occurs in
(a) A MOSFET
(b) A BJT
(c) An IGBT
(d) All the above

23. A diode is properly functioning, when
(a) Conduct current in both directions
(b) Conduct current in reverse biased condition

(c) Not dissipate heat
(d) Conduct current in one direction only

24. Main purpose of diode in power electronics
(a) Conversion of AC to DC
(b) Amplification
(c) Variable speed drive
(d) Converting DC to AC

25. SOA stands for semiconductor device
(a) Semiconductor operating actuator
(b) Semiconductor operational amplifier
(c) Safe operating area
(d) None of the above

26. SOA represents
(a) Maximum capacity that a device can withstand with failure.
(b) Maximum capacity that a device can withstand without failure.
(c) Minimum capacity of the device that withstands with failure
(d) Minimum capacity of the device that a device can withstand without failure

27. A transistor is turned of effectively by
(a) Making base positive
(b) Removing base voltage
(c) Driving base negative
(d) Injecting negative current at the base

28. BJT operates as a switch
(a) Under small signal condition
(b) With no signal condition
(c) In active region of transistor
(d) Under large signal conditions

29. Saturation state of a MOSFET is
(a) A closed switch
(b) An open switch
(c) An amplifier
(d) Pure resistance

30. Forward current gain of a BJT is
(a) Ratio of collector current and base current
(b) Ratio of base current and collector current
(c) Ratio of emitter current and base current
(d) Ratio of emitter and collector current

Answers to multiple-Choice Questions

1. (a) 2.(d) 3. (b) 4. (c) 5. (d) 6. (d) 7. (a) 8. (d) 9. (a) 10. (b) 11. (d) 12. (c) 13. (c) 14. (d) 15.(b) 16. (d) 17. (b) 18. (3) 19. (b) 20. (c) 21. (d) 22. (b) 23. (d) 24. (a) 25. (c) 26. (c) 27. (b),(d) 28. (d) 29.(a) 30.(a)

Short and Long Answer Type Questions

1. What is the difference between power and signal diode?
2. What are the different types of power diodes? Indicate clearly the difference between them.

3. Enumerate the types of power transistors along with their symbols.
4. Classify the power semiconductor devices.
5. What is a bipolar junction transistor? Why it is so called?
6. Describe the types of BJT and with their circuit symbols.
7. What are the different types of BJT configurations? Draw the circuit configuration for each BJT.
8. What is the difference between the current gain (β) and the forced current gain (β_F) of a BJT?
9. Enumerate the conditions at which the transistor operates as a switch.
10. The collector current at saturation is substantially constant even if the bas current is increased. Explain.
11. Explain the switching performance of BJT with necessary waveforms.
12. Explain the constructional details of low power and power MOSFETs and bring the difference between them.
13. What is the difference between current gain (β) and forced current gain (β_F) of a bipolar junction transistor?
14. What are the conditions under which a BJT operates as a switch? Discuss the hard drive and overdrive factors of bipolar junction transistors?
15. A Typical switching waveform of a BJT in CE mode is given in **Figure.1.38.** (a) Show that switch on energy loss is given by $\frac{V_{CC}I_{CSAT}}{6}t_{on}$. (b) obtain the average value of the switch-on loss, (c) derive the expression for switch-off energy loss and average power during switch-off.

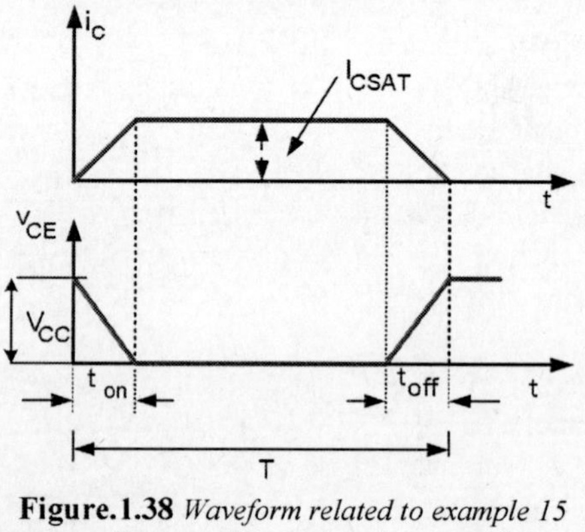

Figure.1.38 *Waveform related to example 15*

16. Describe with the necessary diagram the input and output characteristics of BJT.
17. Classify power MOSFETs.
18. Explain the transfer and output characteristics of power MOSFETs. Use necessary diagrams.
19. Explain the switching performance of power MOSFETs.
20. Compare the power MOSFETs and BJTs.
21. What is IGBT? What are the other names of IGBT?
22. Give the basic structural features and the working of IGBT.
23. Describe the input and transfer characteristics of IGBT.
24. Explain the switching characteristics of IGBT.
25. Why IGBT is popular in converter applications?
26. Enumerate the applications of IGBT.
27. What is single electron transistor (SET)?
28. Describe the construction of SET with the necessary diagram.

29. What is a quantum dot? Explain the quantum energy state of a quantum dot with the necessary diagram.
30. Describe the phenomena of coulomb blockade.
31. Explain the working of SET with the necessary diagram.
32. Draw the circuit equivalent of SET. Explain the circuit.
33. Draw and explain the V-I characteristics of SET.
34. What are the switching limits of BJT? Explain each with the necessary diagrams.

Numerical problems

1. Find the maximum power dissipation and case temperature of a BJT, if the junction and ambient temperature of a BJT are 150^0C and 25^0C respectively. Thermal resistance between junction and case, case and sink, and sink and ambient are 0.3^0C/W, 0.2^0C/W, and 0.5^0c/W respectively.
2. The maximum junction temperature of a BJT is $T_J = 160^0$C and the ambient temperature $T_A = 25^0$C. The maximum power dissipation is 125W. If thermal resistances are R_{JC}, R_{CS} are 0.4^0C and 0.2^0C/W respectively, find the thermal resistance from sink to ambient R_{SA}.
3. The maximum and minimum values of current gain (β) of a BJT (shown in **Figure.1.28**) are 60 and 10 respectively. The value of load resistance connected in the collector circuit is 5Ω. The DC supply voltage in the collector emitter (V_{CC}) is 100V. The base voltage (V_B) is 8V. The saturated value of collector emitter voltage (V_{CESAT}) and that of base-emitter voltage (V_{BESAT}) are given as 2.5V and 1.75V respectively. Determine (i) resistance R_B, due to which saturation occurs with overdrive factor (ODF) equal to 20, (ii) forced current gain (β_F), (iii) BJT power loss.
4. The switching waveforms of BJT connected in CE configuration is shown in Figure.1.29. The various data are as follows. V_{CC}=300V, V_{BESAT} =3V, I_B = 8A; V_{CSAT} =2V, I_{CSAT} = 100A, t_d =0.5μs, t_r =1μs, t_s = 5μs, t_f = 3μs, switching frequency, f_s = 10kHz, duty cycle k =0.5, collector to emitter leakage current, I_{CEO} = 3mA, Find the power loss due to collector current during (i) t_{on}, (ii) t_n, (iii) t_{off}, (iv) t_o, (v) total average power loss.
5. The switching waveforms of BJT connected in CE configuration is shown in **Figure.1.34.** The various data are as follows. V_{CC}=250V, V_{BESAT} =2.3V, I_B = 8A; V_{CESAT} =1.4V, I_{CSAT} = 100A, t_d =0.1μs, t_r =0.45μs, t_s = 3.2μs, t_f = 1.1μs, switching frequency, f_s = 10kHz, duty cycle k =0.5, collector to emitter leakage current, I_{CEO} = 3mA, Find the power loss due to collector current during (i) t_{on}, (ii) t_n, (iii) t_{off}, (iv) t_o, (v) total average power loss.
6. The E-MOSFET used in switching has I_D (ON) = 400 mA at V_{GS} = 12V and V_{GS} (th) = 2V. Determine the drain current for V_{GS} = 4V.
7. A MOSFET switch has the following parameters: I_{DS} = 4 mA, $R_{DS(ON)}$ = 0.4 Ω, duty cycle = 50%, I_D = 8A, V_{DS} =120 V, t_r = 150nS, t_f = 200 nS, switching frequency = 50 kHz, find total power loss through the switch:
8. An IGBT switching circuit is shown in **Figure.1.33.** The supply voltage is 250 V and the load resistance is 20 Ω. The switching frequency of IGBT is 1 KHz. Find the time required for the pulse if the power required to the load is 10 kW.

Practical Experiments

Experiment No.1.1

Title: Testing of the given power transistor

Objectives:

(a) To test the power transistor for determining its type, identification of terminals, and condition (Good or damaged),

(b) To study the output characteristics of the power transistor (BJT)

Resources required:

Table.1.2 Apparatus/ components required

Sl No.	Apparatus/ Component	Nos	Range/ Rating	Remarks
1.	DC Power supply	2	Required range in Volt	The ranges are decided as per the type of transistor available.
2.	Digital Multimeter	1	As available	
3.	Voltmeter	2	Required range in Volt	
4.	Ammeter	2	(μA range for base current measurement, Ma or A range for Collector current measurement)	
5.	Resistor	2	1 for R_B high range say 100kΩ (or decided), 1 for R_C say 1kΩ (as decided)	

Theory: A bipolar junction transistor (BJT) is a three-layer, two-junction semiconductor device. There are two types of transistors. They are *npn* transistor and *pnp* transistor. Three terminals are taken out from each region. The terminals are named emitter, base, and collector. The junction between the emitter and base is called the base-emitter junction (BEJ), and the junction between the collector and base is called the collector-base junction (CBJ). The symbol of *npn* and *pnp* transistors are shown in **Figure.1.39.** Manufacturer have manufactured power transistors of various size and shape. A picture containing various transistors manufactured by the different manufacturers is shown in **Figure.1.40.** The output characteristic is the plot between collector current (I_C) versus collector-emitter voltage (V_{CE}). The output characteristics are shown in **Figure.1.43.** When the base current is zero i.e. $I_B = 0$, with an increase in collector-emitter voltage (V_{CE}), a small leakage current (collector current) is present. With the increase in I_B from 0 to I_{B1}, I_{B2},... collector current also increases as shown in **Figure.1.43.**

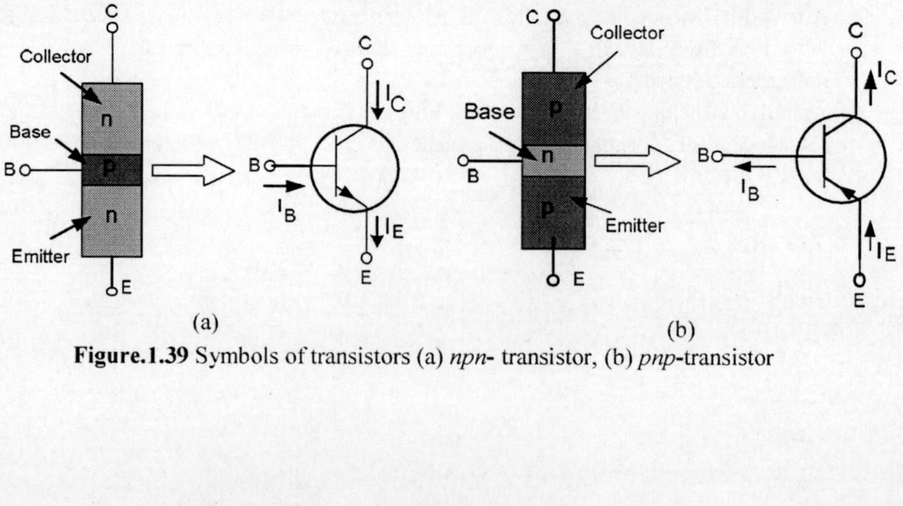

Figure.1.39 Symbols of transistors (a) *npn*- transistor, (b) *pnp*-transistor

(a) To test the power transistor for its type, identification of terminals, and condition (Good or damaged),

To fulfill this objective, the following methods are adopted.

Methods of transistor testing:

Before testing the proper functioning of the transistor, it is required to identify the types and terminals of the particular transistor. In general, the manufacturer data sheets specify the types of transistors by symbol and their nomenclature.

(a) Using Transistor tester

(b) Using Digital Multimeter

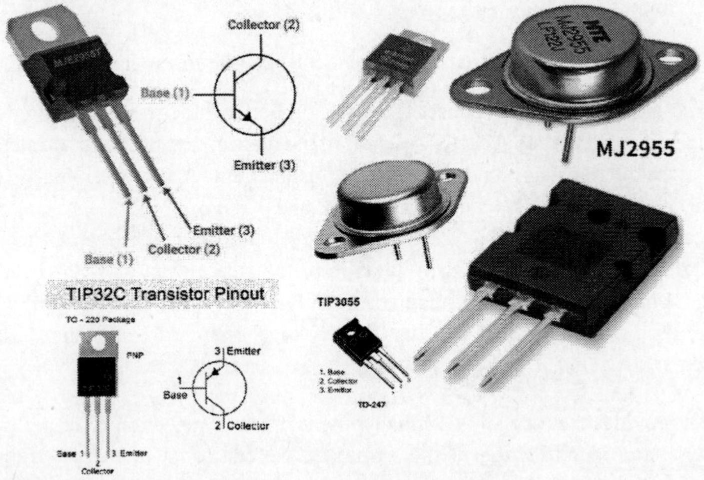

Figure.1.40 Pictures of various transistors manufactured by manufacturers

(a) Using transistor tester

If your laboratory is equipped with a Transistor tester, the transistor can be tested using the same. There are many varieties of testers available in the market. Some of the pictures of such testers are shown in **Figure.1.41**. The detailed procedure of testing is available in the instruction manual.

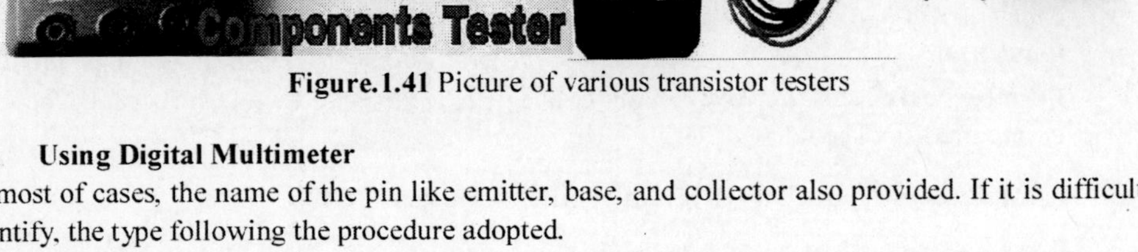

Figure.1.41 Picture of various transistor testers

(b) Using Digital Multimeter

In most of cases, the name of the pin like emitter, base, and collector also provided. If it is difficult to identify, the type following the procedure adopted.

1. Take a digital multimeter, connect the black prob (consider negative) to the com port, red prob (considered positive) connect to V-Ω port of the multimeter. Set the multimeter at diode mode.
2. Now, place the transistor on the wooden table, and numbered the terminals, 1, 2, and 3 from the left.
3. Connect the negative probe to terminal 1 (and 3) and the positive probe to terminal 2 of the transistor. If the multimeter shows a reading, the transistor is a *npn* transistor. If there is no reading shown in the multimeter, the transistor is *pnp* transistor.
4. Connect the negative prob to terminal 2 and the positive prob to terminal 1(and 3) of the transistor. If the multimeter shows no reading, the transistor is *npn* transistor. If the multimeter shows a reading, the transistor is *pnp.*

After knowing the type, if the terminals cannot be identified, the following procedure can be followed.

1. Take a digital multimeter, connect the black probe (consider negative) to com port, red probe (considered positive) connects to V-Ω port of the multimeter. Set the multimeter at diode mode.
2. Now, place the transistor on the wooden table, and numbered the terminals, 1, 2, 3 from left. Say this a *npn* transistor. Hence, 1 is *n*, 2 is P and 3 is *n*
3. Connect the negative prob to terminal 1 (and 3) and the positive probe to terminal 2 of the transistor. In each case note the voltage drop. If voltage drop between 1 and 2 is more than voltage drop between 2 and 3, then terminal 1 is an emitter, and terminal 3 is a collector. If voltage drop between 1 and 2 is less than voltage drop between 2 and 3, then terminal 3 is Emitter, and terminal 1 is the collector.

If the data sheet provides the type and terminal nomenclature, no need to perform the above test. After knowing the type and identification of the terminals, we can test whether the transistor is in good condition or not.

1. Take a digital multimeter,. connect black probe (consider negative) to com port, red probe (considered positive) connects to V-Ω port of the multimeter. Set the multimeter at diode mode.
2. Base to Emitter: Connect the positive prob of the mustimeter to the base (B) and negative probe to the emitter of the transistor. In case of a good *npn* transistor, the multimeter shows a voltage drop between 0.45V and 0.9V. For a good *pnp* transistor, meter will show over limit.
3. Base to Collector: Connect the positive probe to the base (B) and negative probe to the collector (C). In case of good *npn* transistor, the multimeter shows a voltage drop between 0.45V and 0.9V. For good *pnp* transistor, the meter will show over limit.
4. Emitter to Base): Connect the positive probe on the emitter (E) and connect the negative probe to the base (B). For a good *npn* transistor, meter shows over the Limit. For a god *pnp* transistor, the meter should show a voltage drop between 0.45V and 0.9V.
5. Collector to Base: Connect the positive probe to the collector (C) and connect the negative probe to the base (B) of the transistor. For a good *npn* transistor, meter shows Over Limit. For good *pnp* transistor, the meter should show a voltage drop between 0.45V and 0.9V.
6. Collector to Emitter: Connect the positive probe to the collector (C) and connect the negative probe to the emitter (E) – A good *npn* or *pnp* transistor will read over Limit on the meter. Interchange the leads (positive to Emitter and negative to collector) – Once again, a good *npn* or *pnp* transistor should read "OL".

(b) To study the output characteristics of power transistor (BJT)

An experiment for plotting the output characteristics of the transistor is provided here. The circuit diagram is shown in **Figure.1.42**

Circuit diagram:

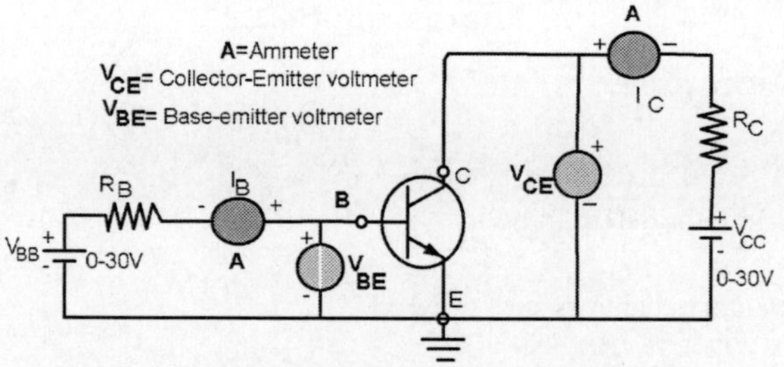

Figure.1.42 Circuit diagram for Experiment No.1.1.

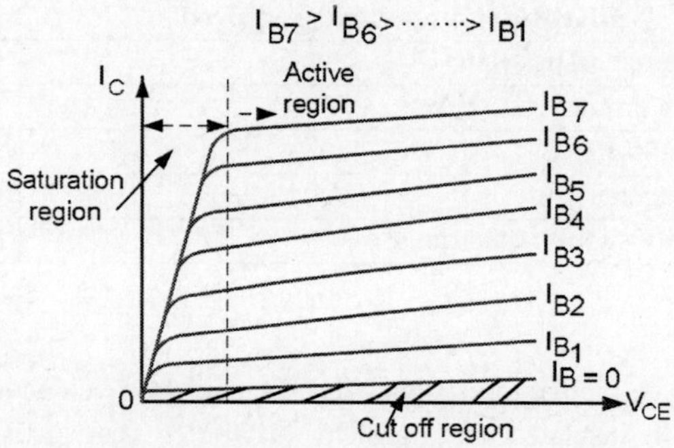

Figure.1.43 V-I characteristics of transistor

Procedure:

1. Make the circuit connection as shown in **Figure1.42**
2. Keep I_B constant at a value (say 10µA (or other value as per data sheet of the transistor), vary V_{CE} (By varying V_{CC}) and note down the collector current I_C. (As shown in **Table.1.3**)
3. Increase the value of I_B in step of some value (say 5µA), vary the V_{CE} (By varying V_{CC}) and note down the collector current I_C. (As shown in **Table.1.3**)
4. Repeat step 3 by keeping the IB to next value in step, vary the VCE (By varying V_{CC}) and note down the collector's current I_C. (As shown in **Table.1.3**)
5. Plot the I_C Versus V_{CE} for each value of I_B.

Result and analysis

Table.1.3

Sl. No	I_B =....		I_B=....		I_B=....	
	V_{CE}	I_C	V_{CE}	I_C	V_{CE}	I_C
1.						
2.						
3.						
4.						
5.						
6.						

Analysis: Using the readings of obtained (shown in Table1.3), three graphs are plotted. The plots are attached. The plots are look like the typical V-I characteristics as shown in Figure.1.43. It shows that the power transistor functioned well as desired.

Conclusion: The shows that the transistor functioned properly.

Experiment No.1.2

Title: Testing of the given IGBT.

Objectives:

(a) To test the IGBT for determining its type, identification of terminals, and condition (Good or damage),

(b) To study the output characteristics of IGBT

Apparatus required:

Sl No.	Apparatus/Components required	Range
1.	IGBT characteristics Kits	As available
2.	CRO with probes	As Available
3.	Digital multimeter	As available
4.	MC Ammeter	0-1A
5.	D.C. Power supply, dual range	0-30V

Resources required:

Theory: The insulated gate bipolar transistor (IGBT) is a newly developed semiconductor device. It is also known as metal oxide insulated gate transistor (MOSIGT), conductively–modulated field effect transistor (COMFET) or gain-modulated field effect transistor (GEMFET). IGBT combines the advantages of both MOSFET and BJT. It has three terminals called gate (G), emitter (E), and collector (C). It has high input resistance like MOSFET and low on-state power loss like BJT. Also, the second breakdown problem is not present in IGBT. An IGBT is a voltage control device like MOSFET. The equivalent circuit and the symbol of IGBT are shown in **Figure.1.44**

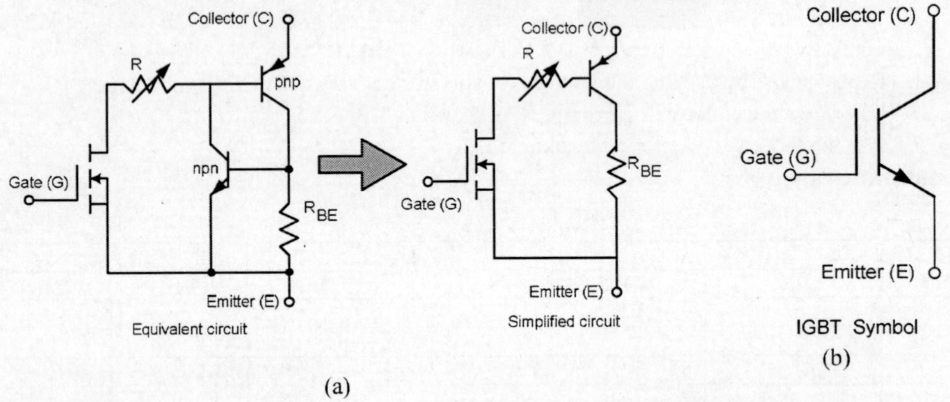

Figure.1.44 Equivalent circuit, and symbol of IGBT (a) equivalent circuit, (b) symbol

The static V-I or output characteristics of an IGBT is the plot of collector current (IC) versus collector-to-emitter voltage (VCE) corresponding to a particular gate-to-emitter voltage. The plot is shown in **Figure.1.45.** In the forward direction, the shape of the plot is similar to the bipolar junction transistor. However, the controlling parameter is the gate to emitter voltage.

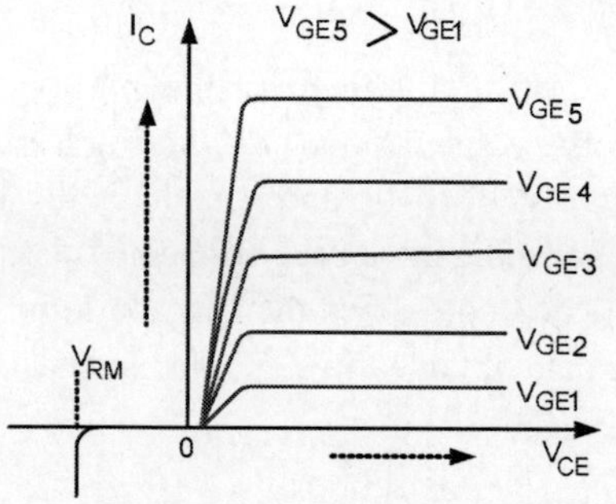

Figure.1.45 Output characteristics of IGBT

(a) To test the IGBT for its type, identification of terminals, and condition (Good or damaged),

Nowadays various IGBT testers are available in the Market. Some pictures of them are shown in **Figure.1.46.** Power electronics laboratories are also in general equipped with the such tester. As per the instruction provided in the manual, the IGBT can be tested with the available instruments.

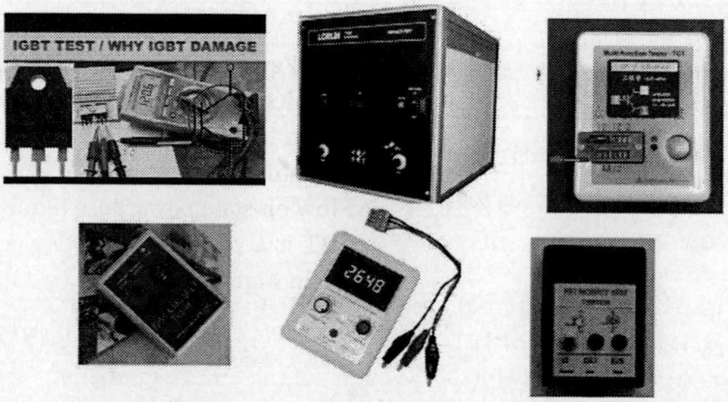

Figure.1.46 IGBT testers

(b) To study the output characteristics of IGBT

Circuit diagram:

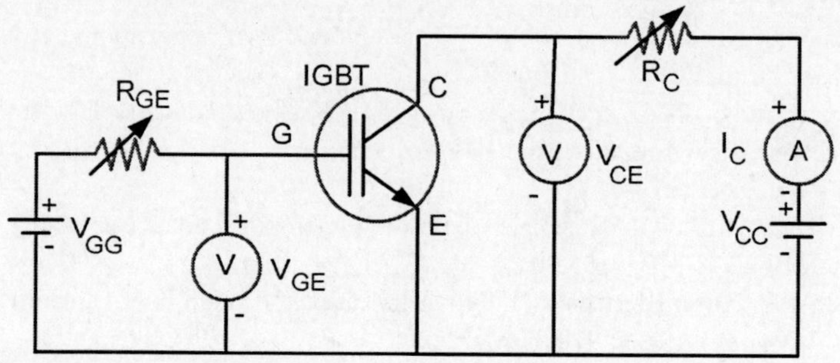

Figure.1.47 Circuit diagram for studying the V-I characteristics of IGBT

Procedure:

1. Make the circuit as per the circuit diagram shown in Figure.1.47.
2. Set the finite gate-source voltage (V_{GE1}) by varying R_{GE} and V_{GG}
3. Vary the voltage source V_{CC} (or R_C) and note down the V_{CE} and I_C in Table1.4
4. Repeat step 2 for the second gate-source voltage (V_{GE2}).
5. Repeat step 3 for V_{GE2} and note down the V_{CE} and I_C in Table1.4
6. Plot the graphs using data obtained in step3 and step5.

Result and analysis:

Table.1.4

Sl. No	V_{GE1} =....		V_{GE2} =.....	
	V_{CE}	I_C	V_{CE}	I_C
6.				
7.				
8.				
9.				
10.				
11.				

Analysis: Using the readings of obtained (shown in **Table1.4**), three graphs are plotted. The plots are attached. The plots are look like the typical V-I characteristics as shown in **Figure.1.43**. It shows that the IGBT functioned well as desired.

Conclusion: The shows that the IGBT functioned properly.

Know More

1. The history of power electronics started with the invention of the mercury arc rectifier in 1900.
2. The bell telephone laboratory, USA is the first organization in which the power electronics revolution started in 1948 with the invention of the Silicon transistor.
3. There are some other power transistors namely static induction transistor (SIT), COOLMOS. SIT is a high-power high frequency device. It is solid-state version of triode vacuum tube. COOLMOS is a new technology for high-voltage power MOSFETs.
4. The power diodes and power transistors are used in power converters like AC to DC converters, Inverters etc.

References and suggested readings

1. Ramamoorty M., An Introduction to Thyristors and their applications, East-West Press Pvt. Ltd., New Delhi, ISBN: 8185336679.
2. Sugandhi, Rajendra Kumar and Sugandhi, Krishna Kumar, Thyristors: Theory and Applications, New Age International (P) ltd. Publishers, New Delhi, ISBN: 978-0-85226- 852-0.
3. Bhattacharya, S.K., Fundamentals of Power Electronics, Vikas Publishing House Pvt. Ltd. Noida. ISBN: 978-8125918530.
4. Jain & Alok, Power Electronics and its Applications, Penram International Publishing (India) Pvt. Ltd, Mumbai, ISBN: 978-8187972228.
5. Rashid, Muhammad, Power Electronics Circuits Devices and Applications, Pearson Education India, Noida, ISBN: 978-0133125900.

6. Singh, M. D. and Khanchandani, K.B., Power Electronics, Tata McGraw Hill Publishing Co. Ltd, New Delhi, 2008 ISBN: 9780070583894.
7. Zbar, Paul B., Industrial Electronics: A Text –Lab Manual, McGraw Hill Publishing Co. Ltd., New Delhi, ISBN: 978-0070728226.
8. Grafham D.R., SCR Manual, General Electric Co., ISBN: 978-0137967711.
9. Mohan Ned, Undeland Tore.M., Robins William P., Power Electronics – Converters, Applications and Design, John Wiley & Sons. Inc,
10. https://www.google.com/search?q=Instruments+for+testing+power+transistor.
11. https://www.researchgate.net/publication/307475653_Single_Electron_Transistor_SET_Operation_and_Application_Perspectives/stats
12. Monika Gupta, "A Study of Single Electron Transistor (SET)", International Journal of Science and Research (IJSR) ISSN (Online),

2 Thyristor Family Devices

UNIT SPECIFICS

This unit covers the following aspects:

- *Introduction to introduction to thyristor family devices.*
- *Silicon Silicon-controlled rectifier (SCR) and its construction, V-I characteristics, and working principle.*
- *Two transistor analogy of Silicon Silicon-controlled rectifier.*
- *A brief description of turn-on methods of SCR.*
- *SCR power loss, cooling, ratings.*
- *Analysis of switching characteristics of SCR.*
- *Classification of thyristors.*
- *Light Activated Silicon Controlled Rectifier (LASCR) and its construction, V-I characteristics, working principle, and application.*
- *Silicon Controlled Switch (SCS) and its construction, V-I characteristics, working principle, and advantages as well as disadvantages.*
- *Gate turn-off (GTO) and its construction, V-I characteristics, working principle, applications, advantages as well as disadvantages.*
- *Unijunction Transistor, or UJT and its construction, V-I characteristics, working principle, applications, types, advantages as well as disadvantages.*
- *Programmable Unijunction Transistor (PUT) and its construction, V-I characteristics, and working principle.*
- *DIAC and its construction, V-I characteristics, working principle, applications, advantages as well as disadvantages.*
- *Triode for alternating current (TRIACs) and its construction, V-I characteristics, working principle, applications, advantages as well as disadvantages.*
- *Protection circuits of SCR and the design of the components of the protection circuits.*
- *Practical experiments on DIAC and SCR.*

The concept of thyristor family members and their construction, V-I characteristics, working, advantages as well as disadvantages, and applications are discussed and analyzed for generating further curiosity and creativity as well as improving problem-solving capacity with some numerical problems.

Besides giving a large number of multiple-choice questions as well as questions of short and long answer types marked in two categories following the lower and higher order of Bloom's taxonomy, assignments through several numerical problems, a list of references, and suggested readings are given in the unit so that one can go through them for practice.

Some practical experiments related to the courses covered in Unit II are also appended at the end of this unit to make the students aware of the hands-on on these topics.

After the related practical experiments on the topic, based on the content, there is a "Know More" section appended. This section has been designed to supplement additional information and higher learning skills on the topic.

RATIONALE

This fundamental unit on thyristor family devices helps students to get a primary idea about the thyristor family devices for application in power modulation and control.

The SCR is the oldest device of the thyristor family. The construction, V-I characteristics, and working principle of thyristor family members such as SCR, LASCR, GTO, SCS, UJT, TRIAC, DIAC, and PUT are discussed.

The physics behind various thyristors' family members and their use are also discussed at length to develop the basic idea about these devices.

Some related problems are pointed out after each section with their solutions which can help further for getting a clear idea of the concerned topics. The mathematics behind the thyristor family members will certainly help students with numerical problem-solving.

As a student in the field of electrical engineering, this unit on thyristor family members helps students to grasp the basic knowledge of various types of power electronic devices.

PRE-REQUISITES

ESC101: Basic Electrical Engineering

UNIT OUTCOMES

After completion of Unit-2 students will be able to:

U2-O1: Identify the application of SCR, LASAR, GTO, UJT, DIAC, TRIAC, SCS, and PUT.
U2-O2: Explain the construction of SCR, LASAR, GTO, UJT, DIAC, TRIAC, SCS, and PUT
U2-O3: Explain the working principle of working with SCR, LASAR, GTO, UJT, DIAC, TRIAC, SCS, and PUT
U2-O4: Analyse the V-I characteristics of SCR, LASAR, GTO, UJT, DIAC, TRIAC, SCS, and PUT

Unit-2 Outcomes	EXPECTED MAPPING WITH COURSE OUTCOMES (1- Weak Correlation; 2- Medium correlation; 3- Strong Correlation)				
	CO-1	CO-2	CO-3	CO-4	CO-5
U2-O1	3	--	1	1	1
U2-O2	-	3	1	1	1
U2-O3	-	--	3	3	1
U2-O4	-	--	--	3	2

2.1 INTRODUCTION TO THYRISTOR FAMILY DEVICES

A semiconductor device called "Thyristor" was fabricated in Bell Laboratories. The prototype of thyristor was introduced by GEC (USA) in 1957. Afterward, similar to thyristor characteristics, many other devices were developed. All such semiconductor devices are included in a common family, named as "Thyristor". The oldest member is the Silicon-controlled rectifier (SCR). Some other family members of thyristor family are TRIAC, MOS-controlled thyristors (MCTs), DIAC, Silicon-controlled switch, Emitter turn-off (ETO) thyristors, programmable unijunction transistor (PUT), static induction thyristor (SITH), gate turn-off thyristor (GTO), reverse conducting thyristor (RCT), MOS-turn off (MTO) thyristor, light-activated silicon-controlled rectifier (LASCR), gate assisted turn-off thyristor (GATT), etc. In most of the books and other literature, the terms Thyristor and SCR are used interchangeably. In this Unit, the construction, working, V-I characteristics, and protection circuits of a few members of the thyristor family are discussed.

2.2 SILICON CONTROLLED RECTIFIER

2.2.1 Introduction

The silicon-controlled rectifier (SCR) is a solid-state semiconductor device having four layers and three terminals. The terminals are called anode, cathode, and gate. It has three *pn* junctions. The schematic diagram and symbol of SCR are shown in Figure.2.1 (a) and (b) respectively. The anode, cathode, and gate are represented as *A, K,* and *G* respectively in Figure.2.1. An actual picture of an SCR is shown in Figure.2.1(c).

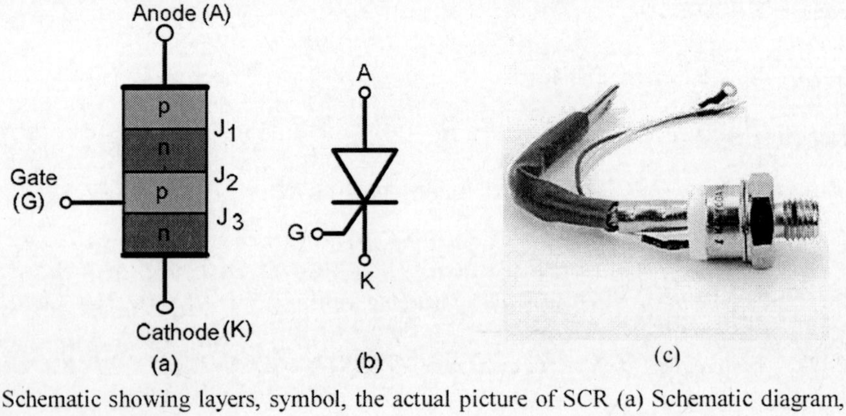

Figure.2.1 Schematic showing layers, symbol, the actual picture of SCR (a) Schematic diagram, (b) Symbol, and (c) Actual picture.

2.2.2 Construction of SCR

Figure.2.2 (a) shows a vertical cross-sectional view of an SCR. The approximate doping level and thickness of the layers are also shown in the figure. The symbol of the SCR is already shown in **Figure.2.1. Figure.2.2 (a)** is similar to BJT but the doping density and thickness are different. The cathode and gate are at the same location as shown for BJT. The n^- layer is for absorbing the depletion layers of the junction if the SCR is in the off state. The purpose of n^- layer is same as that of n^- layer of the collector of BJT.

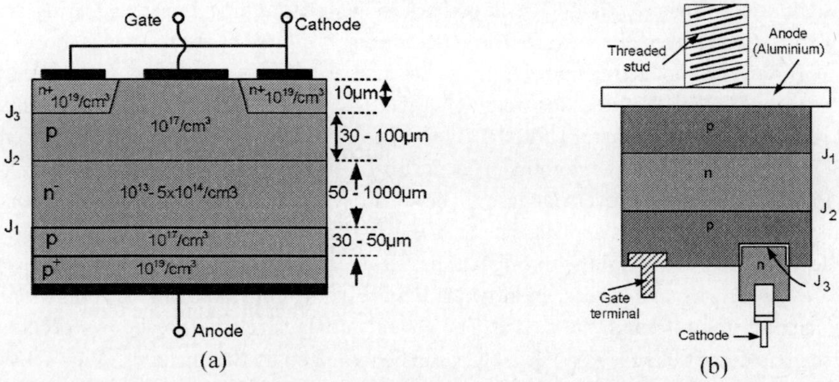

Figure.2.2 Vertical cross section of SCR (a) Vertical cross section showing all layers showing the thickness and doping density for typical SCR, (b) Only the layers and terminals.

Figure.2.2(b) displays only the layers *p, n, n, p* along with terminals. The p^+ substrate n^- layers are not shown separately. It comprises alternate *p* and *n*-type semiconductor layers i.e., *p*-type, *n*-type, *p*-type, and *n*-type. These four layers form three junctions, and the junctions are shown as J_1, J_2, and J_3 in **Figure.2.1(a)** and **(b).** Normally gate is placed nearer to the cathode. A treaded stud is also shown. This is required to fix the SCR with the heat sink or the frame. The anode (*A*), gate (*G*), and cathode (*K*) terminals are taken out from the outer *p*-type layer, inner *p*-type layer, and outer *n*-type layer respectively. The heat sink is used to remove heat from the device. Mainly silicon is used in SCR construction. The operation of the SCR as a rectifier can be controlled. It is unidirectional like a diode.

2.2.3 Working principle of SCR

The different terminals are shown in **Figure.2.1.** A low gate current flows from the gate to the cathode. The *A* and *K* terminals are connected to the main power circuit. When anode, *A* is made positive with respect to (w.r.t) cathode (*K*), the junctions J_1 and J_3 are forward biased and junction J_2 is reversed biased. Because of the existence of the depletion layer, junction J_2 prevents the current from flowing through the SCR. Only a small leakage current called forward leakage current flows through the device due to insufficient drift of mobile charges to make the device on. The SCR under this condition is called the forward blocking state or off-state of the device. If terminal *A* is made more positive w.r.t terminal *K*, the depletion layer across the J_2 junction starts to decrease. When *A*-to-*K* voltage goes on increasing, a stage will be reached at which the depletion layer vanishes and the device starts conducting.

2.2.4 V-I characteristics of SCR

Figure.2.3 (b) displays a typical V-I characteristic of an SCR. It is nothing but a plot between the anode to cathode voltage (V_{AK}) and the anode current (i_A) of the SCR. Let us consider Figure.2.1(a) and the circuit of Figure.2.3(a). When the anode is positive w.r.t cathode J_1 and J_3 junctions are forward biased, J_2 junction is reversed biased. Only a small leakage current flows through the device. The SCR is now called to be in forward blocking mode and the current through the device is called off-state current. When V_{AK} increases, a stage will reach at which its reversed biased J_2 junction breaks down. This breakdown phenomenon is named avalanche breakdown. The value of V_{AK} at which this breakdown occurs is called the forward breakdown voltage (V_{BO}). Since J_1 and J_2 are already forward-biased and J_2 breaks down, the large anode current (I_A) flows due to the movement of carriers across the junctions. This is the forward anode current. The SCR is in a state called conducting state or on-state. The ohmic drop in the junctions is very small in on-state and it is of the order of 1V. The anode current should be more than a minimum anode current, called latching current (I_L) to maintain conduction in the on-state.

In case, if it is less than I_L, the SCR will go back to the forward blocking state. During the conduction period, the SCR behaves like a diode. The anode current is limited by the resistance R_L in **Figure.2.3(a).** The SCR remains conducting because of the absence of a depletion layer at J_2 junction. If the anode current is reduced to a value called holding current (I_H), the depletion layers across J_2 are formed and SCR reaches the blocking state. Thus, the holding current is always less than the latching current. When the SCR is reversed biased i.e. terminal K is positive w.r.t terminal A, the junction J_2 is forward biased but junctions J_1 and J_3 are reversed biased. The situation is like a circuit of two series-connected diodes in which reversed voltage is applied. The SCR is now in a reverse blocking state. A small current called reverse leakage current (I_R) flows through the device due to minority carriers. When the reversed voltage increases further, a stage is reached at which the breakdown of reversed biased junctions occurs and heavy current flows through the device. The voltage at which reverse breakdown occurs is called the reverse breakdown voltage (V_{RBV}). The SCR can be turned on by two methods. They are as follows.

(a) By increasing the forward voltage greater than the forward break-over voltage (V_{BO}). But this method is destructive.

(b) By maintaining a forward voltage below V_{BO}, and applying a positive voltage across gate (G) and cathode (K). The situation is shown by the dashed line in the V-I characteristic. This type of turn-on method is called gate triggering.

After the SCR is turned on by gate triggering and the anode current is above the holding current, the SCR remains conducting because of positive feedback even after the gate signal is removed. Thus, the SCR is a latching device.

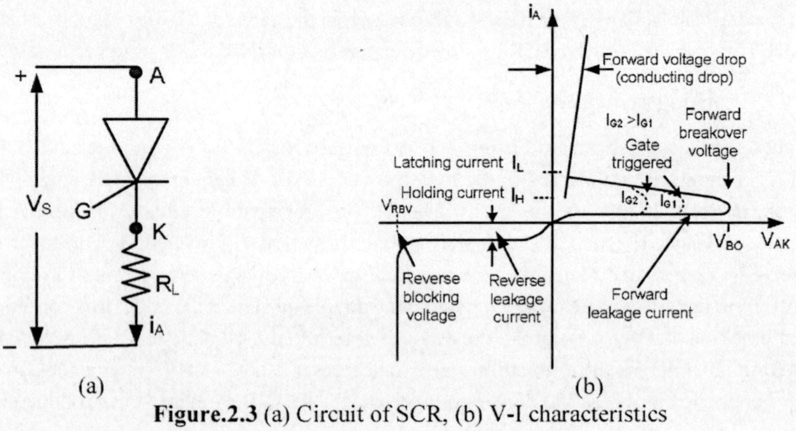

Figure.2.3 (a) Circuit of SCR, (b) V-I characteristics

2.2.5 Two transistor analogy of SCR

The latching action of a SCR can be demonstrated using a two-transistor model of SCR. An SCR can be assumed to comprise two transistors. One of them is *pnp* type (T_1) and the other is *npn* type (T_2) as shown in **Figure.2.4.** The basic structure of the two-transistor model and the equivalent circuit of the same is shown in **Figure2.4(a)** and **(b)** respectively.

The collector current of a transistor is given by equation (2.1), where I_{CBO} is the leakage current of the collector–base junction, I_C is the collector current, α is the forward current gain, and I_E is the emitter current.

$$I_C = \beta I_E + I_{CBO} \tag{2.1}$$

The base-emitter current gain α is defined as equation (2.2)

$$\alpha = \frac{I_C}{I_E} \tag{2.2}$$

In the case of T_1, I_E is the anode current, I_A, hence collector current of T_1 is given by equation (2.3) in which α_1 is the current gain of T_1 and I_{CBO1} is the leakage current of T_1.

$$I_{C1} = \alpha_1 I_A + I_{CBO1} \tag{2.3}$$

Similarly, for transistor T_2, the collector current, I_{C2} is given by equation (2.4), in which α_2 is the current gain of T_2 and I_{CBO2} is the leakage current of T_2.

$$I_{C2} = \alpha_2 I_k + I_{CBO2} \tag{2.4}$$

Combining I_{C1} and I_{C2}, we will get the anode current of the SCR, which is given by equation (2.5).

$$I_A = I_{C1} + I_{C2} = \alpha_1 I_A + \alpha_2 I_K + I_{CBO1} + I_{CBO2} \tag{2.5}$$

Considering the gate current, I_G, the cathode current is given by equation (2.6)

$$I_K = I_A + I_G \tag{2.6}$$

After solving equation (2.6) for I_A, we will get the equation (2.7)

$$I_A = \frac{\alpha_2 I_G + I_{CBO1} + I_{CBO2}}{1-(\alpha_1 + \alpha_2)} \tag{2.7}$$

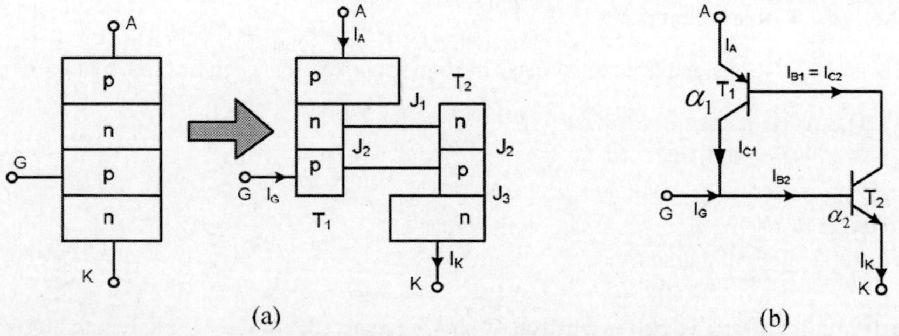

Figure.2.4 Basic structure and equivalent circuit of two transistor model (a) basic structure, (b) equivalent circuit

From the above, it is seen that the current gain α_1 varies with anode current ($I_A = I_E$) and α_2 varies with cathode current ($I_K = I_A + I_G$). It is seen from equation (2.7) that with the sudden increase in gate current, the anode current increases, and both α_1 and α_2 will also increase. An increase in the value of α_1 and α_2 further increases the anode current. It is just like a positive feedback effect or regenerative effect. When the sum of α_1 and α_2 is nearly equal to 1, the denominator of equation (2.7) tends to zero and the anode current will be very high. Hence, the SCR turns on by the application of a small gate current.

The two-transistor model of an SCR showing the capacitances of *pn* junctions is shown in **Figure.2.5.** This helps in deciding the characteristics of the SCR. At the blocking state, the application of rising voltage across the SCR causes high current to flow through the capacitances. Let the voltage across the junction J_2 is V_{j2}, and the capacitance across J_2 is C_{j2}, the charge across J_2 is q_{j2}, and the current through the capacitance is i_{j2}. The current i_{j2} is expressed as equation (2.8).

$$i_{j2} = \frac{d(q_{j2})}{dt} = \frac{d(C_{j2}V_{j2})}{dt} = V_{j2}\frac{dC_{j2}}{dt} + C_{j2}\frac{dV_{j2}}{dt} \tag{2.8}$$

From equation (2.8), it is seen that when the rate of rise of voltage $\left(\frac{dV_{j2}}{dt}\right)$ is large, the current i_{j2} is large and as a result the leakage current I_{CBO1} and I_{CBO2} also rises. Thus, from equation (2.7), it can be concluded that with large values of I_{CBO1} and I_{CBO2}, the sum of α_1 and α_2 becomes nearly equal to 1 and the denominator tends to zero and hence anode current is very large. This large current through the capacitor C_{j2} may damage the SCR. Thus, the applied $\frac{dV_{j2}}{dt}$ must be less than the specified value.

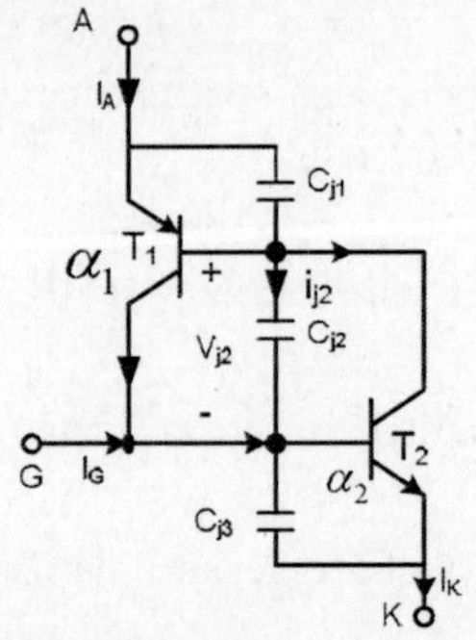

Figure.2.5 Two transistor model of an SCR

2.2.6 SCR turn on methods

To turn on the SCR, it is required to increase the anode current. This can be done by following methods.

(a) Thermal runway
(b) Application of light
(c) Application of high voltage
(d) High *dv/dt*
(e) Application of gate current

(a) **Thermal runway:** The pairs electron - holes are generated to increase in temperature of the SCR. Thus, the leakage currents are increases. When SCR leakage current increases, the sum of α_1 and α_2 approaches unity and from equation (2.7), it is seen that the anode current is very high. Thus, the SCR will turn on. This method is the thermal runway and in general not use because, there is a chance of damage to the SCR.

(b) **Application of light:** When the light is permitted to fall on the junctions, pairs of electrons and holes are generated and this in turn makes the sum of α_1 and α_2 tend to unity and from equation (2.7), it is seen that anode current increases to very high value. The light can be injected using light-activated thyristors

(c) **Application of high voltage:** When the applied voltage between the anode and cathode increases beyond break-over voltage, a significant leakage current flows through the device. This results in an increase in anode current due to positive feedback.

(d) **High dv/dt:** When the rate of rise of voltage across the anode and cathode increases, it is seen from equation (2.8) that junction current increases, and thus anode current increases.

(e) **Application of gate current:** By application of a +ve gate voltage w.r.t the cathode, the forward blocking voltage decreases. This SCR is turned on.

2.2.7 SCR switching characteristics

The static characteristics of SCR are already provided in section 2.2.4. In this section, the switching characteristics will be discussed. Both characteristics are important for the design of converters using SCR. The switching characteristic is also called dynamic or transient characteristics. At the time of turn-on and turn-off, the voltage across, and current through the SCR undergoes changes. The plots of these variations versus time give the dynamic characteristics of the SCR or thyristor. The dynamic characteristics at the time of the turn-on and turn-off processes are discussed separately.

2.2.7.1 Switching characteristics while turn-on

A forward-biased SCR (thyristor) is turned on normally by applying a positive gate pulse to the gate terminal with respect to the cathode. There is a transition state between the off-state and the on-state of the SCR. The time for this transition is called on turn-on time (t_{on}). The t_{on} is the time taken by the SCR to bring it from forward blocking state to final on the state. The t_{on} has three parts. The first part is called delay time (t_d), the second part is called rise time (t_r), and the third part is called spread time (t_p). These time periods are shown in **Figure.2.6.**

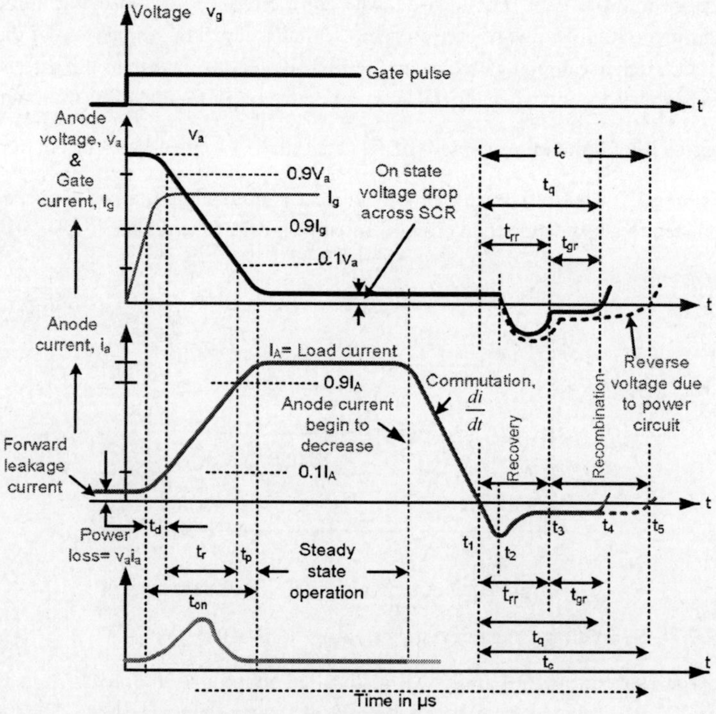

Figure.2.6 Switching characteristics of SCR

The delay time (t_d) is measured from the time when gate current attains 0.9 times of the final gate current (I_g) to the time when the anode current attains 0.1 times the final anode current (I_A). (i.e. 0.9I_g to 0.1I_A) Sometimes t_d is defined as the time period in which the anode voltage falls from the initial anode voltage (V_A) to 0.1 times of initial anode voltage (V_A) (i.e. V_A to 0.1V_A). Let the SCR be in a forward blocking state and at this stage the anode to cathode voltage is V_A, and there is a small leakage current. The turn-on process starts with an increase in anode current from the small leakage current and falls in the anode to cathode voltage from V_A. When a positive gate signal is applied w.r.t the cathode, gate current start flowing from the gate to the cathode.

The rise time (t_r) is measured from the instant at which the anode current rises from 0.1 times of final anode current (i.e. 0.1I_A) to 0.9 times I_A (i.e. 0.1I_A to 0.9I_A). t_r is also defined as the duration of time taken by the anode to cathode voltage for falling from 0.9V_A to 0.1V_A. t_r is inversely proportional to the gate's current magnitude as well as its built-up rate. Thus, t_r can be reduced by the application of high and steep current pulses applied to the gate terminal. The nature of the anode current is the factor defining t_r. The rate of rise of the anode current is slow in the case of the R-L circuit and fast in the case of the R-C circuit. Hence, if the anode circuit R-L, t_r is more than that of the R-C circuit.

The spread time (t_p) is the time taken by the anode current to rise from $0.9I_A$ to I_A. The time taken by forward blocking voltage to fall from $0.1V_A$ to on-state voltage (may be 1 to 1.5V) is also defined as spread time. During t_d, the conduction is spread over the entire cross-section of the SCR. t_p is dependent on the area of the cathode and gate structure. After t_p, the anode current reaches its steady state value, and the voltage drop across SCR is the on-state voltage drop.

The SCR can be considered as a charge control device during turn on. The charge to be injected is proportional to gate current. Thus, if value of gate current is higher, lesser will be the time required to inject charge into the gate. The turn-on time can be reduced by applying higher values of gate current than the minimum value of gate current. Normally it is 3 to 5 times that of the minimum value of gate current. Higher value of gate current facilitates injecting charge to the gate. An SCR is said to be hard firing or overdriving when the gate current is several times that of its minimum value required. Hard driving facilitates the improvement of $\frac{di}{dt}$ capability. A typical gate current versus time curve is shown in **Figure.2.7.** The initial value of gate current is high and sustained for several microseconds and then reduced to a low value to avoid unwanted turn-off of the device.

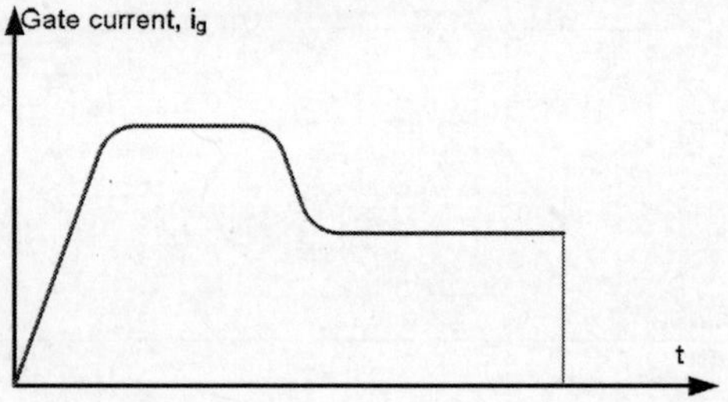

Figure.2.7 A typical waveform of gate current

2.2.7.2 Switching characteristics during turn-off

The process of turn-off of an SCR is also called commutation process. It is a dynamic process in which the device is brought from the on-state to the forward blocking state. The turn-off of the SCR means the change from on to off condition and the SCR is able to block the forward voltage. Once, the SCR is turned-on, the gate lost its control. Thus, no gate signal is required to sustain the conduction of the SCR. The SCR may be turned off by decreasing the anode current below the holding current and this job is accompanied by either natural commutation or forced commutation. When the anode current goes to zero to turn off, the SCR will be unable to block the forward voltage because of the presence of holes and electrons in the semiconductor layers of the SCR. Thus, the situation is favourable for the conduction of the SCR and it will go to conduction though there is no gate signal. To avoid such a situation, the SCR is reversed biased for a duration after the anode current attains zero.

The turn-off time (t_q) is measured from the time when the anode current is zero to the time at which the SCR is in forward blocking state. All the excess carriers (holes from outer p and electrons from n layers) must be removed from the four layers of SCR during t_q. The carrier around the J_2

junction can be removed by recombination. The time t_q has two components. They are reverse recovery time (t_{rr}) and gate recovery time (t_{gr}). The sum of t_{rr} and t_{gr} gives the turn-off time t_q.

The various time instants t_1, t_2, t_3, t_4, t_5 are shown in **Figure.2.6.** At t_1 anode current is zero. After t_1, the anode current increases in the reverse direction. This reversal is due to the existence of minority carriers in semiconductor layers of SCR. The slope $\frac{di}{dt}$ of this anode current is same as whatever before t_1. The surplus carriers from J_1 and J_3 are removed by the reverse recovery current (RRC) during period from t_1 to t_3. At t_2, nearly 60% of storage charge are removed from the outer two layers. The carrier density around J_1 and J_3 start decreasing and RRC starts decaying. The decay of RRC is fast during beginning and gradual afterwards. During fast decay there is a reverse voltage. This voltage is due to the presence of circuit inductance. This reverse voltage charge if appeared across the device may damage it. This is avoided by connecting resistance and capacitance elements across the device. At t_3, the RRC fallen nearer to zero and the junctions J_1 and J_3 are recovered. Thus, SCR is capable of blocking reverse recovery voltage. At the end of period t_1 to t_3, junction J_2 still have some trapped charge due to which SCR is not able to block forward voltage. This trapped charge cannot be eliminated by external circuit. The only way is recombination. The recombination is possible only when reverse voltage is applied. The time taken for recombination is called gate recovery time (t_{gr}) and it is the period between t_3 and t_4 as shown in **Figure.2.6.** At t_4 junction J_2 recovers. Now, forward blocking voltage can be reapplied between anode and cathode. The turn off time t_q is applicable for an individual SCR. The SCR is a part of the power circuit. For practical power circuits, the turn-off time provided to SCR is different and it greater than t_q. In the **Figure.2.6,** it is t_c. The turn-off time t_c is called circuit turn-off time. The turn off time is also dependent of the magnitude of rate of rise anode current $\left(\frac{di}{dt}\right)$ at the time of commutation and junction temperature. The turn-off time increases with increase in the magnitude of these parameters increases. When the forward current is more, the trapped charges across J_2 is more and the time taken for recombination is more. The turn-off time decreases with an increase in reverse voltage. Thyristors or SCR are designed for the slow and fast turn-off. The SCR with slow turn-off (50-100μs) are converter grade SCR. The cost of converter grade thyristor is less. The SCR with fast turn-off (3-50μs) are inverter grade SCR.

2.2.8 SCR Forward gate characteristics

The gate characteristics of an SCR can be shown with the help of graph between gate voltage and the gate current and a such typical characteristic is shown in **Figure.2.8.** In this figure both gate voltage and gate current are DC values and positive. The gate characteristics are spread between curve 1 and curve 2. Curve 1 and curve 2 represent the lowest and highest voltage that can be applied safely to turn on the SCR. Each SCR has maximum and minimum limits of gate voltage and current. In the figure maximum limit of voltage is denoted by V_{gm} and the maximum limit of current is denoted by I_{gm}. The minimum limit of voltage *oy* and minimum limit of current *ox* are also shown. There is also another specification called average gate power dissipation (P_{gav}) which is shown by curve *de*. To avoid permanent damage to junction J_3, the limits should not exceed P_{gav}, V_{gm}, I_{gm}, and for reliable turn on V_g and I_G should not be lower than the minimum limit. It is therefore concluded that the area *abcdefghb* as shown in figure is the preferred area for the gate drive. Manufacturers generally prescribed the non-triggering gate voltage (here denoted by *oa* in the figure). The unwanted signals like noise or spurious signals should be less than *oa*.

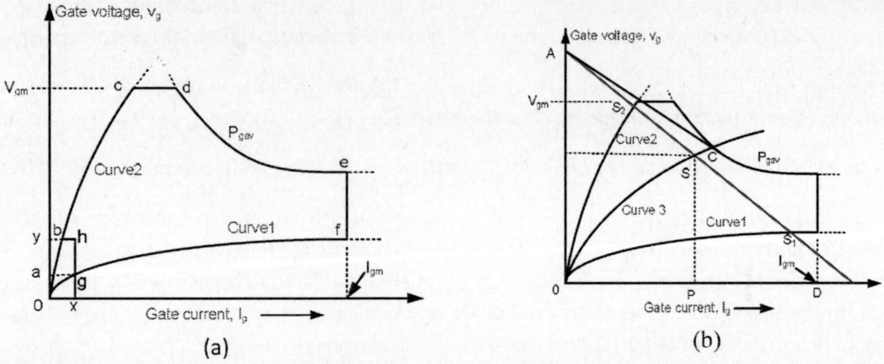

Figure.2.8 (a) SCR forward gate characteristics, (b) SCR forward gate characteristics for selecting gate parameters

2.2.9 Triggering circuits of SCR

The trigging circuits for triggering SCR are shown in Figure.2.9. The gate source voltage, gate source resistance, gate to cathode voltage is denoted by V_S, R_S, V_G respectively. While designing the circuit the ratio $\frac{V_S}{R_s}$ i.e. current should not harm to the source and the gate circuit. In case of low value of R_S, an external resistance required to connect in series with R_S. To provide an easy path for leakage current, a resistance R_l is connected across the gate to cathode. The current through R_l is given by equation (2.9), in which V_{gmn} is the minimum gate to cathode voltage.

$$I_{gmn} = \frac{V_{gmn}}{R_1} \tag{2.9}$$

With resistance R_l, the gate-source voltage is given by equation (2.10).

$$V_S = R_S(I_{gmn} + \frac{V_{gmn}}{R_1}) + V_{gmn} \tag{2.10}$$

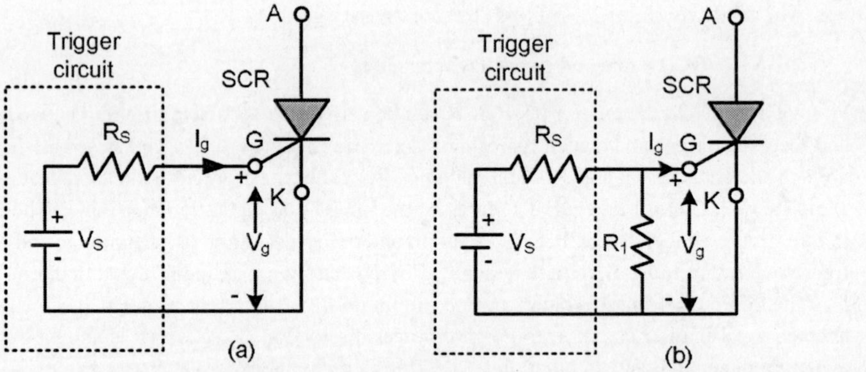

Figure.2.9 (a) Triggering circuit (a) without gate to cathode resistance, (b) with gate to cathode resistance

The operating point of a low power circuit can be obtained by using V-I characteristics of the SCR and gate. To select the operating point for the circuit of **Figure.2.9,** a load line AD is drawn in the **Figure.2.8(b).** Here OD = $I_g = V_S/R_S$, OA = V_S. The curve 3 represent a V_g-I_g characteristic of an SCR and it cuts the load line at point S. Here OP = I_g and PS = V_g. The point S is the operating point. To minimise the turn-on time and unwanted turn-on (called jitter), the point S may vary from point S_1 to S_2 (within curve 1 and curve 2) and it must be closer to P_{gav} curve. The required gate source resistance is given by the gradient of AD (i.e. OA/OD). The minim gate to source resistance is obtained from gradient of the tangent to the P_{gav} curve. In this way gate requirement in terms of DC signal can be obtained from **Figure.2.8(b).** In general pulse triggering is done for SCR turn on. In the case of pulse widths greater than 100μs, the DC data as obtained above can be used. But for pulse widths less than 100μs, the value of gate voltage and current can be increased. As we know that SCR is a charge-controlled device and its turn-on time can be minimized by application of bigger gate current pulse. In this case, the pulse wide should be such that anode current is more than the latching current of the SCR. In general gate pulse wide is taken ≥ to the value of turn on time. The power dissipation in pulse triggering should be less than the peak instantaneous gate power dissipation P_{gm}. The frequency of firing (f) can be obtained by taking P_{gm}, pulse width (T), and periodicity (T_1). The pulse gating is shown in **Figure. 2.10(a).** Hence,

$$P_{gm}T.\text{f} \geq P_{gav}, or \ \frac{P_{gav}}{fT} \leq P_{gm} \tag{2.11}$$

where $f = \frac{1}{T_1}, T =$ pulse widths in second.

In limiting case,

$$P_{gm}T.\text{f} = P_{gav}, or \ \frac{P_{gav}}{fT} = P_{gm} \text{ and f} = \frac{P_{gav}}{T.P_{gm}} \tag{2.12}$$

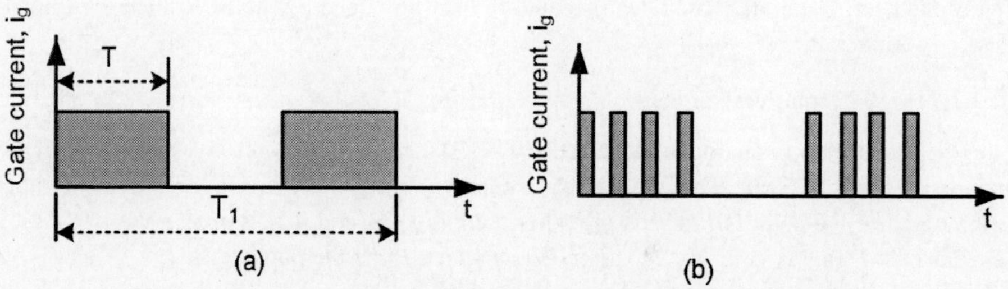

Figure.2.10 (a) Pulse gating, (b) high frequency carrier gating

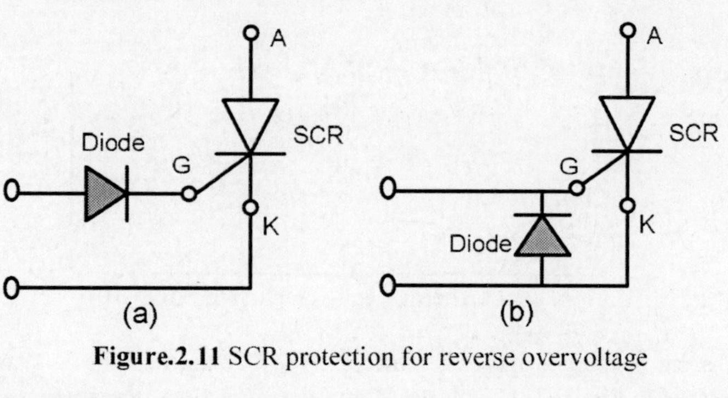

Figure.2.11 SCR protection for reverse overvoltage

The duty cycle (D) is defined as ratio of pulse wide in on period to the periodicity. Thus, duty cycle is given by **equation (2.13).**

$$D = \frac{T}{T_1} = fT \tag{2.13}$$

Thus, the **equation (2.12)** in terms D becomes

$$P_{gm}.D = P_{gav}, or \ \frac{P_{gav}}{D} = P_{gm} \text{ and } P_{gm} = \frac{P_{gav}}{D} \tag{2.14}$$

Sometimes the pulses are modulated and generate train of pulses. This technique is called high-frequency carrier gating. The generated pulses are shown if **Figure.2.10 (b).** There is a prescribed peak reverse voltage of an SCR and if any voltage signals greater than this limit may damage the thyristor. Hence, to prevent this a diode as depicted in **Figure. 2.11.** This diode is called clamping diode.

2.2.10 SCR power loss and cooling

SCR suffers from power losses while it is in operation. These losses occur at various stages. They are as follows.

(a) Forward conduction loss simply on state loss.
(b) Forward and reverse blocking loss because of leakage current.
(c) Switching loss in turn-on and turn off.
(d) Gate loss.

Out of these (a) is most dominant in power frequencies from zero to 400Hz and (b) is significant in high operating frequency. These losses produce thermal heat. The losses as well as temperature are increases with increase in SCR rating. The heat must be removed from the junction region. Hence cooling is extremely necessary. The heat produced in the SCR is dissipated as electrical power to ambient fluid. The ambient fluid may be air or water. This is done by fitting the SCR on a heat sink. The heat sink dissipates the heat. If the dissipated heat is equal to heat produced by the SCR, then the junction will reach steady temperature.

2.2.11 Thermal resistance

The thermal resistance is analogous to electrical resistance. It is denoted by symbol *R*. The unit of thermal resistance is ^{0}C/Watt. The heat energy or thermal energy moves from high to low temperature. If the temperature at a point is $T_1{}^0$C and the other point is $T_2{}^0$C, and if T_1 is greater than T_2, heat energy flows from T_1 to T_2 and is given by equation (2.15), where P_{av} is the power loss.

$$R_{12} = \frac{T_1 - T_2}{P_{av}} \ ^0C/W \tag{2.15}$$

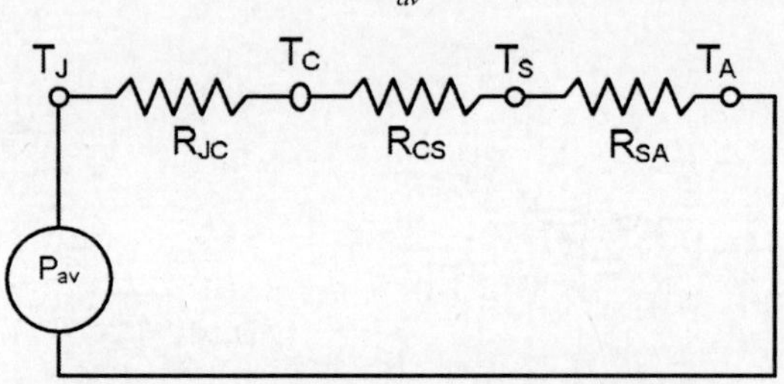

Figure.2.12 Thermal equivalent circuit of SCR

In case of SCR, the heat developed at the junctions of the semiconductor. The thermal equivalent circuit of an SCR is displayed in **Figure.2.12.** The heat flows from Junctions to case, case to sink, and sink to ambient. Let thermal resistances between Junctions and case, case and sink, and sink and ambient are denoted by R_{JC}, R_{CS}, R_{SA}, respectively. The P_{av} is the average rate of heat produced at junction, and it is given by equation (2.16), where $R_{JA} = P_{av}(R_{JC} + R_{CS} + R_{SA})$.

$$P_{av} = \frac{T_J - T_C}{R_{JC}} = \frac{T_C - T_S}{R_{CS}} = \frac{T_S - T_A}{R_{SA}} = \frac{T_J - T_A}{R_{JA}} \tag{2.16}$$

Thus, the difference in temperature between junction and ambient is given by equation (2.17).

$$T_J - T_A = P_{av}(R_{JC} + R_{CS} + R_{SA} + R_{JA}) \tag{2.17}$$

The manufacturer of SCR provides a data sheet or graph in which the temperature difference between junction and ambient (T_J-T_A) versus power loss (P_{av}) for a particular heat sink (say aluminium).

2.2.12 SCR mounting

SCRs are mounted on the case. The case must be designed in such a way that heat flows from junction to the case. Based on the power ratings of the SCR, various techniques of mountings are developed. They are given below.

(a) Lead mounting
(b) Stud mounting
(c) Bolt down mounting
(d) Press fit mounting
(e) Press pack mounting.

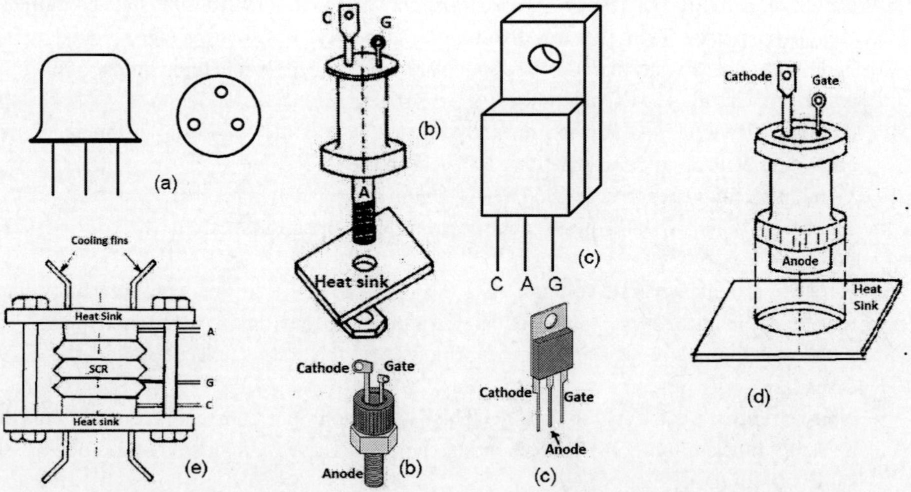

Figure.2.13 Various types of SCR mountings (a) lead, (b) Stud, (c) Bolt down, (d) Press-fit, (e) Press pack

(a) Lead mounting: If the load current is very small of the order of 1A, lead mounted SCR is used. SCR under this category do not require cooling or heat sink. The housing of the SCR dissipates adequate heat via radiation as well as convection.

(b) Stud mounting: Due to the flexibility and ruggedness, the stud mounted SCR is normally used. The SCR is fixed to the sink via threaded stud, bolts, and nuts. The threaded stud is anode of the SCR. Thus, anode is electrically connected to heat sink. If electrical connection in between gate and heat sink undesirable, mica or PTFE washers are used between the joining surfaces.

(c) Bolt down or flat pack mounting: This type of mounting has tabs with one or more holes. Sometimes hole is situated in the middle. Bolts are pushed through these holes to mount the device onto the heat sink. A thin insulating layer of PTFE or mica is provided to insulate the SCR

(d) **Press-fit mounting:** Press-fit or pressure-fit mounting is designed for insertion into appropriate-sized holes in the heat sink. Insertion is done with the help of a vice or by pressing the device using a wooden block. For larger-sized SCR it is carried out using a hydraulic ram.

(e) **Press pack mounting:** It is also called a disc or hockey pack. Here SCR is clamped between two heat sinks. Uniform external pressure is applied for not producing deformation. The heat sink may be air, water or oil cooled. This type of mounting is provided for high current rating SCR.

The above types of mountings are shown in Figure.2.13

2.2.13 SCR ratings

Like other semiconductor devices, SCR also have ratings. These ratings are in terms of voltage, current, power, and temperature limits. The SCR can be used within these limits without any damage. The current and voltage should not exceed the desired limit for reliable operation. The various ratings of SCR are given below.

A. Voltage rating of SCR

(a) **Working peak-off state forward voltage:** This is the maximum instantaneous voltage excluding repetitive and non-repetitive transient voltage at forward OFF state condition.

(b) **Repetitive peak-off state forward voltage:** It is the peak transient voltage blocked by the SCR in OFF state in forward direction and stated at maximum allowable temperature of the junctions when gate circuit is in open condition.

(c) **Non-repetitive peak-off-state forward voltage:** This is the maximum instantaneous voltage across the SCR of any non-repetitive transient in the OFF state.

(d) **Working peak reverse voltage:** The maximum instantaneous reverse voltage across the SCR except all repetitive and non-repetitive transient voltages is called working peak reverse voltage.

(e) **Repetitive peak reverse voltage:** It is the peak reverse transient voltage which may arise repeatedly in the reverse direction at the allowable maximum junction temperature. The junction temperature will increase enormously if this rating exceeded and the SCR may damage.

(f) **Non-repetitive peak reverse voltage:** This is the maximum transient reverse-voltage that can be safety blocked by the SCR. This rating can be increase by inclusion of a diode in series of same current rating that of SCR.

(g) **On state voltage:** The voltage drops between anode and cathode for a specified-on state current and junction temperature is called on state voltage. Typical value is in the range of 1 to 1.5 V

(h) **Gate trigger voltage:** The minimum gate to cathode voltage that is required to produce triggering current.

(i) **Voltage safety factor (V_f):** This factor relates the operating voltage in rms to the peak inverse voltage and is given by

$$\text{Voltage safety factor, } (V_f) = \frac{\text{Peak inverse voltage}}{\sqrt{2}\text{x rms value of input voltage}}$$

The operating voltage in general kept below the rms voltage of the SCR so that the damage of the device avoided. Typical value is in between 2 and 2.5.

(j) **Forward $\frac{dv}{dt}$ rating:** This is the maximum rate of rise of anode voltage with respect to time at which the SCR will nor trigger without firing signal at the gate. If this rating exceeded the SCR may switch from off state to on state. This phenomenon can be explain with the help of charging of the internal capacitances of the SCR. With increase of rate of rise of anode voltage, the charging current also increases, thus sufficient charges flow through the device. These charges help in bringing the SCR from off state to on state like application of gate triggering signal at the gate. The maximum rate of rise of $\frac{dv}{dt}$ can be limited by using snubber circuit across the device.

B. Current ratings

Thermal capacity of thyristor is small because it is made from semiconductor materials. The current carrying capability of an SCR is decided by the junction temperature. The device may get damage because of rise in temperature. The current ratings of SCR accordingly decided. The various current ratings are given below.

(a) Average on-state current: This rating is stated at maximum junction temperature and is varies with case temperature. This rating is repetitive type. As we know that forward voltage drop is low in case of thyristor, hence the power loss mainly depends on forward average current of the device. The rating can be obtained from the manufacturer data sheet in which variation in current with respect to case temperature provided.

(b) RMS on-state current: This rating is a repetitive type and specified at maximum junction temperature. For direct current rms and average current is same. The rms current i.e rms on state current have importance when thyristors are applied for high peak low duty cycle waveforms. The rms value may exceeded the allowable value. To prevent excessive heating in metallic joints, leads, and interfaces due to excessive rms current manufacturer have provided a data sheet specifying the rms values.

(c) Surge current rating or non-repetitive, peak-on state current: When the SCR is operating within the repetitive ratings, its junction temperature not exceeded but during short circuit or faults condition the junction temperature may exceeded which may damage the SCR. This unusual working condition is avoided, the surge current rating is specified. It indicates maximum probable non-repetitive surge current that can be handle by the SCR. This rating is stated in terms of the number of surge cycles with corresponding surge current peak. It is inversely proportional to the duration of the surge. The 1-cycle surge current rating is the peak value of an allowable non-recurrent half-sine wave of 10 ms (50 Hz) duration. If duration less than 1 cycle, a sub cycle surge current is specified. The sub-cycle surge current is given by

$$I_{Sub-cycle} = I_{1-cycle}\sqrt{\frac{T_{1-cycle}}{t}}$$

where, $I_{sub-cycle}$ is the sub cycle current rating in ampere, $T_{1-cycle}$ is the time for 1 cycle, $I_{1-cycle}$ is the surge current in 1- cycle, and t is the duration in seconds of sub cycle surge.

(d) I²t rating: It is the maximum permissible non-recurring value of the square of the instantaneous current integrated with respect to the time (i.e $\int i^2 dt$). This rating is important to decide the rating of the fast-acting fuse used for protection of the device.

(e) $\frac{di}{dt}$ rating: It indicates maximum rate of rise of anode to cathode current of the SCR. The maximum rate of rise of current that can withstand by the SCR is called critical rate of rise of current. The critical rate of rise of current is always specified at the value of the temperature at which the device safely withstand.

(f) Holding current: This is the minimum value of anode to cathode current below which the device stop conducting.

(g) Latching current: It is the minimum value of anode to cathode current that is required to keep the device in on state.

(h) Gate current: This is the current applied to the gate of the SCR for control purpose. It may be minimum gate current and maximum gate current. The minimum value of gate current is the minimum gate current required to fire the SCR, and maximum gate current is maximum value of gate current that can be applied to the gate safely.

C. Power rating

(a) Forward conduction loss: The average on state loss is the SCR is given by

Average power loss = Average anode current x forward voltage drop.

This on state loss is a main source of junction heating. The information related to forward conduction loss are supplied by the manufacturer in terms of graph between the average power loss versus average forward anode current for various conduction angles.

(b) Turn-on losses: During turn on an appreciable amount of power loss taking place. For switching frequency above 400Hz, additional circuits are used to limit the same.

(c) Turn-off losses: These losses are arising during the time of decay of reverse current. The circuit inductance is used to limit the change of current and thus the energy dissipated. However, inductance produce high reverse voltage transients during turn-off. The turn off losses is considered while selecting the device rating.

(d) Forward and reverse blocking losses: In the forward blocking region, anode is made positive w.r.t cathode and the anode current is the small forward leakage current. The forward blocking power loss is the integration of product of the forward blocking voltage and forward leakage current. Similarly, reverse power loss occurs in reverse blocking region.

(e) Gate power loss: The gate power loss is the average power loss due to gate current between the gate and main terminals. Gate losses are negligible for pulse types of triggering signals. Losses may become more significant for gate signals with a high duty cycle or for SCRs in a small package.

D. Thermal ratings

The main thermal ratings are given below.

(a) Junction temperature: The junction temperature of an SCR is specified and the junction temperature should be within the limit. If it exceeds this rating the thyristor may turn on even there is no gate signal. This rating is the deciding factor to control the maximum current that can be carried by the SCR for any significant period of time. The range of junction temperature varies with type of thyristor.

(b) Transient thermal impedance: This impedance (resistance) is the impedance between junctions and the cooling surfaces.

E. Turn-on and turn-off Time Ratings

Thyristors for a specific application is dependent on turn-on and turn -off time. For fast acting thyristors it is of low values. Gold doping in silicon is used to make fast acting thyristors. Life time of minority carriers are reducing with inclusion of gold doping but increases the leakage current.

2.3 TYPES OF THYRISTORS

Based on the physical construction, turn-on and turn-off behaviour, members of thyristors family are classified into the following

(a) Phase controlled thyristors or Silicon controlled rectifiers (SCRs)
(b) Light activated silicon-controlled rectifiers (LASCRs)
(c) Silicon control switches (SCSs)
(d) Gate turn off thyristors (GTOs)
(e) Programable unijunction transistor (PUT)
(f) Unijunction transistor (UJT)
(g) Bidirectional diode (DIAC)
(h) Bidirectional triode thyristors (TRIACs)
(i) Reverse conducting thyristors (RCTs)
(j) FET-controlled thyristors (FET-CTHs)
(k) MOS turn off thyristors (MTOs)
(l) MOS-controlled thyristors (MCTs)
(m) Static induction thyristors (SITHs)
(n) Emitter turn-off thyristors (ETOs)
(o) Integrated gate-commutated thyristors (IGCTs)

Out of all the above SCR have been explain in details in section 2.2. In the preceding sections some of them will be explored with symbol, construction, operating principle and V-I characteristics.

2.4 LIGHT ACTIVATED SILICON CONTROLLED RECTIFIER

2.4.1 Introduction

The Light Activated Silicon Controlled Rectifier (LASCR), which belongs to the photoelectric thyristor family, is widely used as a control element. This device can be activated by optical radiation and has the highest trigger sensitivity to light with a wavelength of 0.8 to 1.5μm. It is commercially available as a 6000 V and 1500 A device with a light triggering power of less than 100 mW, also known as a photo SCR. The LASCR is a semiconductor optoelectronic switch that has a higher capacity for handling power than other similar devices. Furthermore, it requires only a small amount of light energy to control a significantly larger amount of electrical energy. The usual rate of change of current (di/dt) is around 250 A/ms while the rate of change of voltage (dv/dt) can reach up to approximately 200 V/μs. The LASCR provides full electrical separation between switching component of a power converter and light-triggering source.

The LASCR belongs to the category of thyristors that can be activated by photons found in light rays. It is a type of semiconductor optoelectronic switch equipped with a lens that concentrates light onto its Gate. Its conduction is initiated by exposure to light, and it remains in the OFF state if the amount of light received is insufficient.

The LASCR have Anode (A), Cathode (K), and Gate (G) terminals. The gate terminal is utilized for electrical triggering of the LASCR. Optically triggering of the thyristor provides an advantage of protection against electrical noise disturbances. Thus, the LASCR is regarded as one of the most desirable devices. The symbolic illustration of LASCR is shown in Figure.2.14(a).

2.4.2 Construction of LASCR

When light falls on the LASCR, it enters the conduction state. The LASCR functions similarly to a regular SCR, but the difference is that it can be triggered by light through its Gate. The LASCR has a *pnpn* or *npnp* structure and is made up of four semiconductor layers. It contains three junctions, namely J_1, J_2, and J_3. The anode terminal of the LASCR is connected to the *p*-type material of the *pnpn* structure, and the cathode terminal is connected to the *n*-type layer. The gate of the LASCR is connected to the *p*-type layer that is near to the cathode, as displayed in Figure.2.14(b).

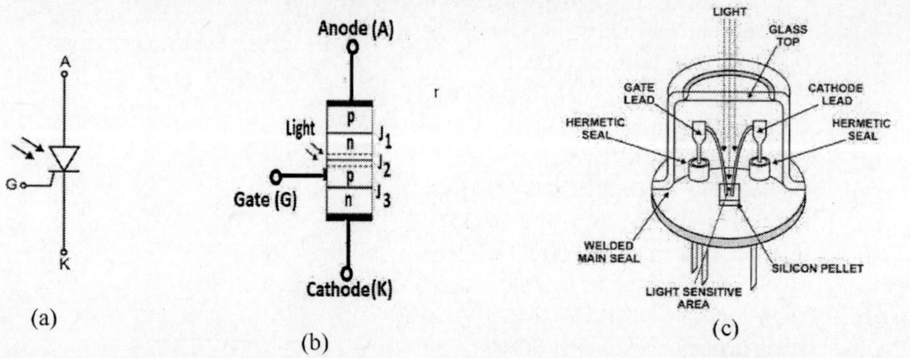

Figure. 2.14 (a) Symbol of LASCR, (b) *pnpn* structure of LASCR, (c) Basic structure of LASCR

The basic structure of LASCR is depicted in Figure.2.14(c). The LASCR consists of a semiconductor material, where the silicon pellet is located at the bottom of the device. To focus the external light source on the semiconductor material, a glass lens is utilized. The semiconductor crystal within the LASCR is designed to have its electrons dislodged by the intensity of the light, thereby contributing to conduction. Due to its high sensitivity, the LASCR is prone to respond to various factors such as temperature, applied voltage, as well as the rate of change of applied voltage. To facilitate its photosensitivity, the LASCR is designed with a glass top to allow light input and a lens to focus on the gate. It can be activated by directing radiation on the silicon wafer of the gate with light. However, its turn-off time is longer compared to a standard SCR. The LASCR pellet is shown in Figure.2.15.

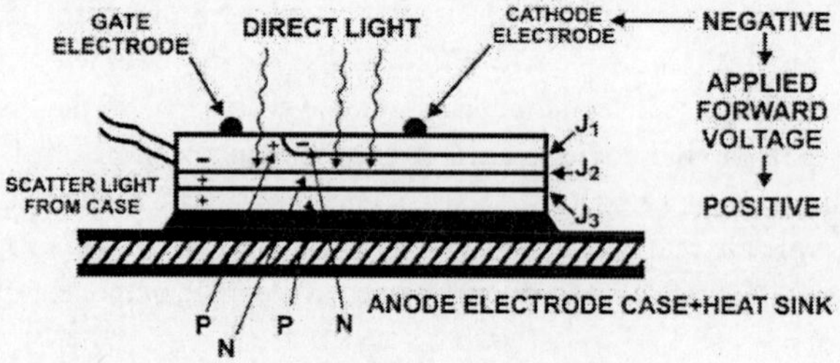

Figure.2.15 LASCR pellet

2.4.3 Operating principle of LASCR

Figure.2.16 (a) illustrates the configuration of a LASCR that is appropriately biased. By applying a supply voltage V_s, the LASCR is forward biased. To achieve optimal sensitivity, it is recommended to leave the gate *G* unconnected. When light does not pass through the transparent window on the p_l base layer, the LASCR remains inactive since the junction J_2 is reverse biased. Therefore, the LASCR is in an OFF state when there is no light, which is known as the forward blocking state. However, if light passes through the transparent window and the light intensity is strong enough, it generates electron-hole pairs in the p_l layer. These pairs then diffuse towards the J_l and J_2 junctions, reducing the width of the depletion layer and the reverse biasing across the junction. The carriers injected into J_2 further decrease the reverse bias, leading to the breakdown of the junction and allowing current to flow through the LASCR from anode to cathode. Hence, when there is light, the LASCR becomes active and is in an ON state or forward conduction state. To activate the LASCR, a positive pulse voltage can be applied to the gate *G* through incident light, which has low sensitivity to light. Additionally, the sensitivity of the circuit to light intensity can be adjusted by inserting a variable resistor in between gate and cathode.

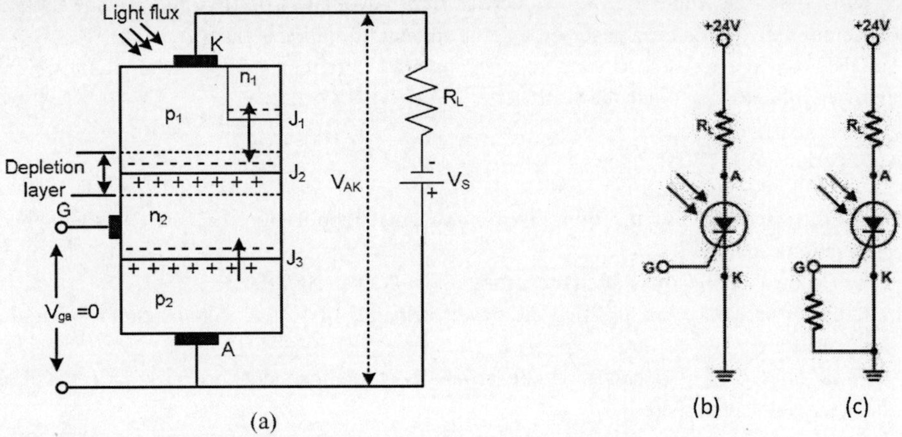

Figure.2.16 (a) Operation of the LASCR, (b) Simple LASCR at maximum sensitivity, (c) Simple LASCR at variable sensitivity

LASCR is utmost responsive to light when the gate terminal is open. To achieve maximum sensitivity, the gate is kept open, as shown in Figure.2.16 (b). However, if the triggering point of the LASCR needs to be adjusted, a variable resistor is installed along with its gate, as depicted in Figure.2.16(c). By applying resistance between the gate and ground, some of the electrons generated by the intensity of light are diverted, altering their path and reducing the circuit's sensitivity to incoming light. This enables the adjustment of the level of light at which the SCR triggers on. It is imperative to note that the LASCR's gate is connected to a lead, similar to a regular SCR, allowing it to be triggered on as needed by providing positive signals to its gate.

2.4.4 Voltage-current (V-I) characteristics

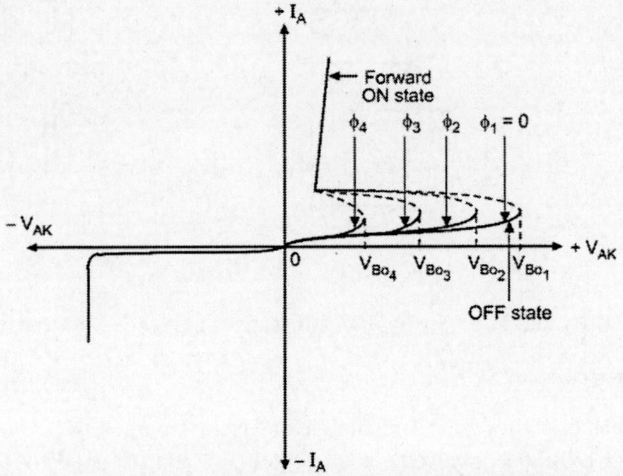

Figure.2.17 V-I Characteristics of the LASCR

Voltage- Current characteristics of a LASCR is as depicted in Figure.2.17. The LASCR is forward biased with open Gate. When there is no light present ($\Phi_1 = 0$), the LASCR will not conduct electricity and will be in a state known as forward blocking (OFF). As light of varying intensities (flux) shines on the LASCR, it gradually becomes conductive at lower breakover voltages, indicating that it has entered a forward conduction state (ON). Once it has been turned on, a small voltage is produced across the LASCR like an SCR. To decrease turn-on voltage, a positive pulse can be applied at gate *G*. Additionally, it has been observed that the amount of luminous flux ($\Phi_4 > \Phi_3 > \Phi_2 > \Phi_1$) incident on the LASCR affects the forward voltage required to turn it on, with larger amounts of light requiring lower voltages to activate the LASCR.

2.4.5 Applications of LASCR

(a) Light activated SCR is typically utilized in low power applications because of its low power generation capacity.
(b) Motor control applications often incorporate light activated SCR.
(c) LASCR is necessary for fulfilling the power demands of various components in computer systems.
(d) Optical light control operates on the principle of photoconduction, making LASCR an indispensable component.
(e) In solid state relays, two LASCR are connected in reverse parallel to produce power during both half cycles of AC.
(f) It can be used to trigger high power SCR.
(g) It also can be used in high voltage DC transmission and static VAR compensation.

2.5 SILICON CONTROLLED SWITCH

2.5.1 Introduction

Like SCR, the Silicon Controlled Switch (SCS) is a silicon-based device with a unilateral structure consisting of four layers and three junctions. The device features four electrodes, namely the cathode (*K*), cathode gate (G_K), anode gate (G_A), and the anode (*A*), which are illustrated in Figure.2.18. The symbol of SCS is shown in Figure.2.18(a).

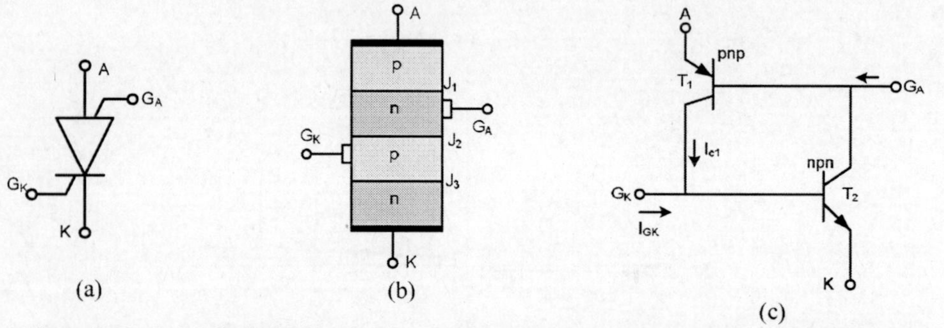

Figure.2.18 (a) Symbol of SCS, (b) *pnpn* structure of SCS, (c) Two transistor model of SCS

2.5.2 Construction of SCS

The SCS is comprised of four semiconductor layers arranged as *pnpn*, and it possesses four terminals: cathode (*K*), cathode gate (G_K), anode gate (G_A), and the anode (*A*). This device has three junctions and is constructed similarly to an SCR. The basic structure for an SCS is shown in

Figure.2.18(b). and it is commonly available in low power ratings. Compared to the SCR, the SCS is a low-power device that can handle currents in the milliampere range rather than the ampere range. The SCS differs from the SCR in several ways, including the presence of an additional gate known as the anode gate, its smaller physical size, lower leakage, and holding currents, and its ability to operate with small triggering signals. Additionally, the SCS exhibits more consistent triggering characteristics from sample to sample. The *pnpn* structure of SCS is shown in Figure.2.18 (b) and the equivalent transistor model is depicted in Figure.2.18(c).

2.5.3 Operating principle of SCS

The SCS is considered as consisting of a complementary pair of transistors with regenerative feedback. The operation of SCS is like an SCR, it can be turned on by either a positive pulse at gate G_K or a negative pulse at gate G_A. Its construction is similar to that of an SCR. The operating principle is illustrated here with the two-transistor model exposed in Figure.2.19(a).

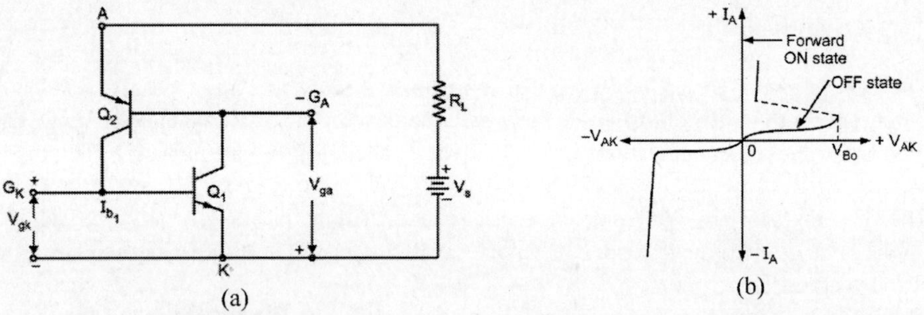

Figure.2.19 (a) Operation of SCS using two transistor model, (b) V-I characteristic

The SCS is made forward biased when a positive voltage is applied at anode (*A*) w.r.t cathode and no voltage is applied at the gates G_K and G_A. As the gate current is zero, both the transistors will be OFF. If a positive pulse is applied at the Gate (G_K), the gate current will flow through the base of transistor Q_1. Due to regenerative feedback action, the transistors, Q_1 and Q_2 are driven into saturation and heavy current will flow through the device from anode to cathode. This is forward ON (conduction) state of a SCS. The SCS can also be turned ON by applying a negative pulse at the anode gate G_A. The SCS can be turned OFF from its ON state by applying a positive pulse at anode G_A or a negative pulse at cathode gate G_K. Once the SCS is turned ON, it behaves like an SCR. The SCS can be turned off like SCR by reducing the anode current below the holding current I_H. Despite, the SCS having the ability to be turned off at either gate, its turn-off gain is relatively low, similar to that of a gate-controlled switch. Nevertheless, since the SCSs are designed for low-current applications, that necessitate low gate currents for turn-off, which means this drawback is not significant. Its versatility stems from its ability to be activated and deactivated by signals of either polarity.

2.5.4 V-I characteristics of SCS

When the gate G_K receives a positive pulse voltage or the gate G_A receives a negative pulse voltage under a forward biased condition that is lower than the breakover voltage, the SCS will not conduct current. However, reverse saturation (leakage) current through the SCS, which will put it in a forward blocking (OFF) state. On the other hand, if the positive pulse voltage at the gate G_K or the negative pulse voltage at the gate G_A exceeds the breakover voltage under the forward biased condition, the SCS will start conducting current from anode to cathode. At this point, the anode current will experience a sudden rise, but the voltage across it will drop to a low voltage of approximately 1V. This state is called the forward conduction (ON) state of the SCS. The forward OFF and ON states of the SCS are illustrated in Figure2.19. Similar to SCR, SCS also displays negative differential resistance in the on state. If there is a sudden application of anode voltage, SCS can unintentionally switch ON due to a phenomenon called **rate effect**. The **transition capacitance** between G_A and G_K electrodes are known as interelectrode capacitance, and is responsible for this effect.

2.5.5 Advantages and disadvantages of SCS

Compared to an SCR, one benefit of using an SCS is its shorter turn-off time, usually falling within the range of 1 to 10μs for the SCS and 5 to 30μs for the SCR. In addition to this advantage, the SCS offers improved control and triggering sensitivity, as well as a more predictable firing situation. Nevertheless, the SCS has its limitations in terms of power, current, and voltage ratings, with maximum anode currents ranging from 100 mA to 300 mA and a dissipation rating of 100 to 500mW.

2.6 GATE TURN-OFF THYRISTOR

2.6.1 Introduction

The SCR is a semi-controlled switch that can be activated by applying a positive gate current, but it is not capable to turned off using the same gate. To turn off the SCR, the main current must be interrupted by means of a commutation circuit. To resolve this issue, the gate turn-off (GTO) thyristor was developed.

The GTO thyristor provides complete control over switching, meaning it can be turned on and off using the same gate terminal. Despite its unique capability, the GTO shares many similarities with a standard thyristor.

The GTO is a semiconductor-based unidirectional switch that is fully controlled and has three terminals named as gate (*G*), cathode (*K*), and anode (*A*). Its ON/OFF state can be controlled through the gate terminal. Application of a positive current pulse to the gate turns ON the GTO, while a negative current pulse switches OFF the GTO. The current flows only in one direction i.e from anode to cathode. A positive current pulse at the gate can also trigger it into conduction mode. Although the GTO has a low on-state voltage drop, it requires a relatively high turn-off current at the gate. In particular, the negative current pulse required to switch it OFF is approximately one-fourth of the anode current. The symbols of GTO are displayed in Figure.2.20 (a) and (b).

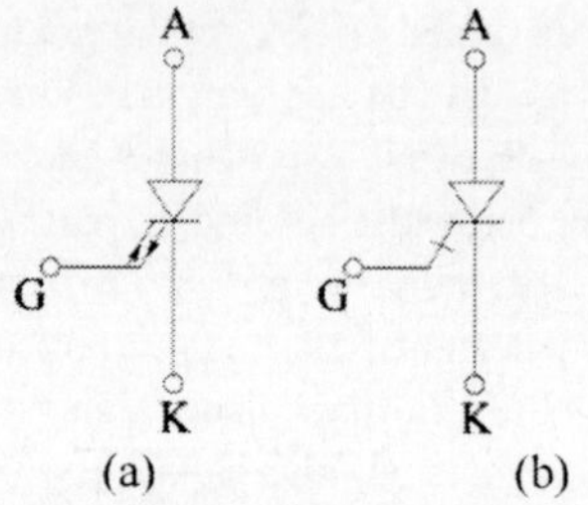

Figure.2.20 Symbols of GTO

2.6.2 Construction of GTO

The GTO shares a similar structure with a regular thyristor in that it is composed of four *pnpn* layers and three junctions. The anode of the GTO is made up of a p^+ layer with n^+ type fingers that are diffused within it, while the highly doped n^+ layer serves as the cathode. This results in a low breakdown voltage for junction J_3, typically ranging from 20 to 40V. To maintain excellent emitter efficiency, the *p*-layer doping level must be kept low, while the region doping must be high for good switch-off properties.

The anode junction is defined as the junction between the p^+ (p^+ heavily doped region for the purpose of maintaining high anode efficiency) anode and n base. A heavily doped p^+ anode region is necessary for achieving good switch-on properties, but this can negatively impact switch-off capabilities. To address this issue, n^+ layers that are heavily doped can be initiated at regular intervals within the p^+ anode layer. This allows electrons to move from the base region to the anode metal contact from the p^+ anode without causing hole-injection, resulting in a GTO structure with anode shorted. However, the reverse blocking capacity of the GTO can be reduced towards the reverse breakdown voltage of the J_3 junction, and the switch-on performance can be degraded with increased anode shorts density. Thus, careful consideration must be taken to strike a balance between anode shorts density and switch-on/off performance. The structure of GTO is depicted in Figure.2.21(a). The two-transistor analogy is depicted in Figure.2.21(b).

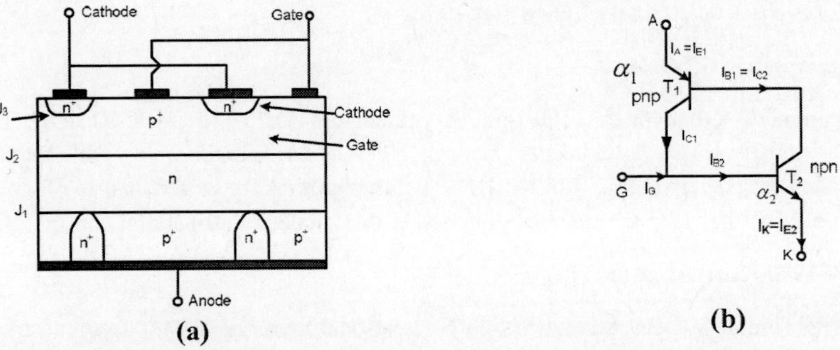

Figure.2.21 (a) Structure of GTO, (b) transistor model

2.6.3 Operating principle

The GTO functions similarly to SCR, but it has the additional feature of being able to turn off.

Turn-on mechanism: The conventional thyristor and GTO have the same turn-on operation. GTO can be activated through two methods: firstly, by raising the forward voltage beyond the break-over voltage and secondly, by application of a positive gate current. When the GTO receives a forward voltage, the anode voltage becomes positive w.r.t cathode, and this results forward biasing J_1 and J_3 junctions and reverse biasing J_2 junction. Thus, junction J_2 prevents the flow of current. However, if the forward voltage is raised beyond the **forward break-over** voltage, an avalanche occurs, and the J_2 turn out to be forward biased, thereby allowing the flow of current. Nonetheless, this type of switching is avoided because it is destructive. To properly turn on a GTO, a positive gate current should be applied while the forward voltage is being applied. The application of a positive gate current injects holes into the p gate region, causing J_3 to become forward biased, which then allows current to pass through the device.

Turn-off mechanism: Providing a negative gate current is necessary to turn off the GTO. When the gate metallization collects holes from the anode, the voltage drop in the p base region is greater than that in the n emitter region. This results in the reverse biasing of junction J_3, which stops the flow of electrons. This occurs at the boundary region between the p base and n emitter layers. At the gate contact anode region, high-density filament current congestion can occur, which may cause device failure if not handled promptly. Once these filaments disappear completely, the flow of electrons stops, and a depletion region is generated at junctions J_2 and J_3. This blocks the forward voltage through the device, but the anode to gate current still flows due to the diffusion of n base carriers towards J_1. This current is known as **tail current** and reduces exponentially when the excess carriers of the n base region recombine completely. To expedite the decay of tail current, the electron from the *n* base region needs an alternative path to reach the anode contact without further emission from the anode.

2.6.4 Types of GTO

There are two types of GTOs based on their structure. They are

(a) Symmetric GTO
(b) Asymmetric GTO

(a) Symmetric GTO

The voltage blocking capabilities of the symmetric GTO are symmetrical, with the reverse blocking voltage (RBV) being just as high as the forward voltage. Unlike a "shorted-anode" structure, the anode of this GTO is constructed solely from a pure p^+ region.

(b) Asymmetric GTO

The most commonly used type of GTOs is the Asymmetric GTOs, which are also referred to as "shorted anode GTO". These GTOs have unequal voltage blocking abilities, meaning their forward blocking voltage differs from their RBV, with the RBV being significantly lower than the forward blocking voltage. Typically, they are utilized in conjunction with a diode in anti-parallel configuration.

2.6.5 V-I characteristics of GTO

GTO exhibits similar V-I characteristics to a conventional thyristor (CT), but with a higher latching current, typically around 2A compared to 100-200 mA for a CT. The V-I characteristics of the GTO include four regions: Forward Blocking Mode (FBM), Forward Conduction Mode (FCM), Reverse Blocking Mode (RBM), and Reverse Conduction Mode (RCM), which are illustrated in the Figure.2.22.

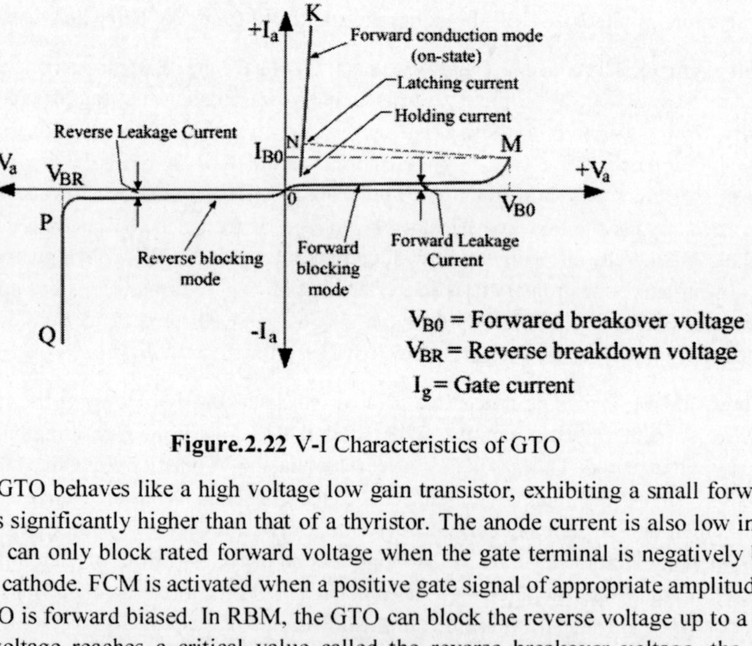

Figure.2.22 V-I Characteristics of GTO

In FBM, the GTO behaves like a high voltage low gain transistor, exhibiting a small forward leakage current that is significantly higher than that of a thyristor. The anode current is also low in this mode, and the GTO can only block rated forward voltage when the gate terminal is negatively biased with respect to the cathode. FCM is activated when a positive gate signal of appropriate amplitude is applied while the GTO is forward biased. In RBM, the GTO can block the reverse voltage up to a limit, but if the reverse voltage reaches a critical value called the reverse breakover voltage, the GTO starts conducting in reverse direction. However, this operation is not destructive to the device if the gate is negatively biased, and the time of operation is short. The reverse blocking capability of the GTO is dependent on the type of GTO. A symmetric GTO has a higher reverse blocking capability than an asymmetric GTO, which typically has a reverse blocking capability of around 20V to 30V.

2.6.6 Advantages and disadvantages of GTO

A. Advantages of GTO

(a) GTO has exponential switching characteristics.
(b) GTO circuit has a smaller size and smaller weight when compared with thyristor circuit unit.
(c) Due to the absence of a commutation circuit, the GTO results in reduced cost, weight, and size.
(d) Compared to the SCR, the GTO has a faster switching speed.
(e) Maintenance requirements are lower for the GTO.
(f) GTO has a current surge capacity similar to that of an SCR.
(g) GTO has a high blocking voltage capacity.
(h) The di/dt ratings are higher during turn ON for the GTO.
(i) GTO has high efficiency.

B. Disadvantages of GTO

(a) The combined loss and ON-state voltage drop are greater in GTOs due to their multi-layered structure, which also results in higher gate triggering currents compared to conventional thyristors.
(b) Gate drive circuits experience significant losses when driving GTOs, and the ON-state voltage drop across the gate turn-off thyristor is also higher.
(c) GTOs require higher magnitudes of latching and holding current as compared to SCRs.
(d) The latching current for GTOs is 2A, while it ranges from 100mA to 500mA for SCRs.
(e) The triggering current required for GTOs is higher than that for SCRs.

2.6.7 Applications of GTO

GTO (Gate Turn-Off) thyristor is favoured over traditional thyristors in several applications due to its exceptional switching characteristics and ability to turn off without requiring a commutation circuit. The following are some examples of the applications where GTO is utilized:

(a) Employed for high-performance motor drives.
(b) Utilized in variable frequency drives (VFD) with regulating frequency.
(c) Used in DC-AC or DC-DC converters, and in inverters.
(d) Used in high-voltage direct current (HVDC) systems.
(e) Utilized in induction heating.
(f) Used in high-power AC/DC power supplies.

2.7 UNIJUNCTION TRANSISTOR

2.7.1 Introduction

Unijunction Transistor, or UJT, is a solid-state device with three terminals. The UJT can function as a gate pulse, timing circuit, or trigger generator for switching and control of thyristors and TRIACs Like diodes, UJTs are made up of distinct *p*-type and *n*-type materials that create a single *pn*-junction within the main conducting *n*-type channel of the device. Despite its name, the UJT has different switching characteristics than BJT or FETS, and cannot amplify signals. Instead, UJTs function as on-off switching transistors with unidirectional conductivity and negative impedance characteristics. It acts like a variable voltage divider during breakdown. The UJT consists of a single solid piece of *n*-type semiconductor forming the main current-carrying channel, with Base 2 (B_2) and Base 1 (B_1) as its outer connections, and the Emitter (*E*) located along the channel. The Emitter terminal is represented by an arrow pointing from the *p*-type emitter to the *n*-type base, and the Emitter rectifying *pn*-junction is formed by fusing the *p*-type material into the *n*-type silicon channel. There are also *p*-channel UJTs with an *n*-type Emitter terminal, but these are not commonly used. The Emitter junction is positioned closer to terminal B_2 than B_1, and the UJT symbol includes an arrow pointing towards the base, indicating that the Emitter terminal is positive and the silicon bar is negative material. The UJT symbol, *p-n* structure, and equivalent structure is shown in Figure.2.23 (a), (b), and (c) respectively. The symbol for the unijunction transistor looks very similar to that of the junction field effect transistor (JFET), except for the bent arrow representing the Emitter (*E*) input. Although the two types of transistors have similar ohmic channels, they operate differently, and one should not confuse them.

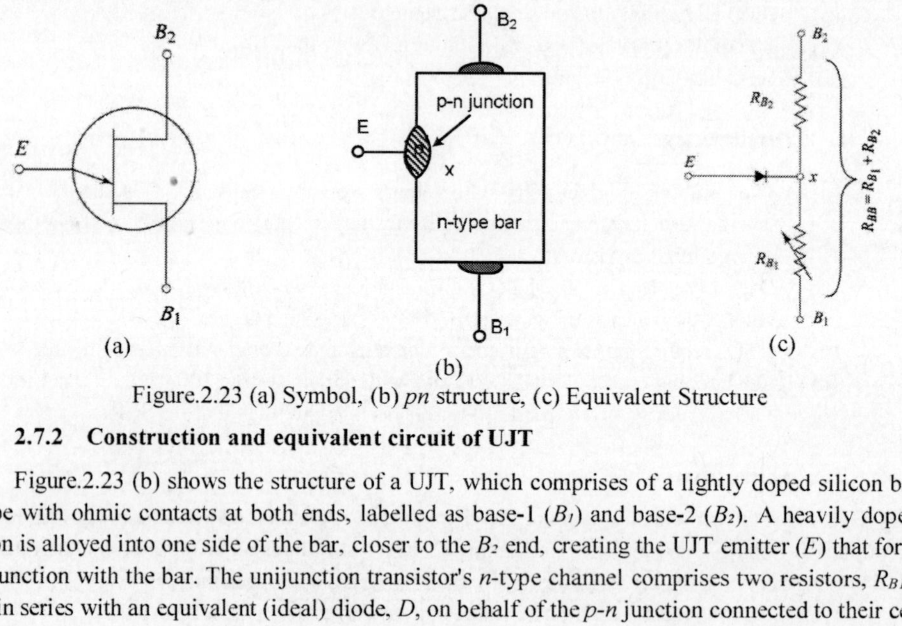

Figure.2.23 (a) Symbol, (b) *pn* structure, (c) Equivalent Structure

2.7.2 Construction and equivalent circuit of UJT

Figure.2.23 (b) shows the structure of a UJT, which comprises of a lightly doped silicon bar of *n*-type with ohmic contacts at both ends, labelled as base-1 (B_1) and base-2 (B_2). A heavily doped *p*-region is alloyed into one side of the bar, closer to the B_2 end, creating the UJT emitter (*E*) that forms a *p-n* junction with the bar. The unijunction transistor's *n*-type channel comprises two resistors, R_{B1} and R_{B2}, in series with an equivalent (ideal) diode, *D*, on behalf of the *p-n* junction connected to their center point.

2.7.3 Operating principle and *V-I* or V_E- I_E characteristics of UJT

During the manufacturing process, the Emitter *p-n* junction is permanently positioned along the ohmic channel and cannot be altered. R_{B1} and R_{B2} are situated between the Emitter (*E*) and terminals B_1 and B_2, respectively. Due to the physical proximity of the *p-n* junction to terminal B_2, R_{B2} has a lower resistive value than R_{B1}. The overall resistance of the silicon bar, also known as its ohmic resistance, depends on the dimensions of the *n*-type silicon channel and the semiconductor's actual doping level. This resistance is represented as R_{BB} and is typically between 4kΩ and 10kΩ when measured with an ohmmeter for most common UJTs, including the 2N1671, 2N2646, and 2N2647. The two series resistances of R_{B1} and R_{B2} create a voltage divider network between the UJT's two base terminals. When a voltage is applied to the device, the potential at any point along the channel is proportional to its position between terminals B_2 and B_1, with the voltage gradient level determined by the supply voltage.

In a circuit, the Emitter serves as the input to the device, and terminal B_1 is connected to ground. If a voltage V_{BB} is applied between B_2 and B_1 while zero Emitter input is given, the voltage developed across R_{B1} of the resistive voltage divider can be calculated as shown below

$$V_x \text{ or } V_{R_{B_1}} = \frac{R_{B_1}}{R_{B_1} + R_{B_2}} \times V_{BB} = \underbrace{\frac{R_{B_1}}{R_{BB}}}_{\eta} \times V_{BB}$$

$$\therefore V_x \text{ or } V_{R_{B_1}} = \eta \cdot V_{BB} \tag{2.18}$$

The intrinsic stand-off ratio, denoted by the Greek letter η (eta), is the resistive ratio of R_{B1} to R_{BB} in a unijunction transistor. Common UJTs typically have η values ranging from 0.5 to 0.8. When a small positive input voltage that is less than $\eta \cdot V_{BB}$ is applied to the Emitter input terminal, the P-N junction of the diode is reverse biased, resulting in a high impedance and no current flow. This turns the UJT "OFF". However, when the Emitter input voltage exceeds $\eta \cdot V_{BB}$ + 0.7V, the *p-n* junction becomes forward biased, and the UJT begins to conduct current ($\eta \cdot I_E$) from the Emitter into the Base region. When additional current is directed towards the Base, the channel between the Emitter junction and the B_1 terminal experiences a decrease in resistive properties, which in turn lowers the resistance value of R_{B1} to a low level. As a result, the Emitter junction becomes more forward-biased, leading to an increase in current flow and generating negative resistance at the Emitter terminal. Conversely, if the voltage between the Emitter and B_1 terminal drops below breakdown, the R_{B1} resistance value increases to a high level, and the UJT acts as a voltage breakdown device. Thus, the resistance of R_{B1} varies depending on the level of Emitter current (I_E). Forward-biasing the Emitter junction causes a reduction in the resistance between the Emitter (*E*) and B_1, leading to a decrease in the voltage drops across R_{B1} (V_{RB1}), an increase in current flow, and the occurrence of negative resistance. In summary, the flow of current into the UJT's Emitter triggers a decrease in the resistive value of R_{B1}, resulting in a negative resistance state.

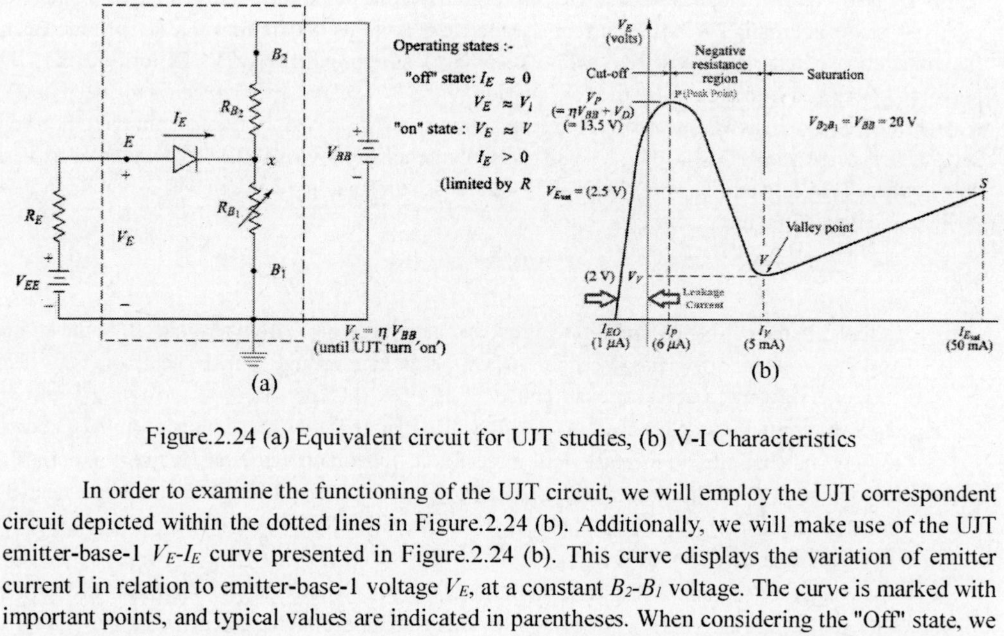

Figure.2.24 (a) Equivalent circuit for UJT studies, (b) V-I Characteristics

In order to examine the functioning of the UJT circuit, we will employ the UJT correspondent circuit depicted within the dotted lines in Figure.2.24 (b). Additionally, we will make use of the UJT emitter-base-1 V_E-I_E curve presented in Figure.2.24 (b). This curve displays the variation of emitter current I in relation to emitter-base-1 voltage V_E, at a constant B_2-B_1 voltage. The curve is marked with important points, and typical values are indicated in parentheses. When considering the "Off" state, we can temporarily disregard the diode and observe that R_{B1} and R_{B2} create a voltage divider, resulting in a voltage V at point *x* in relation to ground, as shown in Figure.2.24(a).

At point x, the voltage corresponds to the voltage on the N-side of the p-n junction. When the V_{EE} source is applied to the emitter, which is on the p-side, the emitter diode becomes reverse-biased. This results in the "off" state, which is represented by a very low current region on the V-I curve. During the "off" state, the UJT displays a high resistance between E and B_1, and the reverse leakage current is usually negligible. Without any I_E, there is zero drop across R_E, and the emitter voltage, V_E, is equal to the source voltage. The UJT's "off" state, depicted on the V_E-I_E curve, continues until the emitter voltage surpasses Vx, augmented by the diode threshold voltage V_D that is required to generate forward current through the diode. V_P, is known as the peak-point voltage and is located at point P. It can be computed as:

$$V_P = V_x + V_D = \eta \cdot V_{BB} + V_D \quad (2.19)$$

with V_D, typically being 0.5V. For instance, when η equals 0.65 and V_{BB} equals 20V, V_P equals 13.5 V. As V_{BB} changes, V_P will also alter. When V_{EE} rises, the UJT remains "off" until V_E approaches the peak-point value V_P. Once V_E approaches V_P, the p-n junction becomes forward-biased and starts conducting in the reverse direction, signifying the "on" state.

On the V_E-I_E curve, it can be observed that the current I_E becomes positive as it approaches the peak point P. When V_E is equal to V_P, the current flowing through the emitter is equal to the peak-point current, I_P. At this juncture, the heavily doped emitter releases holes which move into the n-type bar, particularly in the B_1 area. Due to the lightly doped nature of the bar, the holes find few chances to recombine. Consequently, the lower section of the bar is overwhelmed with additional current carriers, in the form of holes, causing the resistance of R_{B1} to be dramatically reduced. The drop in R_{B1} results in a decline of V_P. Consequently, the diode becomes more forward biased, causing I_E to further increase. The injection of a larger I_E causes more holes to move into B_1, causing a further reduction in R_{B1}, and so on. Eventually, this regenerative process ends with R_{B1} dropping to a very small value (between 2-25 Ω), leading to a very large I_E, limited primarily by the external resistance R_E.

UJT undergoes a transition to the high-current, low-voltage region of its V_E-I_E curve, where the resistance is low and the slope is steep in the "on" region. At this point, the emitter voltage (V_E) remains relatively stable at around 2 V as the current (I_E) increases up to its maximum value ($I_{E(sat)}$). Increasing the voltage V_{EE} results in a boost in I_E while keeping V_E at approximately 2V. To turn off the UJT, reducing V_{EE} causes I_E to drop along the "on" portion of the V_E-I_E curve until it reaches the valley point, V, where I_E equals I_V, the valley current. I_V is the holding current necessary to maintain the UJT's "on" state. If I_E is decreased below I_V, the UJT will turn off and quickly return to the "off" region of its V_E-I_E curve, where I_E is zero and V_E -V_{EE}. The valley current is similar to the holding current found in *pnpn* devices and typically ranges from 1 to 10 mA.

Notes:

- Initially, in the cut-off region of operation, a small current flow from terminal B_2 to the emitter as a result of minority charge carriers when the emitter voltage is increased from zero. This current is referred to as leakage current.
- Once the emitter voltage exceeds a specific value, denoted as V_E, the emitter current (I_E) begins to flow and continues to increase until it reaches its maximum values of V_P and I_P at point P.
- Beyond point P, further increases in V_E lead to a sudden rise in I_E, accompanied by a decrease in V_E. This section of the curve is known as the Negative Resistance Region since an increase in I_E leads to a reduction in V_E.
- The Negative Resistance Region concludes at the valley-point (V), which has a valley-point voltage (V_V) and current (I_V). Following the valley-point, the device enters the saturation region.

2.7.4 Types of UJTs

Below are descriptions of the three main types of unijunction transistors.

(a) **Original UJT:** Essentially, the original unijunction transistor consists of a bar made of *n*-type semiconductor, into which a section of *p*-type semiconductor is diffused at some point along its length.

(b) **Complementary UJT:** The complementary unijunction transistor can be described as a solid piece of semiconductor material that is primarily composed of *p*-type material. However, there is a portion of the bar where *n*-type semiconductor material has been diffused into it.

(c) **Programmable UJT:** The programmable unijunction transistor, which features several junctions and two resistors that are external, is a device that can perform a comparable function to the conventional UJT. However, it is not directly replaceable with the conventional UJT due to its unique structure.

2.7.5 Advantages of UJT

(a) It is affordable
(b) Excellent characteristics
(c) Consumes minimal power during normal operation.

2.7.6 Applications of UJT

(a) Oscillators
(b) Trigger Circuits
(c) Saw tooth generator
(d) Bi-stable networks
(e) Pulse and voltage sensing circuits
(f) UJT relaxation oscillators
(g) Over voltage detectors

2.8 *PROGRAMMABLE UNIJUNCTION TRANSISTOR*

2.8.1 Introduction

The PUT, or Programmable Unijunction Transistor, is a *pnpn* device that shares a striking resemblance to the UJT. Although the two are distinct, the PUT is often discussed in conjunction with the UJT due to their comparable behaviour. Notably, PUT can be viewed as a UJT with a controllable trigger voltage V that can be established by an external voltage divider.

2.8.2 Symbol and construction of PUT

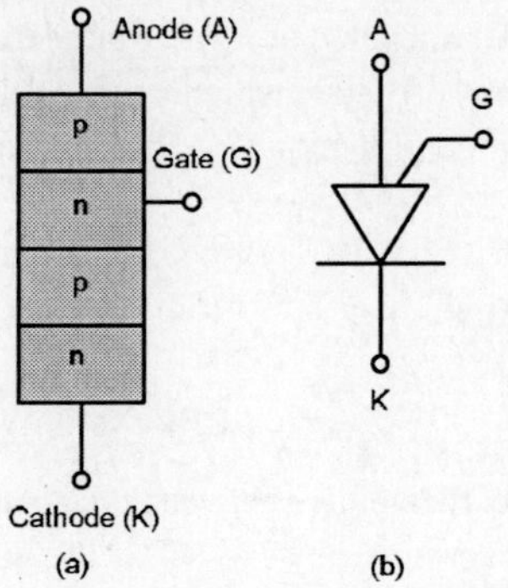

Figure.2.25 (a) *pnpn* structure of PUT, (b) Symbol of PUT

In Figure.2.25, the *pnpn* structure and circuit symbol for the PUT are displayed, with the anode (A) and cathode (K) serving as the same for all *pnpn* devices. The gate (G), located adjacent to the anode, forms a *p-n* junction with it and controls the device's "on" and "off" states. Typically, the gate is positively biased by a certain amount, V_g, with respect to the cathode. When the anode voltage is less than V_g, the anode-gate junction is reverse-biased, and the device remains in the "off" state, behaving like an open-switch between anode and cathode. When the anode voltage exceeds V_g by approximately 0.5 V, the anode-gate junction becomes conductive, causing the device to switch "on" in the same way as forward-biasing the gate-cathode junction of an SCR. In the "on" state, the PUT operates like any other *pnpn* device between the anode and cathode, with low resistance and V_{AK} of 1V. The PUT is sometimes known as a complementary SCR (CSCR).

2.8.3 Operation of PUT

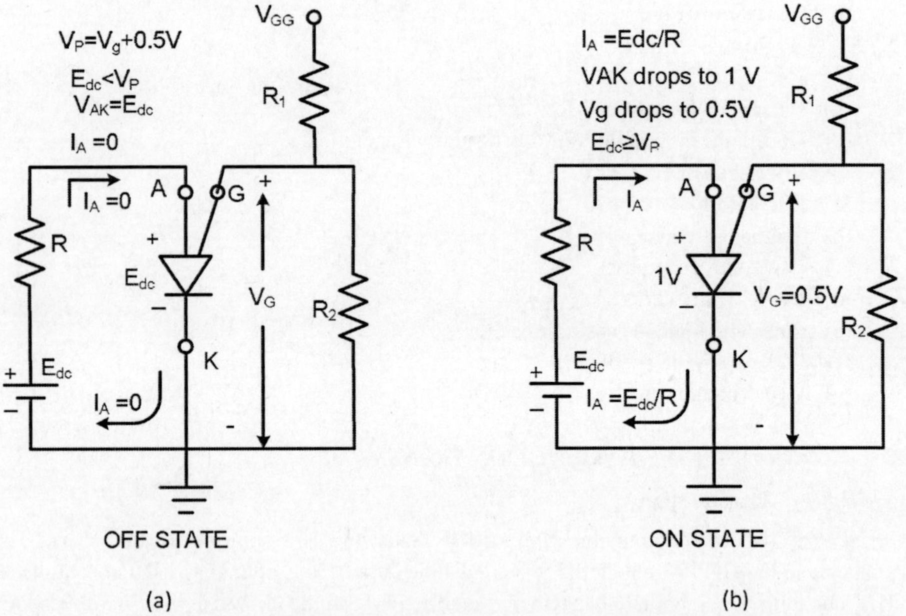

Figure.2.26: (a) off state of PUT, (b) on state of PUT

Figure.2.26 illustrates the typical bias configuration for the PUT. The gate voltage, V_g, is established by the voltage divider consisting of R_1 and R_2. It should be noted that R_1 and R_2 are not part of the device and can be selected to obtain any desired value of V_g. The anode-cathode bias is provided by E_{dc}. When E_{dc} is less than V_g, the device is in an "off" state with no current flowing through it, and the full voltage E_{dc} is applied across the anode-cathode junction ($V_{AK} = E_{dc}$). Part (a) of Figure.2.26 summarizes the "off" state.

Increasing the Emitter-to-Collector voltage (E_{dc}) by approximately 0.5V above the bias value of V_g causes the device to switch to the "on" state. This means that the Peak-Point Voltage (V_p) for the PUT is given as:

$$V_p = V_g = 0.5V \tag{2.20}$$

When the device is in the "on" state, the voltage between the anode and cathode (V_{AK}) drops to 1V, and the anode current (I_A) is mainly determined by E_{dc}/R_1, as the current is limited by R. Moreover, the voltage of V_g decreases significantly to around 0.5V since R_2 is bypassed by the "on" *pnpn* structure.

The device will remain in the "on" state until the anode current drops below the valley current (I_V). The characteristics of the "on" state are described in part (b) of Figure.2.26.

2.8.4 V-I characteristics of PUT

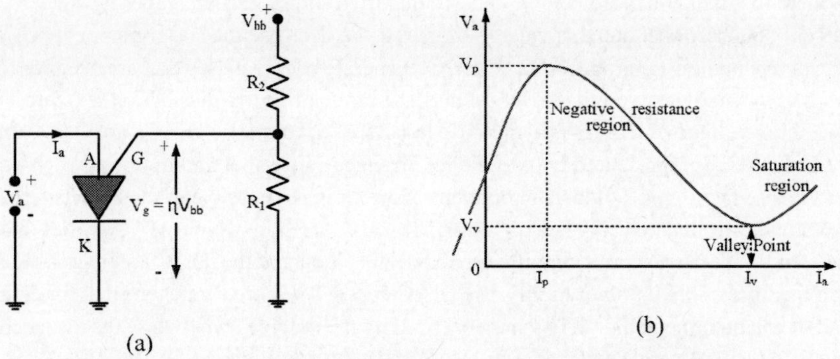

Figure.2.27 (a) Biasing circuit of PUT, (b) V-I characteristics of PUT

PUT (Programmable Unijunction Transistor) characteristics are defined by a plot of anode voltage (V_a) versus anode current (I_a). The biasing circuit and V-I characteristics of PUT are shown in Figure.2.27(a) and Figure.2.27(b) respectively. Typically, the anode is connected to a positive voltage and the cathode to ground. The gate is connected to the junction of two external resistors (R_1 and R_2), which create a voltage divider network that determines the intrinsic standoff ratio (η) and peak voltage (V_p) of the PUT. As the anode-cathode voltage (V_a) is increased, the anode current also increases, and the junction behaves like a standard *p-n* junction. However, there is a limit to how much V_a can be increased, as the junction becomes saturated after a sufficient number of charges are injected. Beyond this point, the anode current (I_a) increases and the anode voltage (V_a) decreases, creating a negative resistance scenario. This negative resistance region in the PUT characteristic is used in relaxation oscillators. When the anode voltage (V_a) is reduced to a specific level known as the "Valley Point," the device becomes fully saturated, and no further decrease in V_a is possible. At this point, the device behaves like a fully saturated *p-n* junction. The point at which the PUT enters the negative resistance region is referred to as the peak voltage (V_p), which is the voltage measured between the anode and cathode.

$$V_p = 0.7V + V_g = 0.7V + V_{R_1} = 0.7V + \eta V_{bb} \tag{2.21}$$

where η represents the intrinsic standoff ratio and V_{bb} represents the total voltage across the external resistor network.

The intrinsic standoff ratio (η) of a PUT is defined as the ratio of external resistor R_1 to the sum of R_1 and R_2. This ratio allows us to predict the amount of voltage that will be dropped across the gate and cathode for a given V_{bb}. Mathematically, the intrinsic standoff ratio can be expressed as

$$\eta = R_1 / (R_1 + R_2) \tag{2.22}$$

2.9 BIDIRECTIONAL THYRISTOR DIODE OR DIODE FOR ALTERNATING CURRENT (DIAC)

2.9.1 Introduction

A DIAC is a type of power semiconductor device comprising of two junctions and three layers. Its name is a combination of two parts: DI, which refers to the diode (such as Di, Tri, Quad, Penta, etc.), and AC, which represents alternating current. Essentially, DIAC is an abbreviation for a diode (member of the thyristor family) designed to handle alternating currents. This is a bidirectional switch with two terminals that conducts electricity in both directions when the applied voltage surpasses its breakover voltage. However, it is not capable of amplifying or providing regulated switching. DIACS are seldom employed individually, but rather combined with other thyristor devices. In phase control circuits, they serve as triggers to provide gate pulses to a TRIAC or SCR. Many gate triggering circuits utilize this component to attain enhanced stability and immunity to noise during triggering These silicon devices are not only bidirectional but also voltage-triggered. To decrease the DC component in the load circuit, DIAC voltage options ranging from 27V to 70V generate trigger pulses that are symmetrically matched at the positive and negative break-over points. The range of power dissipation is 0.5 to 1W. Basically, the switching voltage of DIACs is symmetric in nature. Asymmetric DIACs are also available.

The DIAC is represented by two diodes arranged in parallel and opposing each other, with two terminals. The DIAC can conducts in both directions, it is not possible to identify its terminals. Rather, the terminals are denoted as A_1 and A_2 or MT_1 and MT_2, in which *MT* signifies main terminals. Consequently, like a resistor or ceramic capacitor, the pinouts of the DIAC are interchangeable. Despite being a member of the thyristor family, the DIAC does not feature a gate terminal for control purposes. Instead, it can be turned on or off by merely reducing the voltage level below the avalanche breakdown voltage in either polarity. The symbol of DIAC is depicted in Figure.2.28.

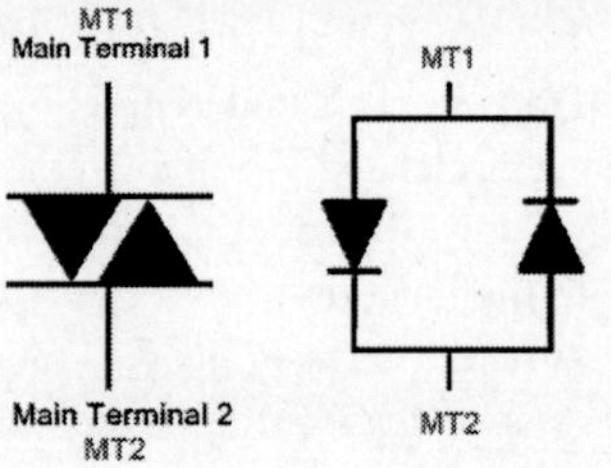

Figure.2.28 Symbol of DIAC

Depending on the polarity of the applied voltage, each terminal has the ability to function as either an anode or a cathode. The end junction that is active and the one that is bypassed are also determined by the polarity of the applied voltage.

2.9.2 PNPN structure or construction of DIAC

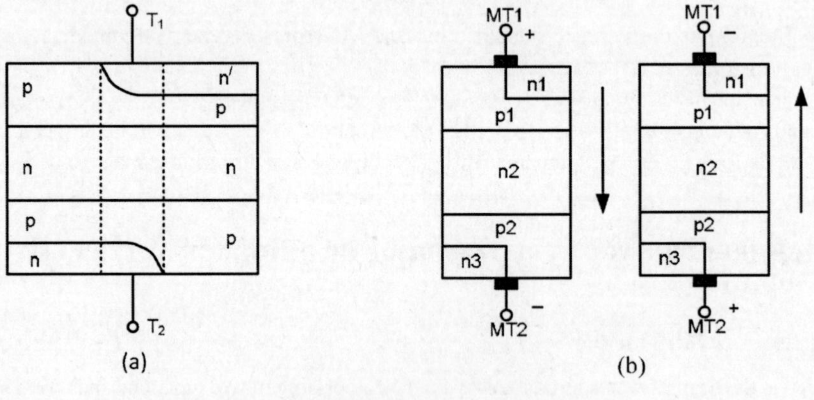

Figure.2.29 (a) *pnpn* construction, (b) Current flow

In essence, the DIAC is a device with two terminals that consists of parallel layers of semiconductor material, which enables it to conduct in one direction. It is commonly used to activate the TRIAC. The DIAC has two terminals - MT_1 (T_1) and MT_2 (T_2). When a positive voltage is applied to MT_1 relative to MT_2, current flows through the four-layer *pnpn* structure, which is another type of diode. The DIAC can conduct in both directions and its symbol looks similar to that of a transistor. It acts as a diode that conducts only after a certain voltage, known as the break-over voltage or V_{BO}, is exceeded. Once this voltage is reached, the diode enters a region of negative dynamic resistance, which causes a decrease in the voltage drop across it with increasing voltage. This results in a rapid increase in current flow, which is controlled by the device. The diode remains in its conducting state until the current falls below a certain value known as the holding current or I_H. Once the holding current is reached, the DIAC reverts back to its non-conducting state. Its behaviour is bidirectional, which means it can conduct in both directions of an alternating cycle. The *pnpn* structure of DIAC is shown in Figure.2.29 (a) and (b).

2.9.3 Operation and V-I characteristics of DIAC

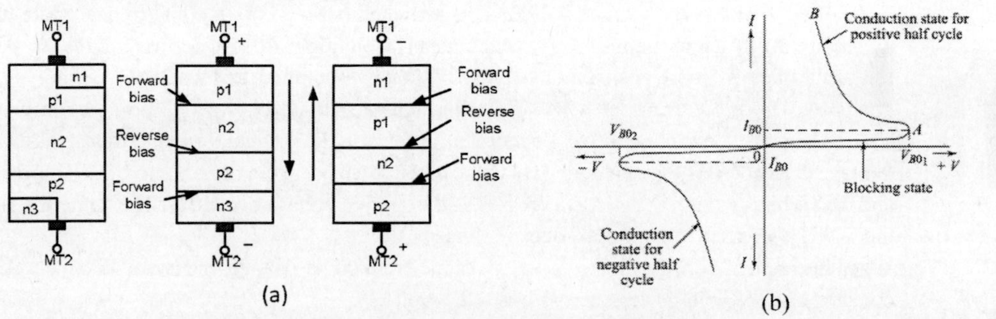

Figure.2.30 (a) Working principle, (b) V-I characteristics of DIAC

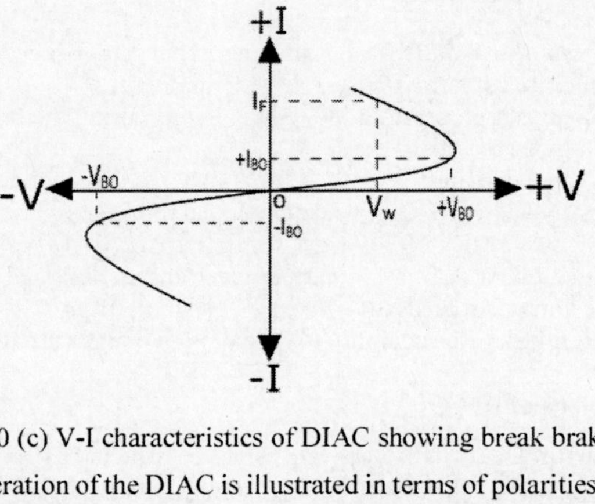

Figure.2.30 (c) V-I characteristics of DIAC showing break brake voltage

In Figure.2.30 (a), the operation of the DIAC is illustrated in terms of polarities. Assuming that the *MT1* terminal is positive, the p_1 layer close to *MT1* will be activated, resulting in conduction in the sequence of p_1-n_2-p_2-n_3. When the current flows from *MT1* to *MT2*, the junctions between p_1-n_2 and p_2-n_3 are forward-biased, while the junction between n_2-p_2 is reverse-biased. Likewise, if we assume that the *MT2* terminal is positive, the p_2 layer close to *MT2* will be activated, and conduction will occur in the order of p_2-n_2-p_1-n_1. The current will flow from *MT2* to *MT1*, and the junctions between p_2-n_2 and p_1-n_1 will be forward-biased, while the junction between n_2-p_1 will be reverse-biased. As a result, conduction will be possible in both directions.

The V-I characteristics of a DIAC is shown in Figure.2.30(b) and (c). The DIAC's V-I characteristic curve takes the form of a Z-shape and spans the first and third quadrants due to its ability to conduct in both positive and negative polarities. In the first quadrant, which corresponds to the positive half cycle, current flows from *MT1* to *MT2*, while in the second quadrant, which corresponds to the negative half cycle, current flows from *MT2* to *MT1*. Initially, the DIAC's resistance is higher due to the Reverse Bias junction between its layers, resulting in a small amount of leakage current flowing through it, which is referred to as the **blocking state** on the curve. When the applied voltage reaches the breakdown voltage, the DIAC's resistance decreases abruptly, causing it to begin conducting, resulting in a sharp decrease in voltage and an increase in current, which is referred to as the **conduction state** on the curve. The breakdown voltage of most DIACs is approximately 30 V, although the exact value is determined by the device type. The DIAC remains in the conducting state until the current reaches the **holding current**, which is the minimum current required to maintain the device in the ON state.

Notes:

(a) The voltage needed to activate the DIAC by passing through its terminals is referred to as the Break-Over Voltage (I_{BO}). Once the DIAC is switched on, the only method to deactivate it is by cutting off the current completely, which means isolating it from the power source.

(b) Initially, when a small positive or negative voltage is applied to the DIAC's terminals, a negligible leakage current (I_{BO}) will flow through the DIAC due to one pn-junction being reverse-biased. The leakage current persists until the voltage gradually rises to the Break-Over Voltage (V_{BO}). At this threshold, an avalanche breakdown occurs in the reverse-biased junction, and the current through the DIAC increases as the applied voltage decreases. The voltage across the DIAC decreases to the Break-Back Voltage (V_W).

(c) When the applied voltage equals or surpasses the breakdown voltage, the DIAC commences to conduct, and the voltage drop across it is a few volts.

2.9.4 Advantages and disadvantages of DIAC

Advantages

(a) DIAC exhibits symmetrical switching properties which facilitate the mitigation of harmonics in a system.

(b) Additionally, it possesses a low voltage drop during the on-state, but this increases as the voltage increases.

(c) DIAC can be readily switched by adjusting the applied voltage and can provide seamless power regulation when employed to trigger other thyristors or TRIACs.

Disadvantages

(a) This device operates at a low power level and is designed to conduct electricity only when the voltage exceeds 30 volts.

(b) However, it lacks the capability to block high voltage currents.

2.9.5 Applications of DIAC

The primary use of DIAC is to activate TRIAC, which has an uneven triggering due to dissimilarities in its two halves. This disparity leads to a dissimilar voltage level for the forward and reverse current, generating harmonics in the system. These harmonics can create various issues, and therefore, it's crucial to minimize them. To solve this issue, the DIAC is connected in series to the gate of the TRIAC. As the DIAC has a similar forward and reverse breakdown voltage, it switches on whenever the voltage exceeds VBO in either direction. This symmetrical triggering causes the TRIAC to trigger at the same voltage level for both halves of the AC cycle, thus preventing asymmetrical switching and minimizing harmonics in the system.

2.10 TRIODE FOR ALTERNATING CURRENT

2.10.1 Introduction

Thyristors, which are semiconductor devices, are extensively utilized for power regulation. They can only conduct in one direction similar to a diode, making them appropriate for DC power regulation. On the other hand, triode for alternating current (TRIACs), is a type of thyristor, can conduct in both directions and provide complete power control. As a result, they are preferred for regulating AC power.

2.10.2 Symbol, pnpn structure and V-I Characteristics of TRIAC

The conventional thyristor, or SCR, is known for its reverse-blocking capability that stops current from flowing from cathode-to-anode. However, in applications such as AC circuits, bidirectional conduction is often necessary. To achieve this, two thyristors can be connected in inverse-parallel, or a single device structure can integrate two antiparallel thyristors, as illustrated in Figure.2.31 (a) and (b).

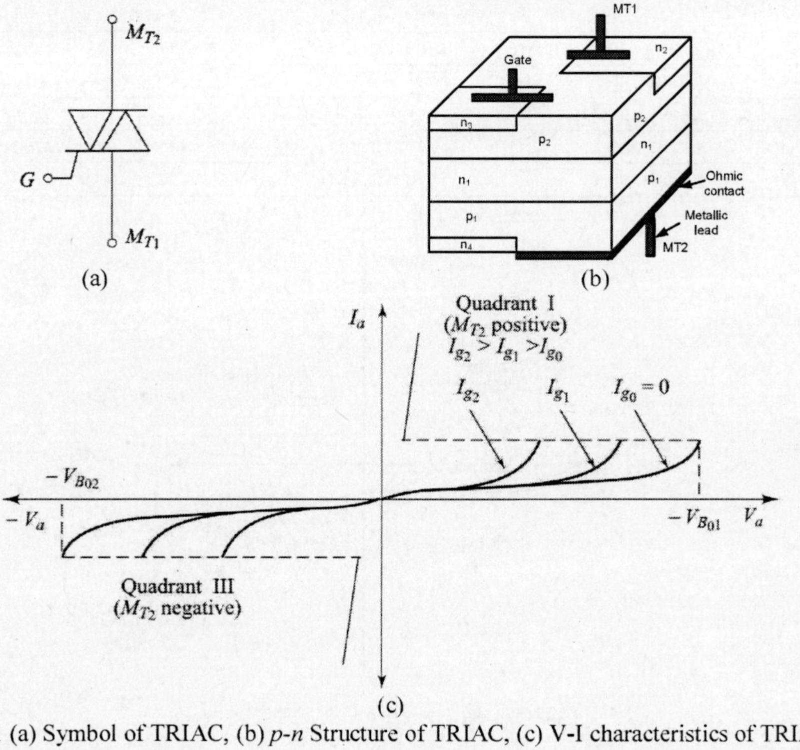

Figure.2.31 (a) Symbol of TRIAC, (b) *p-n* Structure of TRIAC, (c) V-I characteristics of TRIAC

This device is commonly referred to as a TRIAC (triode a.c. switch) and is represented by the circuit symbol shown in Figure.2.31 (a). The term "TRIAC" is formed by combining the capital letters from the words TRIODE and AC, as it can conduct in either direction due to its design, the terms "anode" and "cathode" are not applied to TRIAC. For this reason, its three terminals are usually referred to as "main terminals" MT_1, MT_2, and gate G, similar to a thyristor. The MT_1 terminal serves as the reference point for measuring voltages and currents at the gate terminal and MT_2. The gate is nearer to the MT_1 terminal. The V-I characteristics of a TRIAC is depicted in Figure.2.31 (c) and is based on the MT_1 terminal as the reference point. The first quadrant refers to the region where MT_2 is positive in relation to MT_1, while the third quadrant is the opposite. The maximum voltage applied across the device in either direction must not exceed the breakover voltage to maintain control by the gate. Triggering the TRIAC into conduction in either quadrant requires a gate current of a specified magnitude with either polarity, assuming the device is in a blocking state initially before the gate signal is applied. The TRIAC and SCR share similar characteristics in their blocking and conducting states, with one significant difference: while the SCR conducts in one direction only, but the TRIAC conducts in both directions. The four-layer structure that triggers the regenerative process and turns the device on can be one of p_1 n_1 p_2 n_2, p_1 n_1 p_2 n_3, or p_2 n_1 p_1 n_4, depending on the biasing conditions and polarity of the gate pulse.

2.10.3 Operation of TRIAC

As observed earlier, the TRIAC can be activated by either positive or negative gate current, while maintaining the MT_2 terminal at positive (+ve) or negative (-ve) potential. The TRIAC can be triggered by a variety of sources, including DC, rectified AC, or pulse sources such as UJT, DIAC, SBS, and asymmetrical trigger switch. To comprehend the operation of the TRIAC, one can utilize any of the four methods outlined below and consult Figure.2.32

Mode	Triggering Quadrant	Polarity of MT_2	Polarity of Gate
Mode-1	Quadrant-1	+ve w.r.t MT_1	+ve w.r.t MT_1
Mode-2	Quadrant-2	+ve w.r.t MT_1	-ve w.r.t MT_1
Mode-3	Quadrant-3	-ve w.r.t MT_1	+ve w.r.t MT_1
Mode-4	Quadrant-4	-ve w.r.t MT_1	-ve w.r.t MT_1

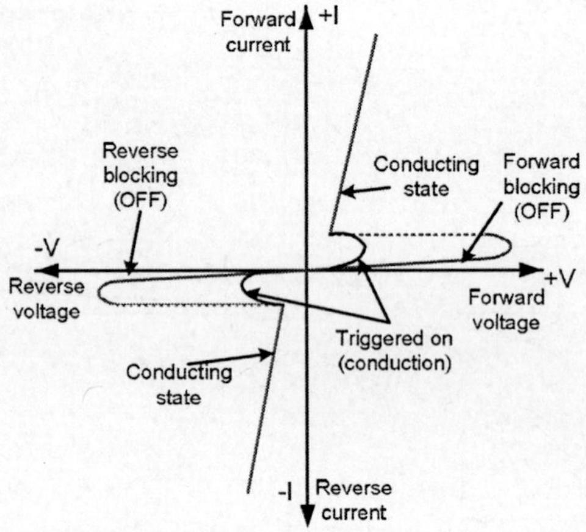

Figure.2.32 V-I Characteristics of TRIAC

Mode-1

When the gate current has a +ve polarity relative to MT_1, it primarily flows from the gate lead towards the MT_1 terminal through the p_2-n_2 junction, as illustrated in Figure.2.33. The device behaves in a conventional manner similar to an SCR when it turns on. However, for a TRIAC, a higher amount of gate current is necessary to achieve turn-on at a specific voltage. This is because of the ohmic contacts between the gate and MT_1 terminals on the p_2-layer, resulting in additional gate current flowing from the gate lead G to the main terminal MT_1 through the p_2 layer of the semiconductor, without having to pass through the p_2-n_2 junction. The p_1 n_1 p_2 n_2 structure can be turned ON through regenerative action. The flow of gate current across the p_2-n_2 junction floods the p_2 layer with electrons. These electrons then diffuse to the junction J_2 and are collected by the n_1 layer, creating a space charge in the n_1 region. To neutralize this negative space charge, more holes from p_1, diffuse into n_1 and attain at junction J_2. This produces a positive space charge in the p_2 region, which in turn results in the injection of more electrons from n_2 into p_2. Ultimately, this positive regeneration causes the p_1 n_1 p_2 n_2 structure to conduct the external current.

Mode-2

In Figure 2.34, the Mode-2 behaviour is illustrated. When the voltage at terminal MT_2 is +ve while the voltage at the gate terminal is -ve with respect to terminal MT_1, a gate current flows through the p_2-n_3 junction. This gate current, also known as I_G, forward biases the gate junction p_2-n_3 of the auxiliary p_1 n_1 p_2 n_3 structure, initiating the conduction of the TRIAC through the p_1 n_1 p_2 n_3 layers. As the p_1 n_1 p_2 n_3 structure conducts, the voltage drops across it reduces, and the potential of the layer between p_2 n_3 rises towards the anode potential of MT_2. Since the right-hand section of p_2 is clamped at the cathode potential of MT_1, there is a potential gradient across layer p_2, with its left-hand region being at a higher potential than its right-hand region. This establishes a current in layer p_2 from left to right, forward biasing p_2 n_2 junction and ultimately leading to the conduction of the main structure p_1 n_1 p_2 n_2. The auxiliary structure p_1 n_1 p_2 n_2 is like a pilot SCR, while the main structure p_1 n_1 p_2 n_2 can be considered as the primary SCR, and both are built into a single common structure. The anode current of the pilot SCR functions as the gate current for the main SCR. The device in which MT_2 is positive but gate current is negative is less sensitive than the turn-on process described in the above section, and thus necessitates more gate current.

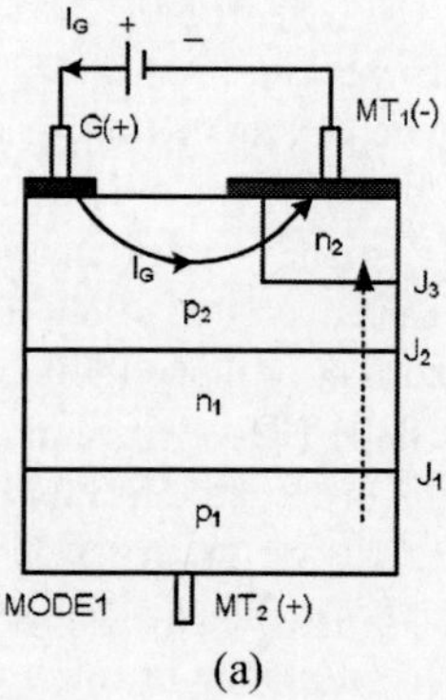

(a)

Figure.2.33 Mode-1

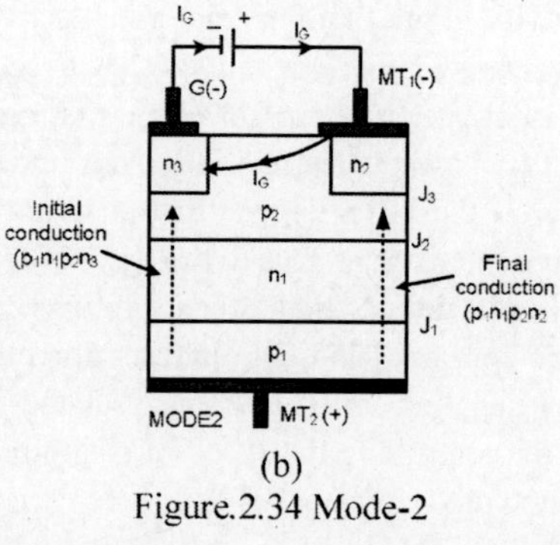

(b)

Figure.2.34 Mode-2

Mode-3

The activation of the device can be achieved by applying a positive voltage between the gate and MT_1 terminal, given that MT_2 terminal is negative and MT_1 terminal is positive. Once triggered, the device will operate in the third quadrant. The p_2 n_1 p_1 n_4 configuration depicted in Figure 2.35 plays a crucial role in the turn-on process, with n_2 serving as a remote gate. By applying a positive voltage, the external gate current I_G will forward bias the p_2 n_2 junction, which enables n_2 to inject electrons into the p_2 layer. The electrons collected will increase the current through the p_2 n_1 junction, while the holes injected from P_2 will diffuse through n_1 and reach p_1, thereby generating a positive space charge in the p_1 region. To counterbalance this positive space charge, more electrons from n_4 will diffuse into p_1, creating a negative space charge in the n_1 region. As a result, more holes will be injected from p_2 into n_1, which continues the regenerative process until the $p_2\ n_1\ p_1\ n_4$ configuration is fully turned ON and conducts the current that is limited by the external load. It's important to note that as the TRIAC is turned ON by the remote gate n_2, the device becomes less sensitive in the third quadrant when a positive gate current is applied.

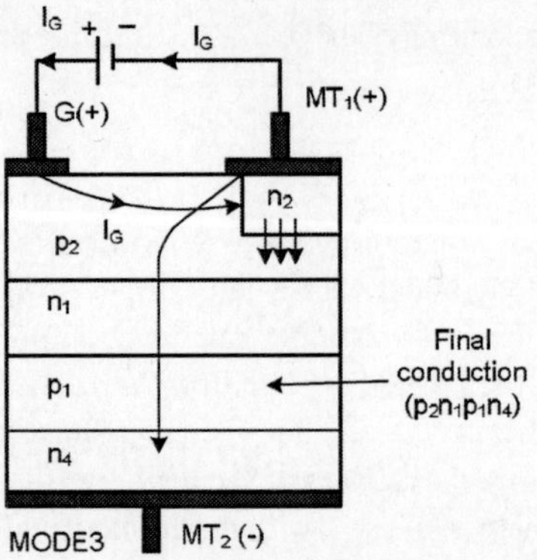

Figure.2.35 Mode-3

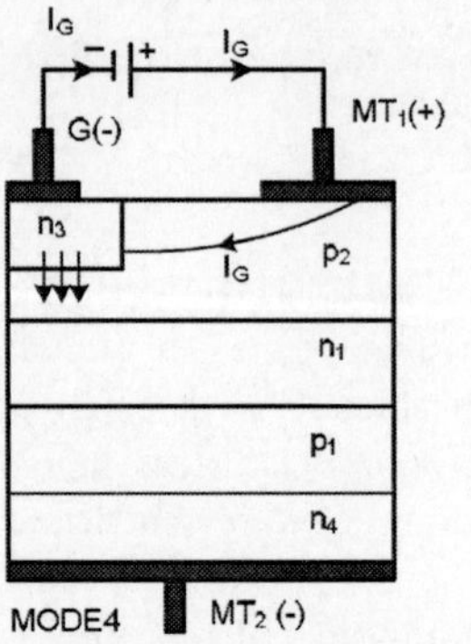

Figure.2.36 Mode-4

Mode-4:

Figure.2.36 presents a cross-sectional view of a structure that employs n_3 as a remote gate. When operating in this mode, $p_2\ n_3$ junction is forward biased by the external gate current I_G, allowing electrons to be injected and collected by $p_2\ n_1$ from n_3, resulting in an increase in current across $p_1\ n_1$. The regenerative action of the $p_2\ n_1\ p_1\ n_4$ structure causes it to turn ON, with the device being more sensitive in this mode than when turned ON by a positive gate current. The TRIAC exhibits its highest sensitivity in the first quadrant when activated with a positive gate current and in the third quadrant when activated with a negative gate current, out of the four available operating modes. When activated with a negative gate current in the first quadrant, its sensitivity is slightly lower, and when activated with a positive gate current in the third quadrant, it is much less sensitive. As a result, it is uncommon to operate the TRIAC in the first quadrant with a negative gate current or in the third quadrant with a positive gate current. Due to the interaction between its two halves, the TRIAC has limitations in terms of voltage, current, and frequency ratings compared to conventional thyristors. Therefore, it is typically used in consumer and light industrial appliances that operate at moderate power levels from 50 or 60 Hz AC supplies. The plastic encapsulated TRIAC is a cost-effective and compact device that is widely used for controlling the speed of single-phase a.c. series or universal motors in consumer appliances like food mixers and portable drills.

2.10.4 Advantages and disadvantages of TRIAC

Advantages of TRIAC

(a) The heat sink used for it is slightly larger in size, whereas for SCR, two smaller heat sinks are required.

(b) Although a secure breakdown in any direction is possible, SCR requires protection through a parallel diode.

(c) In DC applications, to protect SCR from reverse voltage, it needs to be connected with a parallel diode. On the other hand, TRIAC can function alone as it can achieve safe breakdown in any direction.

(d) Once the voltage reaches zero, the TRIAC will turn OFF.

(e) It can be triggered by both positive as well as negative gate signals. It can be protected using a single fuse.

Disadvantages of TRIAC

(a) Compared to SCR, these have lower reliability.

(b) Caution must be taken when turning on the circuit as it can be activated in any direction.

(c) The switching delay is significant.

(d) The dv/dt rating of these is considerably lower than that of SCR.

(e) TRIACs have lower ratings compared to silicon-controlled rectifiers.

(f) These are not suitable for use in DC applications.

2.10.5 Application of TRIAC

TRIACs find extensive application in various fields, including but not limited to, regulating the intensity of light, controlling the speed of fans and motors, and in computerized control circuits of a wide range of household appliances. They are versatile enough to be used in both AC and DC circuits, although their initial purpose was to replace the requirement of using two SCRs in AC circuits. The two primary TRIAC families, namely BT136 and BT139, are widely employed for their specific application-oriented features.

2.11 PROTECTION CIRCUITS

2.11.1 Introduction

To ensure reliable and satisfactory operation of power devices, it should be ensured that the circuit conditions they encounter do not exceed their limited operating capabilities. This can be achieved by using protective components around the device to guard against extreme conditions, allowing for the use of cost-effective and readily available devices. There may be many abnormal conditions like over-voltage, and high device temperature, transients, etc. These conditions can lead to performance degradation or permanent destruction of the power device. Thus, it is crucial to protect the device from such abnormal conditions to maintain its adherence to the manufacturer's specified characteristics and ensure reliable operation.

Figure.2.37 Picture of heat sinks [Courtesy: https://www.google.com]

2.11.2 Cooling and Heat Sink

Power semiconductor devices generate heat due to switching losses that occur during switching on, which must be dissipated to keep the temperature of the device junctions within a definite range. Several methods can be used to transfer heat, including conduction, convection, and radiation, and either natural or forced air can be employed. Industries typically employ the convection method, wherein heat is transferred from the junction to the case and then to the sink. Section 2.2.11 provides the formulas for average power and thermal resistance, which apply to all semiconductor devices, and the device manufacturer specifies the thermal resistance between the junction and case, as well as between the case and sink. The market offers a wide range of heat sinks that employ cooling fins to improve heat transfer, and Figure 2.37 shows images of some of these heat sinks.

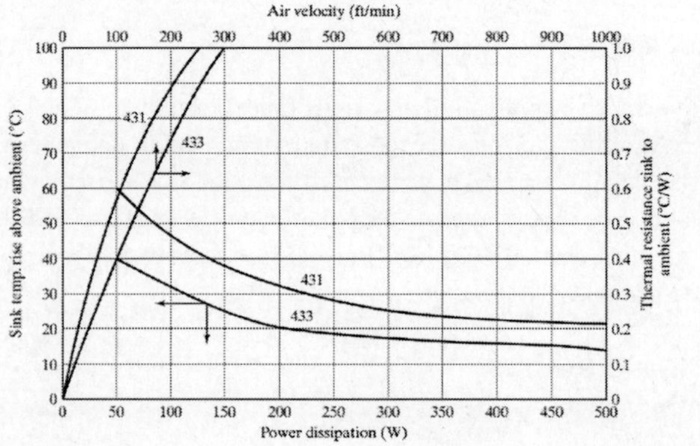

Figure.2.38 Thermal resistance characteristics [Source: EG&G Wakefield engineering]

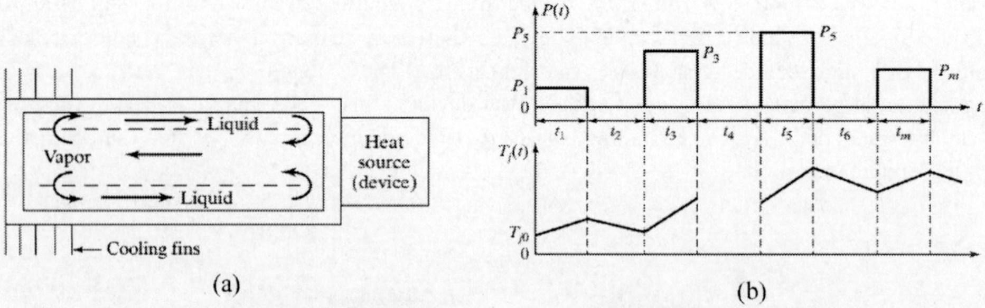

Figure.2.39 (a) Heat pipes, (b) Junction temperature with corresponding rectangular pulses

A typical thermal resistance characteristic is shown in **Figure. 2.38.** In this figure, the plots between power dissipation and heat sink temperature rise at natural cooling is depicted. The device needs to properly mounted on sink so that it maintains the proper mounting pressure between the mating surfaces. Installation guidelines are provided by the manufacturer. The mounting torques must be such that there will not be any damage to the silicon wafer. The stud and nuts must be properly greased or lubricated to increase the tension on the stud in case stud mounted device. Heat pipes (Figure.2.39(a)) are also used for cooling the device. The pipes are partly filled with low vapor-pressure liquid. In high-power applications, the cooing is done by using liquids (may be oil or water). Water cooling is more efficient than oil cooling.

Temperature of the junction varies with instantaneous power loss because of low thermal impedance. The instantaneous power loss should be kept below the acceptable value. The manufacturer generally supplied a data sheet in terms of plots between transient thermal impedance and duration of square wave pulse. The plot of power loss versus time is determined with known value of the current waveform. The step response of the first order system can be used to represent the transient thermal impedance of the device. The instantaneous thermal impedance, $Z(t)$ is given by equation (2.23) in which Z_O = junction to case thermal impedance at steady state, τ_{th}= thermal time constant.

$$Z(t) = Z_0(1 - e^{-t/\tau_{th}}) \tag{2.23}$$

The instantaneous junction temperature (T_J) can be expressed in terms of power loss and $Z(t)$, and given by equation (2.24).

$$T_J = P_d Z(t) \tag{2.24}$$

For a pulsed-type power loss, the corresponding step response by utilizing the equation (2.24) for the junction temperature (T_J) $T_J(t)$ is shown in Figure.2.39(b). Here, P_i .... are power pulses in which i =1, 3, 5 etc. At the end of mth pulse, the device T_J may be represented as equation (2.25) in which T_{JO} = initial junction temperature. The -ve sign on Z_{n-1} (i.e Z_2, Z_4,...etc) represents the T_J fall during the interval t_2, t_4, ...

$$T_J(t) = T_{J0} + \sum_{n=1,3,...}^{m} P_n(Z_n - Z_{n-1}) \tag{2.25}$$

T_J at the end of mth pulse is given by equation (2.26), where Z_n is the impedance of the nth pulse of duration $t_n = \delta t$, P_n is the power loss for n^{th} pulse, $P_O = 0$, and t is the time interval.

$$T_J(t) = T_{J0} + \sum_{n=1,2,3...}^{m} Z_n(P_n - P_{n-1}) \tag{2.26}$$

2.11.3 Overvoltage protection

It is typically essential for power semiconductor equipment to have a mechanism that restricts transient overvoltage to prevent the semiconductor devices from being excessively stressed. To ensure protection against all voltage transients, each device must be protected individually. To limit the number of protective components needed, a safety factor V_f can be implemented, and devices are chosen with maximum voltage ratings of 1.5 to 2.5 times to that of normal peak operating voltage. Protective elements and circuits are commonly used to minimize the impact of overvoltage. The three main protection circuits are as follows.

(a) Snubber Circuits

(b) Crowbar Circuits

2.11.3.1 Snubber circuits (dv/dt suppression)

Protection is typically necessary in power electronic circuits to avoid unintended breakover caused by a high *dv/dt* across devices. If left unchecked, such a condition could cause circuit malfunction and potential device failure. The *dv/dt* that exceeds normal levels can stem either from few external factors for example, the closure of the main power supply contactor or the circuit's inherent operation. To control this excessive *dv/dt*, snubber circuits are employed for damping purposes. As seen in Figure.2.40, the basic components of the snubber circuit are an SCR put-in shunt with a series-connected resistor and capacitor. If there is a capacitor C present across the thyristor, then high dv/dt values that appear at the SCR terminals will cause the capacitor to conduct at the proper current (*C= dv/dt*). The size of the current passing through the capacitor will be severely constrained by the circuit's inductance, which will also limit *dv/dt*. Figure.2.40 (a) shows that the effective inductance, L, restricts the initial *di/dt*, when the device is turned on. To enhance the supply circuit inductance, a linear or saturable series inductor may be added. Usually, the LCR circuit exhibits a slight underdamped behaviour, and when a forward voltage step is applied, it restricts the device's peak voltage and its rate of change to acceptable levels. However, when operating at high-switching frequencies, losses caused by the snubber can become considerable.

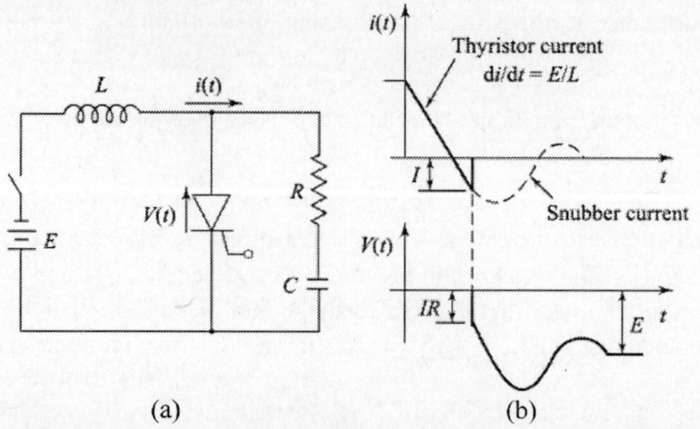

Figure.2.40 (a) Snubber circuit for a thyristor, (b) reverse recovery characteristics

After a thyristor is turned off, a brief burst of reverse recovery current is produced, peaking before the device becomes blocked. Without an RC snubber, the sudden cessation of the reverse recovery current (RRC) flowing through the series inductance, *L*, can lead to transient overvoltage of *Ldi/dt*, which have the potential to harm the thyristor or other semiconductors within the converter circuit.

Connecting an RC snubber to a thyristor can redirect the RRC to the RC path, as demonstrated in Figure.2.40 (b), thereby producing an oscillatory reverse voltage across the semiconductor. By properly designing the snubber, the amplitude and rate of rise of the reverse-recovery voltage can be limited, and the build-up of reverse voltage across the semiconductor can be delayed, resulting in reduced recovery losses in the device. However, excessive ringing may occur in the LCR circuit if the snubber resistance is too small, and the semiconductor's reverse blocking voltage capacity may be exceeded.

When the thyristor T_1 turns on in Figure.2.40 (a), the capacitor *C* in the snubber discharges into the thyristor, with the discharge current being limited by resistor *R* to prevent excessive *di/dt* during turn-on. However, during turn-off, a sudden forward *IR* voltage drop appears across the thyristor due to diode reverse recovery, as shown in Figure.2.40 (b). Thus, the presence of the resistor in the snubber weakens its ability to limit forward *dv/dt* on the thyristor.

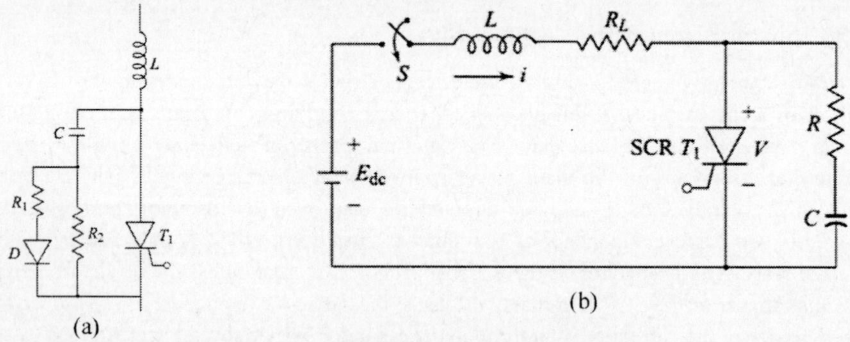

Figure.2.41(a) Polarized snubber, (b) The snubber circuit

To tackle this issue, the polarized snubber in Figure.2.41 is commonly employed, which involves using a small damping resistor R_1 to enhance forward dv/dt protection and a larger value of resistor R_2 to limit snubber discharge current when the thyristor is gated.

In utility-powered circuits, Snubber networks are commonly utilized to restrict incoming transients by connecting them between the AC supply lines. Additionally, RC snubbers are employed across transformer secondary windings and other sources of inductive voltage transients. Nonetheless, snubber networks are only partly successful in mitigating transient over voltages, hence requiring significant safety margins for semiconductor blocking voltages. Compared to RC snubbers, voltage-damping devices are more effective at limiting transient over voltages and have lower losses during normal operating conditions.

Design of snubber network for DC circuit

Figure.2.41(b) illustrates the usage of a snubber circuit in a direct current (DC) circuit, which is crucial when switching a DC voltage across an SCR. It is imperative to ensure that the rate of voltage change across the SCR is lower than its *dv/dt* rating to prevent it from conducting without a gate trigger pulse. When the switch *S* is closed, the capacitor functions as a short circuit while the SCR is in a state of forward blocking, which results in a significantly high resistance. The load resistance in the circuit is indicated as R_L, and the source inductance is denoted as *L* within the depicted diagram. The voltage is given by

$$E_{dc} = (R + R_L)i + L\frac{di}{dt}$$

The solution of the above expression is

$$i = I\left(1 - e^{-t/\tau}\right) \tag{2.27}$$

Where $I = \dfrac{E_{dc}}{R+R_L}$ *and* $\tau = \dfrac{L}{R+R_L}$, in (2.27), *t* is time (s) of closing the switch.

Differentiating (2.27) w.r.t *t*, we get

$$\frac{di}{dt} = I \cdot e^{-t/\tau} \cdot \frac{1}{\tau} = \frac{E_{dc}}{R+R_L} \cdot \frac{R+R_L}{L} e^{-t/\tau} = \frac{E_{dc}}{L} e^{-t/\tau}$$

(2.28)

The di/dt is maximum when $t = 0$.

$$\therefore \left(\frac{di}{dt}\right)_{max} = \frac{E_{dc}}{L} \quad or \quad L = \frac{E_{dc}}{(di/dt)_{max}} \tag{2.29}$$

Now the voltage at the thyristor is

$$V = R \cdot i \quad \therefore \frac{dv}{dt} = R \cdot \frac{di}{dt} \quad or \quad \left(\frac{dv}{dt}\right)_{max} = R\left(\frac{di}{dt}\right)_{max} \tag{2.30}$$

Substituting the value of $(di/dt)_{max}$ from (2.29), then

$$\left(\frac{dv}{dt}\right)_{max} = R\left(\frac{E_{dc}}{L}\right) \quad or \quad R = \frac{L}{E_{dc}}\left(\frac{dv}{dt}\right)_{max} \tag{2.31}$$

To make the circuit critically damped, the parameters *L*, R_L, *R* and *C* should be properly selected.

$$\sqrt{(R_L + R)^2 - \frac{4L}{C}} = 0 \quad or \quad (R_L + R)^2 - \frac{4L}{C} = 0 \tag{2.32}$$

Hence,

$$R_L + R = 2\sqrt{\frac{L}{C}} \tag{2.33}$$

By taking the allowed *dv/dt* value as a reference to prevent any malfunctioning of the SCR, and considering the provided values of *L* and R_L, the snubber components *R* and *C* can be calculated using equations (3) and (4), respectively.

Design of snubber network for AC circuit

The following equation mentions the capacitance requirement for keeping the voltage transients within the device ratings.

$$C = 10 \cdot \frac{VA}{V_s^2} \cdot \frac{60}{f} \tag{2.34}$$

Where, C = minimum capacitance required (μF), VA = transformer Volt-Ampere rating, Vs = transformer secondary RMS voltage, f = operating frequency
The *R* can be calculated as

$$R = 2\rho\sqrt{L/C} \tag{2.35}$$

Where, ρ is the damping factor of 0.65, *L* and *C* are the effective circuit inductance, and minimum capacitance. So, if the maximum *dv/dt* for the thyristor is specified, the *C* can be computed as

$$C = \frac{1}{2L}\left(\frac{0.564E_m}{dv/dt}\right)^2 \tag{2.36}$$

Where E_m is the peak input line to line voltage.

2.11.3.2 Crowbar Circuits

In both AC and DC circuits, a crowbar can serve as protection against overvoltage and/or overcurrent. To safeguard sensitive DC power electronic circuits and loads from faults, an SCR can be utilized, as depicted in Figure.2.42.

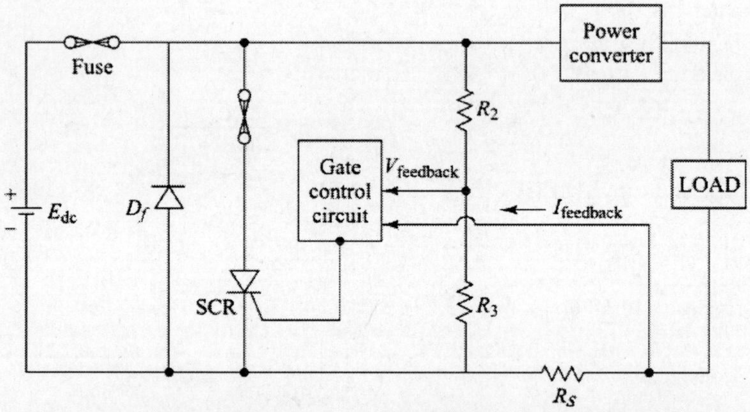

Figure.2.42 SCR crowbar for overvoltage and overcurrent protection

In the event of a fault, the crowbar SCR is activated, which results in a short circuit of the power supply. This creates a high-supply current flow that triggers either the blowing of a fuse or the activation of a fast-acting circuit breaker, which then disconnects the load from the supply, thereby isolating it. Diode D_f provides a current pathway for inductive load energy. The voltage across the sense resistor *R* is used to measure the load current. Once this voltage surpasses a predetermined limit, which indicates that the load current has attained the fault level, the SCR is triggered. The voltage of the load or DC link is measured by the resistor divider R_2 - R_3. If this voltage surpasses the pre-set limit, the SCR is triggered, and the crowbar short-circuit current causes the fuse to blow, thereby separating the sensitive load from the supply.

The Crowbar SCR can handle a significantly higher amount of current than its normal rating. When the fuse is disconnected briefly, the SCR's I^2t feature for surge current can be utilized. For proper operation, the SCR's I^2t rating should exceed the total I^2t rating of the fuse. If a fuse link is used to protect the Crowbar SCR, the total I^2t rating of the dc-link fuse link must be lower than the pre-arcing I^2t rating of the Crowbar SCR's fuse link. An AC Crowbar can consist of either two SCRs connected in antiparallel across the fuse-protected AC line or a single SCR in a four-diode rectifying bridge.

2.11.4 Overcurrent protection

Specially designed fast-acting fuses and circuit breakers are typically employed to provide overcurrent protection for thyristors, as they have a limited overcurrent capacity. These fuses are designed to open and clear fault currents within milliseconds as the fault current rises. Before delving into the details of the fuse, it is important to carefully consider the specific requirements that it must meet, which include:

(a) The fuse must be capable of carrying the device's rated current continuously.

(b) The thermal storage capacity of the fuse should not exceed that of the device it is safeguarding. This implies that the fuse's I^2t let-through value, which is the amount of energy that can pass through the fuse before it clears the fault current, must be lower than the rated I^2t value of the device.

(c) To dissipate the energy in the circuit and reduce the current, it is necessary for the voltage of the fuse to be sufficiently high during arcing.

(d) The fuse should have the ability to endure any restriking voltage that may occur across it once the current is interrupted.

Figure 2.43(a) displays an image depicting a fuse and its components. The fusing component is composed of silver and possesses one or more necks, where fusing occurs due to restricted current flow. The stages of fuse current can be seen in Figure 2.43(b). In the event of a fault, the fault current elevates the fuse temperature until it reaches tm, at which point fuse melt arcs are produced, generating an arc voltage across the fuse. As a result of the heat produced, the fuse element vaporizes, causing an increase in arc length and a decrease in current. This cumulative effect leads to the extinguishing of the arc in a very brief period. The clearing time is determined by the sum of melting time and arc time t_a, with melting time being reliant on load current and arc time being reliant on arc time.

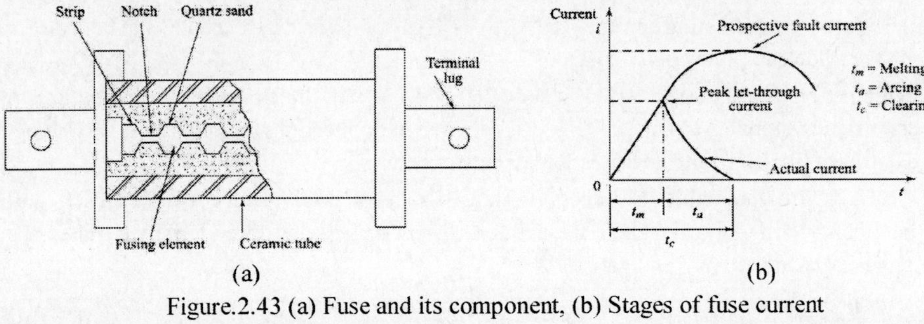

Figure.2.43 (a) Fuse and its component, (b) Stages of fuse current

Unit Summary

This unit explores the construction, operating principle, V-I characteristics, various characteristics of Silicon controlled rectifier (SCR) in details. Also, some important members of thyristor family such as LASCR, SCS, GTO, UJT, PUT, DIAC and TRIAC are presented in terms of symbols, construction, operating principles, V-I characteristics etc. The overvoltage and overcurrent protection circuits for thyristors are presented.

1. Thyristors are four-layer semiconductor device. The doping density and thickness of different layers are different. The oldest member is the Silicon-controlled rectifier (SCR). In most of the books and other literature, the term Thyristor and SCR are used interchangeably.
2. SCR has three terminals. They are called anode, cathode, and gate. The anode, cathode, and gate are represented as *A, K,* and *G* respectively in most of the books and other literature.
3. There are many members in thyristor family. Some of them are unidirectional and some are bi-directional. Some of them are self-turned-off devices and some are forced turned-off.
4. No commutation circuits required for self-turned off devices. On the other hand, commutation circuits are required for forced commutated thyristors.
5. SCR, LASCR, GTOs, SCS etc. are unidirectional and TRIACs, bidirectional phase-controlled thyristors (BCTs) etc are bi-directional. The SCR, LASCR etc. are not gate turn of devices but GTOs, static induction thyristors (SITHs) are gate turned-off thyristors.
6. SCR is a simple turn-on latching device with high turn-on gain. It is low-cost high voltage and high current device.
7. Light Activated Silicon Controlled Rectifier (LASCR) belongs to the photoelectric thyristor family, is widely used as a control element. The LASCR is a device with three terminals - the Anode (*A*), Cathode (*K*), and Gate (*G*) terminal. The gate of the LASCR is isolated and it can be operated separately. This device can be activated by optical radiation and has the highest trigger sensitivity to light with a wavelength of 0.8 to 1.5μm. It is also known as a photo SCR.
8. Similar to the SCR, the Silicon Controlled Switch (SCS) is a silicon-based device with a unilateral structure consisting of four layers and three junctions. The device features four electrodes, namely the cathode (*K*), cathode gate (G_K), anode gate (G_A), and the anode (*A*). Compared to an SCR, one benefit of using an SCS is its shorter turn-off time, usually falling within the range of 1 to 10μs for the SCS and 5 to 30μs for the SCR.
9. GTO is a semiconductor-based unidirectional switch that is fully controlled and has three terminals. They are Gate (*G*), Cathode (*K*), and Anode (*A*).
10. DIAC is represented by two diodes arranged in parallel and opposing each other, with two terminals. The DIAC has two terminals - *MT1* (*T1*) and *MT2* (*T2*). The primary use of DIAC is to activate TRIAC, which has an uneven triggering due to dissimilarities in its two halves.
11. The triode for alternating current (TRIACs), is a type of thyristor, can conduct in both directions and provide complete power control.
12. Unijunction Transistor, or UJT, is a solid-state device with three terminals that can function as a gate pulse, timing circuit, or trigger generator in applications involving the switching and control of thyristors and TRIACs for AC power control.
13. PUT, or Programmable Unijunction Transistor, is a *pnpn* device that shares a striking resemblance to the UJT. The PUT has four layers (*pnpn*) and three terminals. They are the anode (*A*), cathode (*K*), and gate (*G*).

14. The minimum value of the anode current to maintain the conduction of the device is called latching current (I_L). The value of the anode current at which the SCR reached the blocking state is called the holding current (I_H).
15. The value of anode to cathode voltage (V_{AK}) at which this breakdown of the reverse biased junction occurs is called the forward breakdown voltage (V_{BO}).
16. To turn on the SCR it is required to increase the anode current. This can be done by (a)thermal runway, (b) application of light, (c) application of high voltage, (d) high *dv/dt*, and (e) application of gate current
17. SCR are mounted on the case. The case must be designed in such a way that heat flows from the junction to the case. Based on the power ratings of the SCR, various techniques of mountings are developed. They are (a) Lead mounting, (b) stud mounting, (c) bolt-down mounting, (d) press fit mounting, and (c) press pack mounting.

Exercises

Example.2.1

A thyristor circuit comprises a 100V source, the thyristor and a R-L series impedance of R = 20Ω, L = 0.5H. The duration of triggering pulse is 50μs. The latching current of the SCR is 50mA. Determine whether the SCR get fired.

Solution:

The circuit diagram is shown in the Figure.2.44

When the SCR is triggered, the anode to cathode current will rise exponentially, and it is given by

$$i(t) = \frac{V}{R}(1 - e^{-t/(L/R)}) = \frac{100}{20}(1 - e^{-t/(0.5/20)})$$

At t = 50μs, $i(t) = \frac{100}{20}(1 - e^{-50x10^{50x10^{-6}}/0.025}) = 9.99mA$

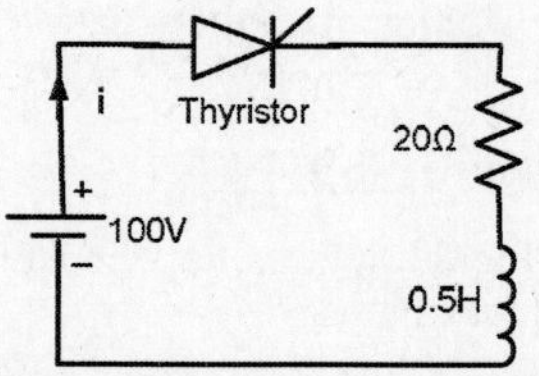

Figure.2.44 Circuit related to example 2.1

The calculated value of circuit current is less than latching current, hence the thyristor will not be triggered.

Example.2.2

An SCR circuit is shown in the Figure.2.45 The latching current is 4mA. Determine the minimum value of gating pulse required to turn on the device.

Solution: The equation of the circuit is given by

$V = L\frac{di}{dt}$ Or $dt = \frac{L}{V}di$, i = latching current

Integrating, $t = \frac{L}{V}i$ or $t_{min} = \frac{0.1}{100}x4x10^{-3} = 4\mu s$

Thus, the minimum width of the gate pulse is 4μs.

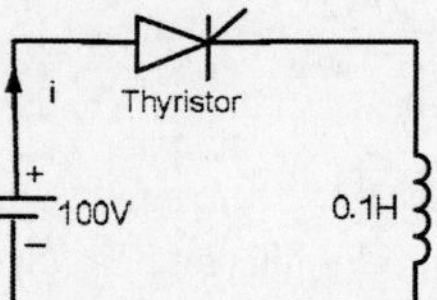

Figure.2.45 Circuit related to example 2.2

Example.2.3

The gate cathode characteristics of an SCR has straight-line with slope of 150. For trigger source voltage of 20V and gate power dissipation of 0.6 watt, Find out the value of gate-source resistance in (Ω).

Solution:

$V_g I_g = 0.6$ W Slope of the line is , $\frac{V_g}{I_g} = 150$

Thus, $150 I_g^2 = 0.6$ from which

$$I_g = \left(\frac{0.6}{150}\right)^{1/2} = 63.24 \text{ mA}$$

Hence, the gate voltage, $V_g = 150 \times 63.24 \times 10^{-3} = 9.486$ V

For the gate circuit, $E_s = I_g R_g + V_g = 63.24 \times 10^{-3} \times R_g + 9.486 = 20$

Thus, gate to source resistance is $R_g = 166.25\ \Omega$

Example.2.4

The latching current rating of thyristor used in the following circuit is 50 mA. Compute the minimum width of gate pulse to be applied for thyristor to turn on.

Solution:

$$Vs = Ri(t) + L\frac{di}{dt}$$

The solution of above equation can be written as,

$$i = \frac{V}{R}\left(1 - e^{-\frac{R}{L}t}\right)$$

Putting the values of given parameters, the solution of equation for t as follows.

$$50 \times 10^{-3} = \frac{100}{120}\left(1 - e^{-\frac{120}{0.2}t}\right)$$

Thus, $t = = 103.12 \mu s$

Thus, width of minimum gate pulse is 103.12µsecond.

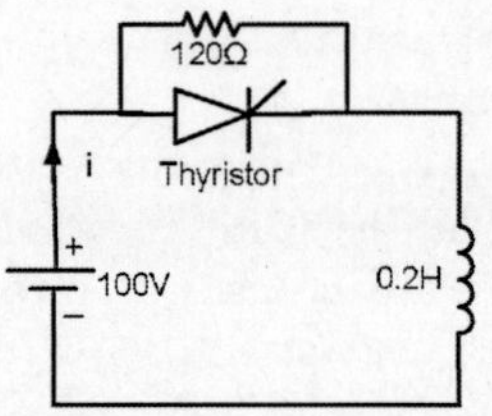

Figure.2.46 Circuit related to example 2.4

Example.2.5

A trigger circuit of a thyristor is supplied by 20 V and the slope of load line is -130 V/ampere. The SCR take 20 mA minimum gate current to turn on. Then find out, (a) Source resistance requires in the gate circuit, (b) the trigger voltage and trigger current for an average gate power loss of 0.6 watt.

Solution:

The necessary gate source resistance is determined by the slope of a load line. Hence, series resistance is required in the gate circuit is 120 Ω

It is given that $V_g I_g = 0.6$

From the gate circuit, $E_g = R_g I_g + V_g$

$$20 = 130 I_g + \frac{0.6}{I_g} \quad \text{or} \quad 130 I_g^2 - 20Ig + 0.6 = 0$$

I_g = 113mA or 40.84 mA

$$V_g = \frac{0.6}{113\times10^{-3}} = 5.309V \text{ and } V_g = \frac{0.6}{40.84\times10^{-3}} = 14.691 \text{ V}$$

Choosing, $V_g = 5.309$ and $I_g = 40.84$ mA for minimum gate current of 20 mA

Example.2.6

The gate-cathode characteristics of a thyristor is given by $V_g = 2+20I_g$, Gate source voltage is a rectangular pulse of 20V with 25μs duration. If the average power dissipation is 0.4W, and a peak gate drive power is 5W, find (a) the resistance to be connected in series with SCR gate, (b) the triggered frequency, (c) the duty cycle of gate pulse.

Solution:

(a) Here, $V_g = 2 + 20I_g$

For pulse- triggering of SCRs,

(Peak gate voltage) × (peak gate current) = Peak gate drive power (P_{gm})

Hence, $(2 + 20I_g)I_g = 5$ or $20Ig^2 + 2Ig - 5 = 0$

Solving for I_g, Ig = 0.4524 A

The amplitude of current pulse = 0.4524A

During the pulse on period, $E_s = R_sI_g + V_g$, or $20 = R_s I_g + (2+20I_g)$

Putting the values of, I_g, and solving for series resistance, R_s, we will get

$$R_s = \frac{20\text{-}(2 + 20\times0.4524)}{0.4524} = 11.78\ \Omega$$

(b) $P_{gm} = \frac{P_{avg}}{fT}$, the triggering frequency, f is given by $f = \frac{P_{gav}}{PgmT} = \frac{0.4 \times 10^6}{5 \times 25} = 3.2$ kHz

(c) Duty cycle (D) = f x T = $3.2 \times10^3\times25\times10^{-6} = 0.08$

Example.2.7

The latching for an SCR inserted in between a dc voltage source of 400 V and the load current is 200 mA. Calculate the minimum width of gate pulse current required to turn on the SCR in case the load consist of (a) L= 0.25H, (b) Series com bination of R = 40Ω and L= 0.25H

Solution:

(a) For purely inductive (L) load, as, we know,

$$E = L\frac{di}{dt} \text{ or } di = \frac{E}{L}t \text{ or } i = \frac{E}{L}t$$

$$0.200 = \frac{400}{0.25}t \text{ or } t = \frac{0.200\times0.25}{400} = 125\ \mu \text{ sec}$$

Therefore, the minimum gate pulse is required 125 μsec.

(b) For R-L series load, from the circuit equation, $E = Ri + L\frac{di}{dt}$

Solution for I, is given by $i = \frac{E}{R}\left(1\text{-}e^{\frac{-R}{L}t}\right)$

Putting the given values in $i = \frac{E}{R}\left(1\text{-}e^{\frac{-R}{L}t}\right)$,

$$\text{or}\quad 0.200 = \frac{400}{40}\left(1 - e^{\frac{-40}{0.25}t}\right), \text{ and and solving for t,}\quad t = 126.26\ \mu s$$

Hence, the minimum gate pulse is required 126.26 μsec.

Example.2.8

The data sheet of an SCR provides maximum rms on-state current as 45A. If the SCR is used in a resistive circuit, find out the average on-state current rating for the half sine wave current for condition for conduction angle (a) 30°, (b) 90°, and (c) 180°

Solution:

The half wave sine wave for current depicted in Figure.2.47

From this figure

$$I_{avg} = \frac{1}{2\pi}\int_{\theta 1}^{\pi} I_m \sin\theta\, d\theta = \frac{I_m}{2\pi}(1 + \cos\theta_1)$$

$$I_{rms} = \left[\frac{1}{2\pi}\int_{\theta_1}^{\pi} I_m^2 \sin^2\theta d\theta\right]^{1/2} = \left[\frac{I_m^2}{2\pi}\int_{\theta_1}^{\pi}\left\{\frac{1}{2} - \frac{\cos 2\theta}{2}\right\}\right]^{1/2}$$

$$= \left[\frac{I_m^2}{2\pi}\left(\frac{\theta}{2} - \frac{\sin 2\theta}{4}\right)_{\theta_1}^{\pi}\right]^{1/2}$$

$$= \left[\frac{I_m^2}{2\pi}\left\{\frac{\pi - \theta_1}{2} + \frac{1}{4}\sin 2\theta_1\right\}\right]^{1/2}$$

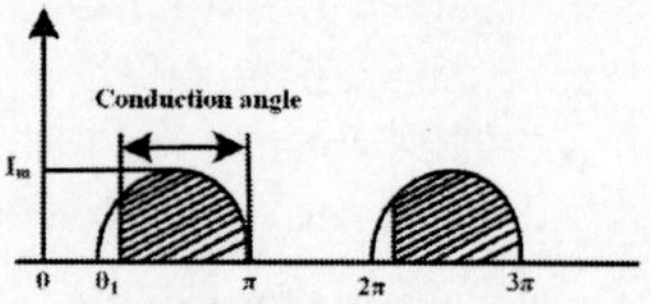

Figure.2.47 *Waveform related to example2.8*

(a) For conduction angle 30°, $\theta_1 = 150^\circ = \frac{5\pi}{6}$

$$I_{avg} = \frac{I_m}{2\pi}\left(1 + \cos\left(\frac{5\pi}{6}\right)\right) = 0.0213 I_m$$

$$I_{rms} = \left[\frac{I_m^2}{2\pi}\left\{\frac{\pi - \frac{5\pi}{6}}{2} + \frac{1}{4}\sin\left(2\times\frac{5\pi}{6}\right)\right\}\right]^{1/2} = \frac{I_m}{2\sqrt{2}}$$

$$\text{Form factor (FF)} = \frac{I_m}{2\sqrt{2}} \times \frac{1}{0.0213 I_m} = 3.9818$$

$$I_{TAVG} = \frac{I_{rms}}{FF} = \frac{45}{3.9818} = 11.30\ A$$

(b) For conduction angle 90°, $\theta_1 = 90^\circ$

$$I_{avg} = \frac{I_m}{2\pi}\left(1 + \cos 90^\circ\right) = \frac{I_m}{2\pi}$$

$$I_{rms} = \left[\frac{I_m^2}{2\pi}\left\{\frac{\pi - 0^\circ}{2} + \frac{1}{4}\sin(2\times 90^\circ)\right\}\right]^{1/2} = \frac{I_m}{2\sqrt{2}}$$

$$\text{From factor (FF)} = \frac{I_{rms}}{I_{avg}} = \frac{I_m}{2\sqrt{2}} \times \frac{2\pi}{I_m} = \frac{\pi}{\sqrt{2}}$$

$$I_{TAVG} = \frac{I_{rms}}{FF} = \frac{45 \times \sqrt{2}}{\pi} = 20.25 \text{ A}$$

(c) For conduction angle 180°, $\theta_1 = 0°$

$$I_{avg} = \frac{I_m}{2\pi}(1+\cos\theta_1) = \frac{I_m}{2\pi}(1+\cos 0°) = \frac{I_m}{\pi}$$

$$I_{rms} = \left[\frac{I_m^2}{2\pi}\left\{\frac{\pi - \theta_1}{2} + \frac{1}{4}\sin 2\theta_1\right\}\right]^{1/2} = \left[\frac{I_m^2}{2\pi}\left\{\frac{\pi - 0°}{2} + \frac{1}{4}\sin(2\times 0°)\right\}\right]^{1/2} = \frac{I_m}{2}$$

$$\text{Form factor (FF)} = \frac{I_{rms}}{I_{avg}} = \frac{I_m}{2} \times \frac{\pi}{I_m} = \frac{\pi}{2}$$

$$I_{TAVG} = \frac{I_{rms}}{FF} = \frac{45 \times 2}{\pi} = 28.64 \text{ A}$$

Example.2.9

A thyristor circuit is shown in Figure.2.48 The width of gate pulse is 60µs. The latching current of thyristor is 40mA. For a series R-L load having R = 40Ω and L= 4H, will the thyristor get turned on? Check, if the answer is negative, how this difficulty can be overcome for a given load. Find the maximum value of the remedial parameter shown dotted.

Solution:

The current thought load and thyristor is

$$i_t = \frac{Vs}{R}\left(1 - e^{-\frac{R}{L}t}\right)$$

From the given circuit,

$$i_t = \frac{400}{40}\left(1 - e^{-\frac{40}{4}\times 60\times 10^{-6}}\right) = 5.998 \text{ mA}$$

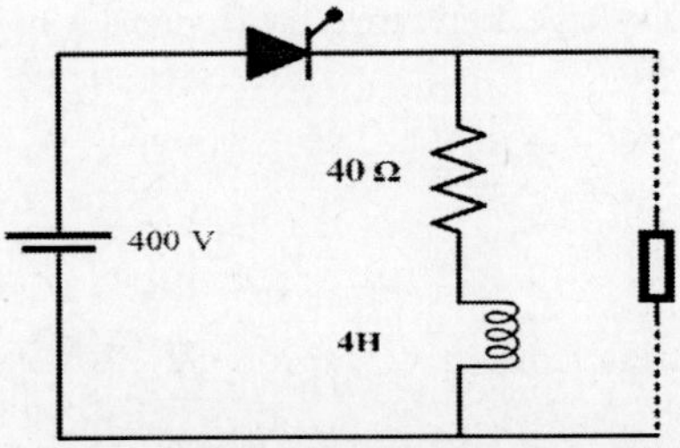

Figure.2.48 Circuit of example 2.9

This shows that for a pulse width of 60µs, the anode current rises to 5.998 mA, which is far less than 40 mA. So thyristor will not get turn on.

The remedial parameter, shown in dotted line in given circuit diagram should be resistance say R, because current can rise in the resistance without any time delay. The value of R can be obtained s under,

$$i_t = 40 \times 10^{-3} = \frac{400}{R} + \frac{400}{40}\left(1 - e^{-\frac{40}{4}\times 60 \times 10^{-6}}\right)$$, from which we can find the value of R,

$$\text{i.e } R = \frac{400 \times 10^{-3}}{34.002} = 11.764 \text{ k}\Omega$$

Example.2.10

The static V-I characteristics of a forward conduction thyristor is given in Figure.2.49 which is a straight line. Obtain average power loss in the thyristor and its rms current rating for (a) A constant current of 100 A for one half cycle, (b) A constant current of 40 A for one-third cycle.

Solution:

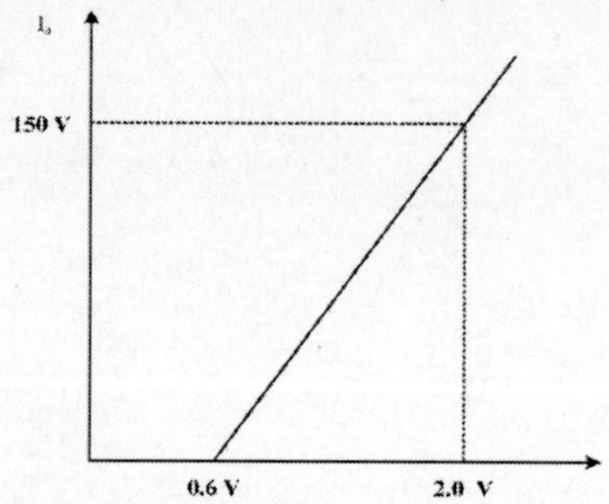

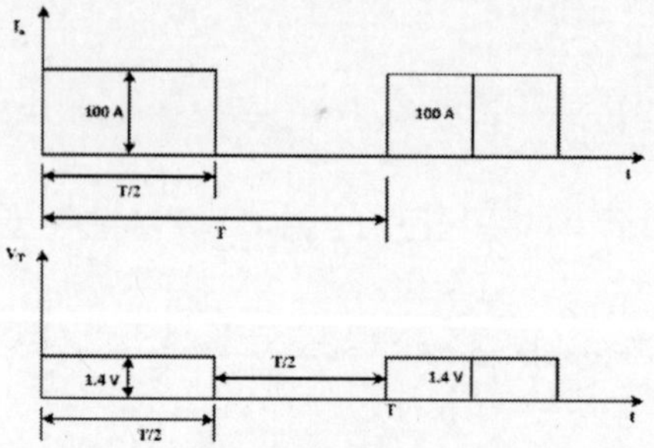

Figure.2.49 (a) V-I characteristics **Figure.2.49 (b)** Wave form

As per Figure.2.49 (a) for any current i_a, the voltage drop across thyristor is

$$V_T = 0.6 + \frac{2.0 - 0.6}{150} \times i_a = 0.6 + 0.008\ i_a$$

(a) A constant current i_a =100A for $\frac{1}{2}$ cycle is shown in figure).

The voltage drop across the device due to i_a =100A flowing is given by

$$V_t = 0.6 + 0.008 \times 100 = 1.4\ \text{V}.$$

The average on state power dissipation in thyristor is

$$P_{avg} = \frac{1}{T}\int_0^{T/2} 1.4 \times 100\ dt = \frac{1.4 \times 100 \times T}{2T} = 70\ \text{W}$$

Waveform of i_a gives the rms current rating of thyristor as

$$\sqrt{\frac{100^2 \times T}{2T}} = 70.71\ \text{A}$$

(b) When a constant current of 30A flowing through the device, *the voltage drop across the device is*

$$V_t = 0.6 + 0.008 \times 40 = 0.92$$

The average on state power dissipation in thyristor in one third cycle is

$$P_{avg} = \frac{0.92 \times 40 \times T}{3T} = 12.26\text{W}$$

Rms rating of current is $= 40 \times \frac{1}{\sqrt{3}} = 23.09$ A

Example.2.11

The peak supply voltage at which an SCR operates is 600V. The specifications of the device are given as follows. Repetitive peak current I_P = 200A, $(di/dt)_{max}$ = 60A/µs, $(dv/dt)_{max}$ = 300 V/µs, safety factor = 2, Design a suitable snubber circuit. Consider minimum value of load resistor is 20 Ω, and damping factor (ξ) is equal to 0.65.

Solution:

Given specifications are

The repetitive peak current, I_P = 200Aξ,

$$\left(\frac{di}{dt}\right)_{max} = 60A/\mu s$$

$$\left(\frac{dv}{dt}\right)_{max} = 300V/\mu s$$

safety factor = 2,

If we consider safety factor = 2, the specifications become

$$I_P = \frac{200}{2} = 100A, \left(\frac{di}{dt}\right)_{max} = \frac{60}{2} = 30\ A/\mu\ sec, \left(\frac{dv}{dt}\right)_{max} = \frac{300}{2} = 150\ V/\mu\ sec$$

In order to restrict the rate, rise of current beyond specification value of (di/dt), inductor must be inserted in series with thyristor. Hence,

$$L = \frac{Vs}{(di/dt)_{max}} = \frac{600 \times 10^{-6}}{30} = 20\ \mu H$$

The value of resistance in series, is given by

$$R_s = \frac{L}{Vs}\left(\frac{dv}{dt}\right)_{max} = \frac{20 \times 10^{-6}}{600} \times \frac{150}{10^{-6}} = 5\ \Omega$$

Before thyristor is tuned on, Cs is charge to 600 V, when thyristor is tuned on, the peak current through the thyristor is

$$\frac{600}{20} + \frac{600}{5} = 150\ A$$

As this peak current through SCR is more than the permissible peak of 100 A. The value of R_s must be increased.

Taking $R_s = 10\ \Omega$, The peak current through SCR $= \frac{600}{20} + \frac{600}{10} = 90$ A (which is less than the permissible peak current)

Also, $$C_s = \left(\frac{2\xi}{R_s}\right)^2 L = \left(\frac{2 \times 0.65}{10}\right)^2 \times 20 \times 10^{-6} = 0.338\ \mu F$$

The value of C_s may be considered as lower, therefore, $C_s = 0.3\ \mu F$
At the instant switch S is closed, the thyristor is open circuited and current through C_s is given by

$$C_s\frac{dv}{dt} = \frac{V_s}{R_s + R_L}$$

Or $$0.3 \times 10^{-6}\frac{dv}{dt} = \frac{600}{10 + 20}$$

Or $$\frac{dv}{dt} = \frac{600}{10+20} \times \frac{1}{0.3 \times 10^{-6}} = 66.66V/\mu sec$$

Since, the value of $\frac{dv}{dt}$ is less than the specified maximum value of 150V/µ sec. The value of C_s is correct. Therefore, L=20µH, $R_s = 10\ \Omega$, $C_s = 0.3\mu F$.

Example.2.12

The thyristor shown in Figure.2.50 has I^2t rating of $40A^2s$. If terminal P get short-circuited to ground, find the fault clearance time so that thyristor is not damage.

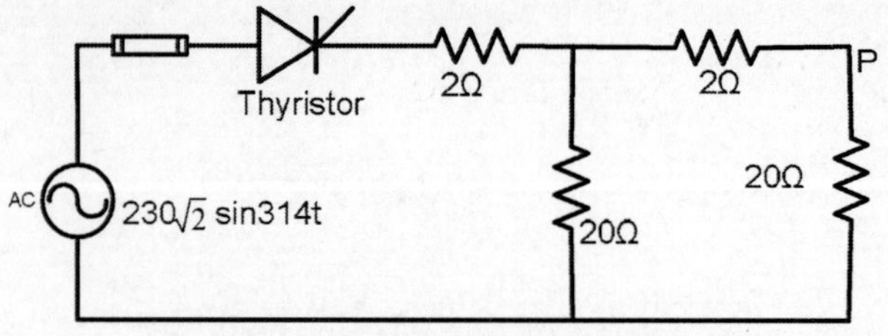

Figure.2.50 Circuit of example 2.12

Solution:

The worst possible fault current should be considered for calculating the fault clearance time.
The maximum fault current occurs when source voltage at its peak =230√2 V.
When terminal P gets short-circuited to ground,

the resistance offers to source = $2+\dfrac{20\times 2}{20+2}=2+\dfrac{10}{11}=\dfrac{31}{11}\Omega$

Assuming maximum fault current = $\dfrac{230\sqrt{2}x11}{31}$ A to remain constant during the short-circuit clearance time t_c, we will get

$$\int_0^{t_c} i^2dt=\int_0^{t_c}\left(\frac{230\sqrt{2}\times 11}{31}\right)^2 dt = 40A^2s$$

$$t_c=40\left[\frac{31}{230\sqrt{2}\times 11}\right]^2\times 1000\text{ ms}=2.99\text{ ms}$$

Example.2.13

The maximum junction temperature of an SCR is 130°C. The thermal resistance for the SCR- sink combination is $\theta_{jr}=0.15$ and $\theta_{cs}=0.06^{\circ}C/W$. The heat sink temperature is 80°C, find the total average power loss in the SCR-sink combination. In case the heat sink temperature bought sown to 70°C by forced cooling. Find the percentage increases in the device rating.

Solution:

As, we know that $$T_j=T_s+P_{avg}\left(\theta_{jc}-\theta_{cs}\right)$$

$$P_{avg}=\frac{130\text{-}80}{0.15\text{-}0.06}=238.095\text{ W}$$

Thus, total average power loss in the thyristor-sink combination as 238.095 W.

With improved cooling, $$P_{avg}=\frac{130\text{ -}70}{0.15+0.06}=285.714\text{ W}$$

Thyristor rating is proportional to the quire root of average power loss.

$\therefore$ % increases in the thyristor rating = $\dfrac{\sqrt{285.714}\text{ - }\sqrt{238.095}}{\sqrt{238.095}}\times 100=9.544\ \%$

Multiple Choice Questions

1. An SCR contains….. pn junctions
   (a) 2
   (b) 3
   (c) 4
   (d) None of the above
2. The name of three terminals of SCR is:
   (a) Emitter, Base and Collector
   (b) Drain, Gate, and Source
   (c) MT_1, Gate, and MT_2
   (d) Anode, Gate, and Cathode
3. An SCR acts like ……………. Switch
   (a) Bidirectional
   (b) Unidirectional
   (c) Mechanical
   (d) Neither of the above
4. Features that an SCR combines are
   (a) A rectifier as well as resistance
   (b) A rectifier as well as transistor
   (c) A rectifier as well as capacitor
   (d) None of the above
5. …… is the control terminal of an SCR
   (a) Cathode
   (b) Anode
   (c) Anode supply
   (d) Gate
6. Which of the following is NOT an advantage of employing an SCR switch?
   (a) Very high switching speed.
   (b) The operation does not produce harmonics.
   (c) It provides noiseless operation while at high efficiency.
   (d) It has no moving parts.
7. In the SCR structure the gate terminal is located
   (a) Near the anode terminal
   (b) Near the cathode terminal
   (c) In between the anode and cathode terminal
   (d) Non
8. In conventional reverse blocking thyristor
   (a) External layer are lightly dopped and internal layer are heavily doped
   (b) External layer are heavily dopped and internal layer are lightly doped
   (c) The p- layer are heavily dopped and the n-layer are lightly doped
   (d) The p- layer are lightly dopped and internal n-layer are heavily doped
9. Which of the following does not cause permanent damage of an SCR?
   (a) High current
   (b) High rate of rise of current
   (c) High temperature rise
   (d) High rate of rise of voltage

10. Surge current rating of an SCR specified the maximum
    - (a) Repetitive current with sine wave
    - (b) Non-repetitive current with rectangular wave
    - (c) Non-repetitive current with sine wave
    - (d) Repetitive current with rectangular wave
11. A forward voltage can be applied to an SCR after its
    - (a) Anode current reduce to zero
    - (b) Gate recovery time
    - (c) Reverse recovery time
    - (d) Anode voltage reduces to zero
12. In order to get best result per unit cost heat sinks on which the thyristor mounted are made of
    - (a) Aluminium
    - (b) Copper
    - (c) Nickel
    - (d) Stainless steel
13. The SCR is turned OFF when the anode current is .... holding current
    - (a) below
    - (b) above
    - (c) equal to
    - (d) twice the
14. The value of holding current of SCR is
    - (a) Equal to latching current
    - (b) Equal to one-third of latching current
    - (c) Two time of the value of latching current
    - (d) Slightly less than the value of latching current
15. In SCR, the turn off time
    - (a) Increases with increases of anode current I_A
    - (b) Is independent of I_A
    - (c) Varies with 1/IA
    - (d) Varies with $1/I_A^2$
16. For normal SCR, turn on tine is
    - (a) < turn off time
    - (b) > turn off time
    - (c) = turn off time
    - (d) 0.5 times the turn off time
17. What is the total anode current of SCR in the equivalent circuit from the two-transistor (T_1 & T_2) analogy of SCR?
    - (a) The sum of both the base currents
    - (b) The sum of both the collector current
    - (c) The sum of base current of T_2 & collector current of T_1
    - (d) The sum of base current of T_1 & collector current of T_2
18. The anode of SCR is always maintained at .... potential w.r.t cathode
    - (a) high
    - (b) low
    - (c) positive
    - (d) negative
19. Normally followed way for turning on a SCR is
    - (a) Breakover voltage

(b) Appropriate anode current
(c) Appropriate gate current
(d) None of the above

20. In forward bias, an SCR
(a) 2 junctions reverse biased
(b) 3 junctions reverse biased
(c) 1 junction reverse biased
(d) All 3 junctions are forward biased

21. The reverse bias of an SCR has
(a) 1 junction reverse biased
(b) 2 junctions reverse biased
(c) 3 junctions reverse biased
(d) 3 junctions forward biased

22. The instant at which an SCR starts conducting
(a) Small current starts flowing
(b) Its act like short circuit device
(c) Its act line an open circuit device
(d) Heavy voltage drop takes place across the device

23. The maximum anode current upon which an SCR is turn off is called
(a) Holding current
(b) Break down voltage
(c) Latching current
(d) peak reverse voltage

24. A conducting thyristor may be turned off by (increasing/decreasing) the anode current (below/above) the (latching/holding) current and applying a (reverse/forward/zero) voltage across the SCR for a short interval.
(a) decreasing, below, latching, reverse
(b) decreasing, below, holding, reverse
(c) decreasing, below, latching, forward
(d) increasing, above, holding, zero

25. Forward voltage triggering is not generally used for turn on thyristor since
(a) In this method losses are increases.
(b) It produces noise
(c) Junctions may be damaged and may be destroyed the device.
(d) It is not an efficient method of triggering.

26. A resistor connected across the gate and cathodes of an SCR increase its:
(a) holding current, noise immunity and turn off time
(b) dv/dt rating, holding current, noise immunity and turn off time
(c) dv/dt rating, holding current, and turn off time
(d) dv/dt rating, holding current, and noise immunity

27. The dynamic equalizing circuit for series SCRs comprising of
(a) R and C in series and a diode across C
(b) R and C in series circuit and with diode across R
(c) C and Diode series and with R and C parallel
(d) R and Diode in series but with C and R parallel.

28. To obtained static voltage equalization in series SCRs, circuit is made of
(a) 1 resistor across the string
(b) Same value of resistor across each SCR

(c) Different resistor values across each SCR
(d) 1 resistor in series with string

29. The magnitude of anode current will increase when the gate current of the thyristor is
    (a) Decreases
    (b) Increases
    (c) Replace by a voltage spike
    (d) None
30. The angle of conduction for an SCR can be changed by ....
    (a) Varying anode voltage
    (b) Varying gate voltage
    (c) Reverse biasing the gate
    (d) None of all the above
31. The most common method to fire an SCR is
    (a) Radiation firing
    (b) Thermal firing
    (c) Voltage firing
    (d) Gate firing
32. Increased in triggering angle of an SCR, the output
    (a) is same as before
    (b) is increased
    (c) is decreased
    (d) None of the above
33. An SCR can exhibit control over ....
    (a) Positive half-cycles ac supply
    (b) Negative half-cycles ac supply
    (c) Both positive and negative half-cycles of ac supply
    (d) Positive or negative half-cycles of ac supply
34. For a series string of SCR, the sharing of the transient voltage is balance by means of
    (a) Big inductor
    (b) Balancing resistor
    (c) Shunt capacitor
    (d) Auxiliary thyristor
35. di/dt protection is offered to SCR by means of
    (a) connecting an inductor in parallel across the load
    (b) connecting an inductor in series with the load
    (c) connecting an inductor in parallel across the gate terminal
    (d) connecting an inductor in series with load
36. dv/dt protection is offered to SCR by means of
    (a) connecting an capacitor in parallel across the load
    (b) connecting an inductor in series with the load
    (c) connecting a capacitor and resistance in parallel with the SCR
    (d) connecting an inductor in series with gate
37. The effect of over voltage on SCR are minimize by means of
    (a) Circuit breaker
    (b) RK circuit
    (c) Varistor
    (d) Inductor

38. Thyristor are used in electronics crowbar protection circuit since thyristor exhibits
    (a) High surge current capability
    (b) Less switching loss
    (c) Voltage clamping properties
    (d) High voltage rating
39. dv/dt protection is provided to
    (a) Limit the power loss
    (b) Reduction in junction temperature
    (c) Avoid accidental turn-on of the device
    (d) Avoid sudden large voltage across the load
40. In SCR, the turn ON-time
    (a) Is independent of V_g
    (b) Decreases with increases of V_g
    (c) Varies as V_g
    (d) Varies as V_A
41. In a SCR
    (a) Gate current is directly proportional forward break over voltage
    (b) As gate current is raised, forward breakover voltage reduces
    (c) Gate current has to kept on continuously for conduction
    (d) Forward breakover voltage is low in forward blocking state
42. After firing an SCR, if the gate pulse is detached, the SCR current
    (a) Remain the same
    (b) Reduces to zero
    (c) Rise up
    (d) Rises a little and then falls to zero
43. An SCR is fired by a pulse train of 5 KHz. The duty ratio is 0.4. if permissible average power is 100 W, the maximum permissible gate drive power is
    (a) $100\sqrt{2}$ W
    (b) 50W
    (c) 150W
    (d) 250 W
44. A thyristor comparable of thyratron tube is a
    (a) SCR
    (b) UJT
    (c) DIAC
    (d) TRIAC
45. Which of the following device provides complete isolation between triggering circuit and power circuit
    (a) PUT
    (b) LASCR
    (c) SUS
    (d) DIAC
46. Calculate the Derating factor if the sting efficiency is 90%
    (a) four
    (b) one
    (c) six
    (d) five

47. Calculate the sting efficiency if the derating factor is 22.
    (a) 78 %
    (b) 42 %
    (c) 20 %
    (d) 90 %
48. The absolute maximum operating frequency in kHz of a converter grade SCR of which turn on time and turn off time are 3μs and 200μs respectively is:
    (a) 9.4
    (b) 4.9
    (c) 5.2
    (d) 300
49. An SCR can withstand maximum temperature of 390°K at the junction with ambient temperature of 345°K. The thermal resistance of the SCR between junction to ambient is 1.5°C/W, the maximum internal power loss in watt allowed will be:
    (a) Twenty-five
    (b) sixty
    (c) Thirty
    (d) Fifty
50. The source voltage is 100 V and the load resistance is 10 Ω. The SCR can withstand a di/dt 50 V/μsec. if the snubber discharge current must be limited to 2A. Compute the snubber circuit resistor and capacitor:
    (a) 150 Ω, 0.6μF
    (b) 50 Ω, 0.2μF
    (c) 100 Ω, 0.4μF
    (d) 130 Ω, 0.8μF

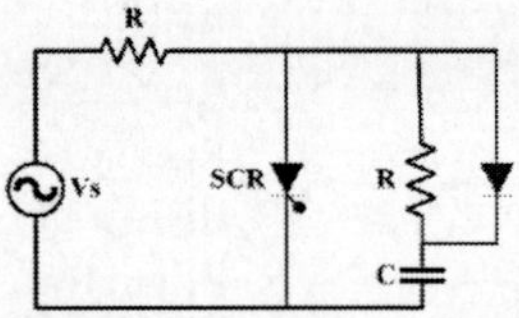

51. Which is a four-electrode thyristor?
    (a) Silicon Unilateral Switch
    (b) Silicon Controlled Rectifier
    (c) Programmable Unijunction Transistor
    (d) Silicon Controlled Switch
52. The SCS can be turned on by application of
    (a) Positive pulse to the anode gate, Negative pulse to the cathode gate
    (b) Positive pulse to the anode gate, positive pulse to the cathode gate
    (c) Negative pulse to the anode gate, positive pulse to the cathode gate
    (d) Negative pulse to the anode gate, Negative pulse to the cathode gate
53. The SCS can be turned on by application of
    (a) The anode current
    (b) The cathode gates
    (c) Either of the gates
    (d) Both the gate together

54. Which power switches has only two terminals?
    (a) DIAC
    (b) TRIAC
    (c) GTO

(d) SCR

55. Which power switches has only two terminals?
    (a) TRAIC
    (b) SCR.
    (c) 4-layer diode.
    (d) DIAC
56. Which one is a current controlled device?
    (a) MOSFET
    (b) SIT
    (c) MCT
    (d) GTO
57. A device with combined properties of thyristor and transistor is
    (a) SCR
    (b) DIAC
    (c) TRIAC
    (d) GTO
58. Gate turn off thyristor is most suitable for
    (a) DC to DC converter
    (b) Voltage source inverter
    (c) AC to AC converter
    (d) DC to AC inverter
59. Turn-on and turn-off times of a thyristor based on
    (a) Voltage gain
    (b) temperature of the device Junction
    (c) Current gain
    (d) Junction capacitances
60. Due to low internal generation in GTO, the GTO has
    (a) Lower latching current
    (b) Lower holding current
    (c) Lower latching and holding current
    (d) Higher latching and holding current
61. To turn-off a GTO what is essential at the gate?
    (a) A high amplitude (but low energy) negative current
    (b) A low amplitude negative current
    (c) A high amplitude negative voltage
    (d) A low amplitude negative voltage
62. Turn off gain of GTO, compared to its turn on gain is
    (a) equal
    (b) greater
    (c) less
    (d) independent
63. A unijunction transistor (UJT) is used to generate a ____ wave
    (a) square
    (b) sine
    (c) Parabolic
    (d) Circular
    (e) sawtooth

64. The number of terminals in an UJT is...
    (a) one
    (b) two
    (c) three
    (d) four
65. The UJT used for...
    (a) Amplifying a circuit
    (b) Circuit breaker
    (c) Splitting device
    (d) On-Off switching device
66. The material used to make the channel of a UJT is ...
    (a) pn type
    (b) It doesn't affect the working
    (c) p type
    (d) n type
67. The working regions of a UJT is..
    (a) Linear region
    (b) Negative Resistance region
    (c) Saturation region
    (d) Cut-off region
68. .... device from the thyristor family has its gate terminal connected to the n-type material near the anode.
    (a) SCR
    (b) RCT
    (c) PUT
    (d) SUT
69. The Programmable Unijunction Transistor (PUT) turns on & starts conducting when the
    (a) gate voltage exceeds anode voltage by a certain value
    (b) anode voltage exceeds gate voltage by a certain value
    (c) gate voltage equals the anode voltage
    (d) gate is given negative pulse w.r.t to cathode
70. Which of the following devices provides complete isolation between triggering circuit and power circuit?
    (a) PUT
    (b) LASCR
    (c) SUS
    (d) DIAC
71. The TRIAC is represented by
    (a) 2 SCRs in anti-parallel
    (b) 2 SCRs in parallel
    (c) 2 diodes in anti-parallel
    (d) 2 diodes in parallel
72. The DIAC is represented by
    (a) 2 SCRs in anti-parallel
    (b) 2 SCRs in parallel
    (c) 2 diodes in anti-parallel
    (d) 2 diodes in parallel
73. A crowbar is used to

(a) Protect a voltage sensitive load from enormous dc power supply output voltages
(b) Protect a current sensitive load from enormous dc power supply output voltages
(c) Protect a voltage sensitive load from enormous ac power supply output voltages
(d) Protect a current sensitive load from enormous ac power supply output voltages

74. The application of snubber circuit for SCR is to
    (a) Suppress dV/dt
    (b) Increase dV/ dt
    (c) Decrease dV/ dt
    (d) Decrease di/dt

Answers to multiple-Choice Questions

1. (b)	11. (b)	21. (b)	31. (d)	41. (b)	51.(d)	61(a)	71.(a)
2. (d)	12. (a)	22. (a)	32. (c)	42. (a)	52.(c)	62.(c)	72.(c)
3. (b)	13.(a)	23. (a)	33. (d)	43. (d)	53.(c)	63.(e)	73.(a)
4. (b)	14. (d)	24. (b)	34. (c)	44. (a)	54.(a)	64.(c)	74.(a)
5. (d)	15. (a)	25. (c)	35. (b)	45. (b)	55.(b)	65.(d)	
6. (a)	16. (a)	26. (b)	36. (c)	46. (a)	56.(d)	66.(d)	
7. (a)	17. (b)	27. (b)	37. (c)	47. (a)	57.(d)	67.(b)	
8. (b)	18. (c)	28. (b)	38. (a)	48. (b)	58.(a)	68.(c)	
9. (a)	19. (c)	29. (d)	39. (c)	49. (c)	59.(d)	69.(b)	
10. (c)	20. (c)	30. (b)	40. (b)	50. (b)	60.(d)	70.(b)	

Short and Long Answer Type Questions

1. What is an SCR?
2. What is the difference between SCR and Thyristor?
3. Provide the list of family members of Thyristor.
4. Provide schematic diagram, symbol and the constructional details of an SCR.
5. Explain the different operating modes of an SCR with the help of V-I characteristics.
6. Differentiate between SCR and Thyristor?
7. What are the different triggering methods used to bring SCR into conduction? Describe briefly each method.
8. What do you mean by latching current?
9. Describe with necessary diagram, the effect of gate current on the forward breakover voltage of an SCR.
10. Discuss the necessary conditions to turn on an SCR.
11. Define turn on and turn off time of an SCR.
12. Draw the dynamic characteristics of an SCR during turn on and turn off processes. Label all the intervals into which the SCR turn on and turn off.
13. Discuss the conditions to be satisfied to turn on an SCR with a gate signal.
14. Discuss the requirements for commutating an SCR from its forward conducting state.
15. Justify the statement, "Higher the gate current, lower is the forward breakover voltage".
16. What is on state condition and off state condition of SCR?
17. What is the difference between SCR and TRIAC?
18. What are the advantages and disadvantages of GTO?

19. Discuss with necessary circuit of two transistor model of SCR? Using two transistor model discuss the mechanisms of SCR for turning on.
20. Describe the electronic crowbar protection scheme employed for overcurrent protection of converters.
21. What are the methods adopted for protection of SCR? Explain each.
22. Describe the significance of $\frac{di}{dt}$ and $\frac{dv}{dt}$ in SCRs.
23. Explain about the thermal resistance of power semiconductor devices.
24. Snubber circuit of SCR mainly consist of a capacitor only. But in practice one resistor also used in series with the capacitor. Explain why?
25. Describe the LASCR. Provide the V-I characteristic of the same and explain the working of it.
26. Give the working principle, V-I characteristics of GTO.
27. Discuss briefly about the working, construction, V-I characteristics of PUT, SUS, SCS, DIAC, TRIAC, and UJT.
28. Draw the cross-sectional view of a DIAC and explain its working.
29. Discuss how a TRIAC may sometimes operate in the rectifier mode?
30. How GTO is differed from SCR?

Numerical problems

1. The gate cathode characteristics of an SCR is a straight-line. The slope of the line is130. The trigger source voltage is 15 V. Power dissipation at gate is 0.5 watt. Find gate-source resistance.
2. The gate cathode characteristics of an SCR is a straight-line. The slope of the line is 160. The trigger source voltage is 15 V. Power dissipation at gate is 0.8 watt, Find gate-source resistance.
3. A trigger circuit of a thyristor is supplied by 15 V and the slope of load line is -120 V/ampere. The SCR take 25 mA minimum gate current to turn on. Then find out, (a) source resistance requires in the gate circuit, (b) the triggering voltage and triggering current for an average gate power loss of 0.4 watt.
4. A trigger circuit of a thyristor is supplied by 25 V and the slope of load line is -150 V/ampere. The SCR take 25 mA minimum gate current to turn on. Then find out (i) trigger voltage, (ii) trigger current if average gate power loss is 0.8 watt.
5. The gate cathode characteristics of an SCR is represented by Vg = 1+10Ig, Gate source voltage is a rectangular pulse of 15V with 20μs duration. If the average power dissipation is 0.3W, and a peak gate drive power is 5W, find (a) the resistance to be connected in series with SCR gate, (b) the triggered frequency, (c) the duty cycle of gate pulse.
6. The gate cathode characteristics of an SCR is represented by $V_g = 2 + 40\ I_g$. A rectangular pulse of 25 V with 25 μ sec duration is used as gate source voltage. The average power dissipation is 0.4 W, and peak gate drive power is 8 W, find
   (a) The resistance to be connected in series with SCR gate
   (b) The triggered frequency
   (c) The duty cycle of triggering pulse.
7. The latching for an SCR inserted in between a dc voltage source of 200 V and the load current is 100 mA. Calculate the minimum width of gate pulse current required to turn on the SCR in case the load consist of (a) L= 0.2H, (b) Series com bination of R = 20Ω and L= 0.2H
8. The latching for an SCR inserted in between a dc voltage source of 500 V and the load current is 300 mA. Calculate the minimum width of gate pulse current required to turn on the SCR in case the load consist of (a) L=0.40 H, (b) R= 50 Ω in series with L= 0.40 H

9. The data sheet of an SCR provides maximum rms on-state current as 35A. If the SCR is used in a resistive circuit, find out the average on-state current rating for the half sine wave current for condition for conduction angle (a) 180°, (b) 90°, and (c) 30°
10. The static V-I characteristics of a forward conduction thyristor is given in Figure.2.51 which is a straight line. Obtain average power loss in the thyristor and its rms current rating for (a) A constant current of 80 A for one half cycle, (b) A constant current of 30 A for one-third cycle.

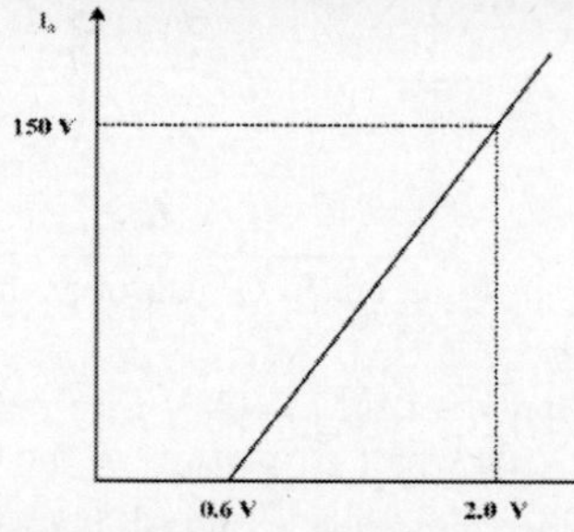

Figure.2.51 Figure related to example 10

11. The peak supply voltage at which an SCR operates is 400V. The specification of the device are given as follows. Repetitive peak current I_P = 200A, $(di/dt)_{max}$ = 50A/μs, $(dv/dt)_{max}$ = 200 V/μs, safety factor = 2, Design a suitable snubber circuit. Consider minimum value of load resistor is 10 Ω, and damping factor (ξ) is equal to 0.65.
12. The thyristor shown in Figure.2.52 has I^2t rating of 20A^2s. If terminal P get short-circuited to ground, find the fault clearance time so that thyristor is not damage.

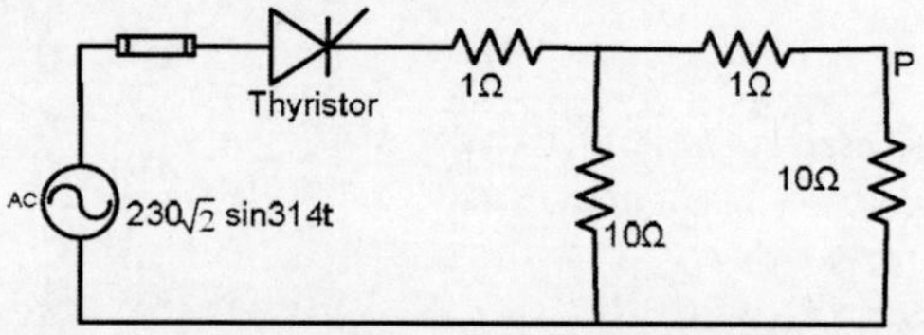

Figure.2.52 Circuit of example 12

13. The maximum junction temperature of an SCR is 125°C. The thermal resistance for the SCR- sink combination is θ_{jr} = 0.16 and θ_{cs} = 0.08°C/W. The heat sink temperature is 70°C, find the total average power loss in the SCR-sink combination. In case the heat sink temperature bought sown to 60°C by forced cooling. Find the percentage increases in the device rating.
14. For a thyristor, Maximum junction temperature is 120°C. The thermal resistance for the thyristor - sink combination is θ_{jr} = 0.18 and θ_{cs} = 0.10°C/W. For, a heat sink temperature of 90°C, find the total average power loss in the thyristor-sink combination. In case the heat sink temperature bought sown to 80°C by forced cooling. Find the percentage increases in the device rating.
15. The specification sheet for an SCR gives maximum rms on-state current as 45A. If this SCR is used for resistive circuit. Then find out the average on-state current rating for the rectangular wave form of current **(Figure.2.53)**, for condition for conduction angle (a) 180°, (b) 90°,(c) 30°

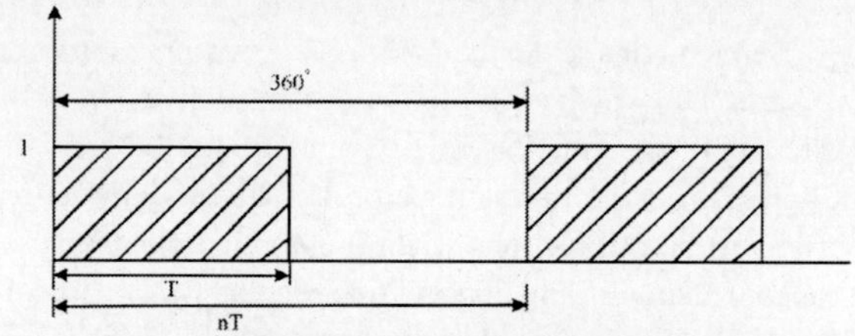

Figure.2.53 Figure related to example 15

16. The circuit shown in Figure.2.54, the thyristor is gated with a pulse width of 50 μ sec. The latching current of thyristor is 40 mA. For a load of 40 ohm and 4 H, will the thyristor get turned on? Check, if the answer is negative, how this difficulty can be overcome for a given load. Find the maximum value of the remedial parameter shown dotted.

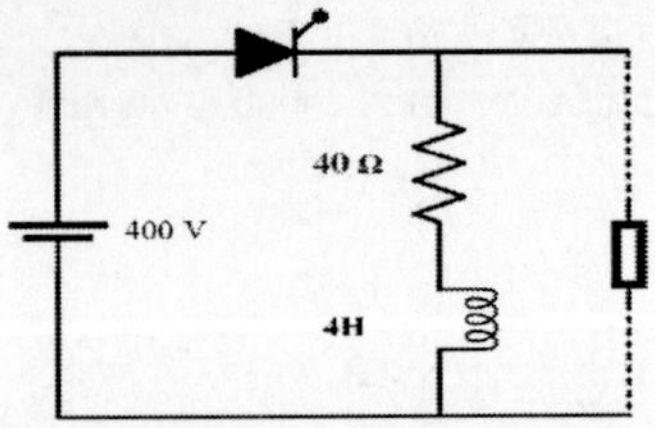

Figure.2.54 Circuit of example 16

17. An SCR is operated at peak supply voltage of 600 V with following specification: Repetitive peak current IP = 200A, $(di/dt)_{max}$ = 60A/ μ sec, $(dv/dt)_{max}$ = 300 V/μ sec, safety factor = 2, Design a suitable snubber circuit. Considered minimum value of load resistor is 30 Ω

Practical Experiments

Experiment No.2.1

Title: Test the proper functioning of DIAC and determine the break-over voltage

Objectives:

(a) To plot the V-I characteristics of DIAC
(b) To find the breakover voltage of the DIAC

Resources required:

Table.2.1 Apparatus/ component required

Sl No.	Apparatus/ Component	Specification Range/ Rating	Nos/Quantity
1.	DIAC Kit		1
2.	D C Power supply		1
3.	Digital Multimeter		2
4.	Voltmeter	0-30V	1
5.	Ammeter	0-1A	1
6.	Connecting leads		

Theory:

A DIAC is a type of semiconductor device comprising two junctions and three layers. Its name is a combination of two parts: DI, which refers to the diode (such as Di, Tri, Quad, Penta, etc.), and AC, which represents alternating current. Essentially, DIAC is an abbreviation for a diode (member of thyristor family) designed to handle alternating current. This is a bidirectional switch with two terminals that conducts electricity in both directions when the applied voltage surpasses its breakover voltage. However, it is not capable of amplifying or providing regulated switching. DIACS are seldom employed individually, but rather combined with other thyristor devices. In phase control circuits, they serve as triggers to provide gate pulses to a TRIAC or SCR. Many gate triggering circuits utilize this component to attain enhanced stability and immunity to noise during triggering. These silicon devices are not only bidirectional but also voltage-triggered. To decrease the DC component in the load circuit, DIAC voltage options ranging from 27V to 70V generate trigger pulses that are symmetrically matched at the positive and negative break-over points. Basically, DIACs have symmetric switching voltages. Asymmetric DIACs are also available. Its Power dissipation ranges from 0.5 to 1 watt.

Despite being a member of the thyristor family, the DIAC does not feature a gate terminal for control purposes. Instead, it can be turned on or off by merely reducing the voltage level below the avalanche breakdown voltage in either polarity. The symbol of DIAC is depicted in Figure.2.55.

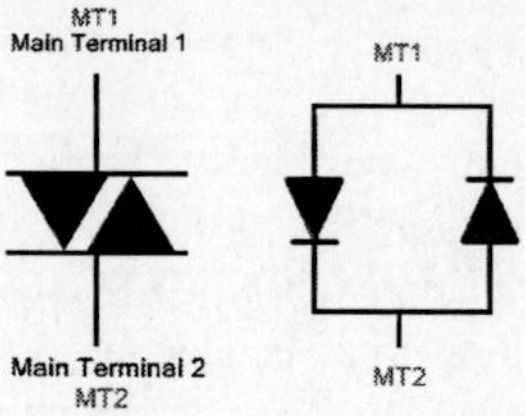

Figure.2.55 Symbol of DIAC

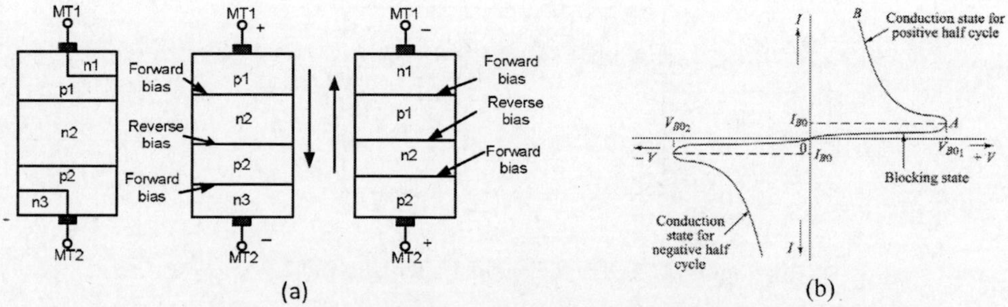

Figure.2.56 (a) Working principle, (b) V-I characteristics of DIAC

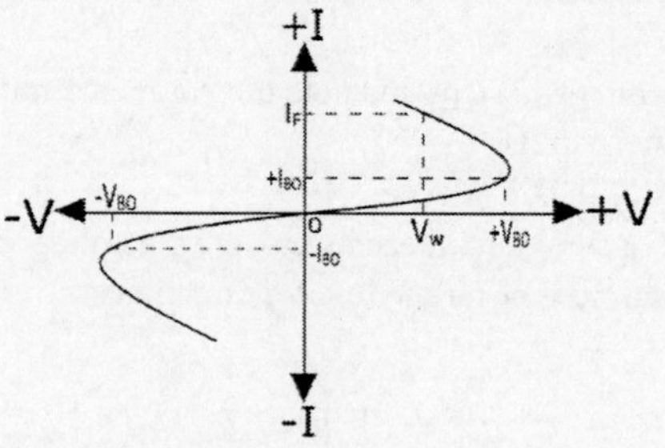

Figure.2.56 (c) V-I characteristics of DIAC showing break brake voltage

Depending on the polarity of the applied voltage, each terminal has the ability to function as either an anode or a cathode. The end junction that is active and the one that is bypassed are also determined by the polarity of the applied voltage.

In Figure.2.56 (a), the operation of the DIAC is illustrated in terms of polarities. Assuming that the MT1 terminal is positive, the P_1 layer close to MT1 will be activated, resulting in conduction in the sequence of p_1-n_2-p_2-n_3. When the current flows from MT1 to MT2, the junctions between p_1-n_2 and p_2-n_3 are forward-biased, while the junction between n_2-p_2 is reverse-biased. Likewise, if we assume that the

MT2 terminal is positive, the p_2 layer close to MT2 will be activated, and conduction will occur in the order of p_2-n_2-p_1-n_1. The current will flow from MT2 to MT1, and the junctions between p_2-n_2 and P_1-N_1 will be forward-biased, while the junction between n_2-p_1 will be reverse-biased. As a result, conduction will be possible in both directions.

The V-I characteristics of a DIAC is shown in Figure.2.56(b) and (c). The DIAC's V-I characteristic curve takes the form of a Z-shape and spans the first and third quadrants due to its ability to conduct in both positive and negative polarities. In the first quadrant, which corresponds to the positive half cycle, current flows from MT1 to MT2, while in the second quadrant, which corresponds to the negative half cycle, current flows from MT2 to MT1. Initially, the DIAC's resistance is higher due to the reverse bias junction between its layers, resulting in a small amount of leakage current flowing through it, which is referred to as the **blocking state** on the curve. When the applied voltage reaches the breakdown voltage, the DIAC's resistance decreases abruptly, causing it to begin conducting, resulting in a sharp decrease in voltage and an increase in current, which is referred to as the **conduction state** on the curve. The breakdown voltage of most DIACs is approximately 30 volts, although the exact value is determined by the device type. The DIAC remains in the conducting state until the current reaches the **holding current**, which is the minimum current required to maintain the device in the ON state.

Circuit diagram:

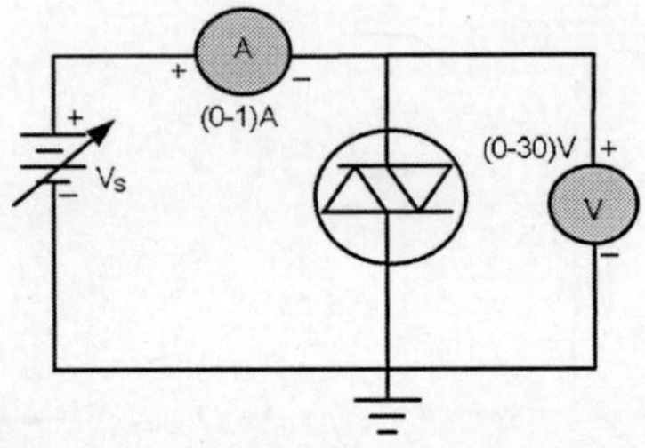

Figure.2.57 Circuit diagram Experiment No.2.1

Procedure:

1. Connect the millimeter, voltmeter, DIAC as per the circuit diagram of **Figure.2.57.**
2. Switch on the power supply.
3. Increase the supply voltage in steps and note the voltmeter, milliammeter readings in each step.
4. Plot the voltage versus current.
5. Reverse the terminal of the DIAC.
6. Increase the voltage in steps and in each step note the voltage and current in **Table 2.2**
7. Plot the voltage versus current for the reverse direction.

Result and discussion

Table.2.2

Forward bias			Reverse bias		
SL. No	Voltage (V)	Current (mA)	SL. No	Voltage (V)	Current (mA)
1.			1.		
2.			2.		
3.			3.		
4.			4.		
5.			5.		
6.			6.		

Discussion: The plots in both the forward and reverse bias have been plotted from the data obtained from test. It is seen that the plot is similar to that of typical plots as shown in Figure. The breakover or breakdown voltage are noted.

Break down voltage: .....

Conclusion: From this experiment it is concluded that the DIAC is function properly as its V-I characteristics resemble to that of typical characteristics.

Experiment No.2.2

Title: Determine the latching current and holding currents using V-I characteristics of SCR.

Objectives:

(a) To plot the V-I characteristics of SCR
(b) To find the breakover voltage, latching current, and holding current of the SCR

Resources required:

Table.2.3 Apparatus/ component required

Sl No.	Apparatus/ Component	Specification Range/ Rating	Nos/Quantity
1.	SCR Kit		1
2.	D C Power supply	0-30V	2
3.	Digital Multimeter		1
4.	Voltmeter	0-30V	1
5.	Milli ammeter	0-30mA	1
6.	Microammeter	0-30μA	1
7.	Resistance	1kΩ, 1W	1
8.	Resistance	100Ω, 20W or 30W	1
9.	Connecting leads		

Theory:

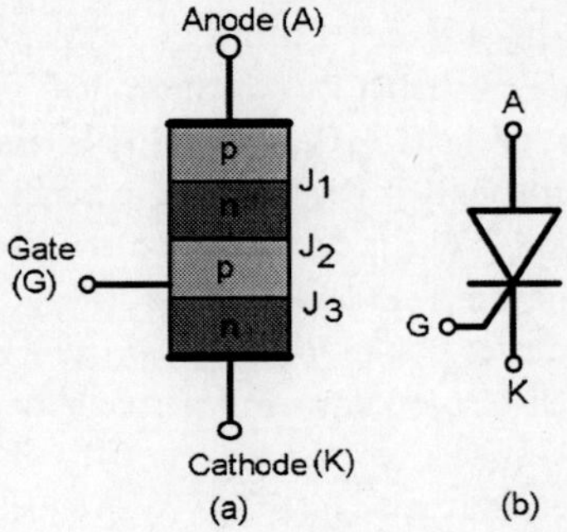

Figure.2.58 Schematic showing layers, symbol, the actual picture of SCR (a) Schematic diagram, and (b) Symbol.

The silicon-controlled rectifier (SCR) is a solid-state semiconductor device having three layers and three terminals. The terminals are called anode, cathode, and gate. It has three *pn* junctions. The schematic diagram and symbol of SCR are shown in Figure.2.58 (a) and (b) respectively. The anode, cathode, and gate are represented as A, K, and G respectively in Figure.2.58.

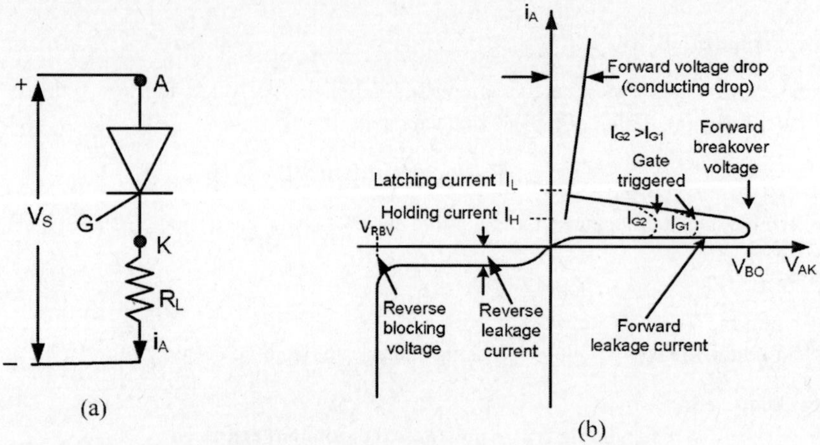

Figure.2.59 (a) Circuit of SCR, (b) V-I characteristics

Figure.2.59 (b) displays a typical V-I characteristic of an SCR. It is nothing but a plot between anode to cathode voltage (V_{AK}) and anode current (i_A) of the SCR. Let us Consider Figure.2.58 (a) and the circuit of Figure.2.59(a). When the anode is positive w.r.t cathode J_1 and J_3 junctions forward biased, J_2 junction reversed biased. Only a small leakage current flows through the device. The SCR is now called forward blocking and the current through the device is called off-state current. When V_{AK} goes on increasing, a stage will reach at which it reversed biased J_2 junction breaks down. This breakdown phenomenon is named as avalanche breakdown. The value of V_{AK} at which this breakdown occurs is called the forward breakdown voltage (V_{BO}). Since J_1 and J_2 are already forward-biased and J_2 breaks down, the large anode current (I_A) flow due to the movement of carriers across the junctions. This is the forward anode current. The SCR is in a state called conducting state or on-state. The ohmic drop in the junctions is very small in on-state and it is of the order of 1V. The anode current should be more than a minimum anode current, called latching current (I_L) to maintain conduction in the on-state. In case, if it is less than I_L, the SCR will go back to the forward blocking state. During the conduction period, the SCR behaves like a diode. The anode current is limited by the resistance R_L in **Figure.2.59(a).** The SCR remain conducting because of the absence of depletion layer at J_2 junction. If the anode current is reduced to a value called holding current (I_H) at which the depletion layers across J_2 formed and SCR reached the blocking state. Thus, the holding current is always less than the latching current. When the SCR is in reversed biased i.e terminal K is positive w.r.t terminal A, J_2 is forward biased but J_1 and J_3 are reversed biased. The situation is like a circuit of two series-connected diodes in which reversed voltage is applied. The SCR is now in a reverse-blocking state. A small current called reverse leakage current (I_R) flows through the device due to minority carriers. When the reversed voltage increased further a stage is reached at which the breakdown of reversed biased junctions occurs and heavy current flows through the device. The voltage at which reverse breakdown occurs is called the reverse breakdown voltage (V_{RWM}). The SCR can be turned on by two methods. They are as follows.

1. Increased the forward voltage greater than the forward break-over voltage (V_{BO}). But this method is destructive.
2. Maintain a forward voltage below VBO, and apply a positive voltage across gate (G) and cathode (K). The situation is shown by the dashed line in the V-I characteristic. This type of turn on method is called gate triggering.

After the SCR is turned on by gate triggering and the anode current is above the holding current, the SCR remains conducting because of +ve feedback even after the gate signal is removed. Thus, a SCR is a latching device.

Circuit diagram:

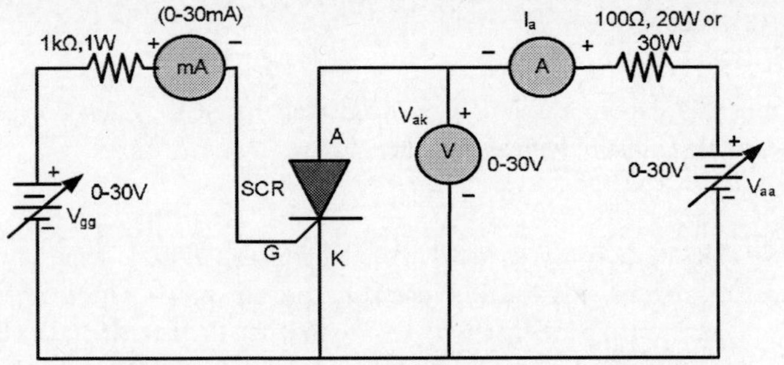

Figure.2.60 Circuit diagram Experiment No.2.2

Procedure:

1. Make the circuit as shown in **Figure.2.60**
2. Set the value of the gate current, I_G to a convenient value by adjusting the V_{gg}
3. Vary the anode to cathode voltage, V_{aa} in steps and in each step note down the voltage V_{ak} and current I_a. Note down the value of V_{ak} and I_a at the instant of firing of the SCR and after the firing (reducing the voltmeter and ammeter range) and then increasing the V_{aa}. Note the values of V_{ak} and I_a
4. The point at which SCR fired is the breakover voltage and the point at which Ia start increases suddenly gives the latching current.
5. Plot the graph between V_{ak} and I_a
6. Switch off V_{gg}
7. Observe the ammeter reading by reducing the supply V_{aa}. The point at which ammeter reading fall to zero is the holding current.
8. Repeat the steps 1 to 8 for another value of I_g
9. Interchange the of the supply voltage V_{aa}, and repeat the experiment. This will give the reverse characteristic.

Result and discussion

Table.2.4

Forward bias			Reverse bias		
SL. No	Voltage (V)	Current (mA)	SL. No	Voltage (V)	Current (μA)
1.			1.		
2.			2.		
3.			3.		
4.			4.		
5.			5.		
6.			6.		
7.			7.		

Discussion: The plots in both the forward and reverse bias have been plotted from the data obtained from test. It is seen that the plot is similar to that of typical plots as shown in Figure. The breakover or breakdown voltage, latching current, and holding currents are noted.

Break down voltage: ....

Holding current:.......

Latching current: ....

Conclusion: From this experiment it is concluded that the SCR is function properly as its V-I characteristics resemble to that of typical characteristics.

Know More

There are different types of thyristors as listed in section 2.3. Out of these some have been discussed in this unit in terms of symbols, construction, operating principles, V-I characteristics etc. It is also required to study the other members such as reverse conducting thyristors (RCTs), bi-directional phase controlled thyristors (BCTs), FET-controlled thyristors (FET-CTHs), MOS turn-off thyristors (MTOs), MOS-controlled thyristors (MCTs), Static induction thyristors (SITHs), Emitter turn-off thyristors (ETOs), and Integrated gate-commutated thyristors (IGCTs). The SCR has controlled turn-on and uncontrolled turn-off capabilities. Like TRIAC, RCTs also have the bidirectional capability. Like GTO, MCT and SITH are also unidirectional current capabilities. t can be considered as a thyristor with built in anti-parallel diodes. BCT is a new semiconductor device and it combines advantages of two having two thyristors in one pack. Electrical behaviours of BCT is similar to two anti-parallel thyristors. FET-controlled thyristors (FET-CTHs) combines the features of MOSFET and a thyristor and connected in parallel. MTOs are new power semiconductor device and it combines the good features of MOSFET and GTO and eliminate the limitations of GTO turn of ability. ETO is a hybrid device that combines the advantages of GTO and MOSFET. The IGCT integrates a gate-commutated thyristor with a multi-layered printed circuit board drive. The SITH is also known as field-controlled diode. It was invented in 1960s.

References and suggested readings

1. Ramamoorty M., An Introduction to Thyristors and their applications, East-West Press Pvt. Ltd., New Delhi, ISBN: 8185336679.
2. Sugandhi, Rajendra Kumar and Sugandhi, Krishna Kumar, Thyristors: Theory and Applications, New Age International (P) ltd. Publishers, New Delhi, ISBN: 978-0-85226- 852-0.
3. Bhattacharya, S.K., Fundamentals of Power Electronics, Vikas Publishing House Pvt. Ltd. Noida. ISBN: 978-8125918530.
4. Jain & Alok, Power Electronics and its Applications, Penram International Publishing (India) Pvt. Ltd, Mumbai, ISBN: 978-8187972228.
5. Rashid, Muhammad, Power Electronics Circuits Devices and Applications, Pearson Education India, Noida, ISBN: 978-0133125900.
6. Singh, M. D. and Khanchandani, K.B., Power Electronics, Tata McGraw Hill Publishing Co. Ltd, New Delhi, 2008 ISBN: 9780070583894.
7. Zbar, Paul B., Industrial Electronics: A Text –Lab Manual, McGraw Hill Publishing Co. Ltd., New Delhi, ISBN: 978-0070728226.
8. Grafham D.R., SCR Manual, General Electric Co., ISBN: 978-0137967711.
9. Mohan Ned, Undeland Tore.M., Robins William P., Power Electronics – Converters, Applications and Design, John Wiley & Sons. Inc,

10. https://www.google.com/search?q=picture+of+heat+sink+for+power+device.
11. Rashid, M. H. (2014). Power Electronics: Devices, Circuits, and Applications: International Edition. United Kingdom: Pearson Education.
12. Undeland, T. M., Mohan, N., Robbins, W. P. (2003). Power Electronics: Converters, Applications, and Design. India: John Wiley & Sons.
13. Fewson, D. (1998). Introduction to Power Electronics. United Kingdom: Elsevier Science.
14. Williams, B. W. (1992). Power electronics: devices, drivers, applications, and passive components. United Kingdom: McGraw-Hill.
15. Robbins, W. P., Mohan, N., Undeland, T. M. (2003). Power Electronics: Converters, Applications, and Design. India: John Wiley & Sons.

3 Turn-on and Turn-off Methods of Thyristors

UNIT SPECIFICS

This unit covers the following aspects:

- *Silicon-controlled rectifier (SCR) turn-on methods.*
- *Basic requirement of the firing of SCR.*
- *Pulse transformer and its application in SCR firing.*
- *Optical isolator for SCR firing.*
- *The resistance gate triggering circuits for SCR and its analysis.*
- *Resistance – capacitance gate triggering circuits for SCR and its analysis.*
- *The programmable unijunction transistor (PUT).*
- *UJT oscillator Triggering circuits for SCR and its analysis.*
- *Pulse transformer and opto - coupler based triggering for SCR.*
- *Classification of SCR turn-off methods.*
- *Analysis of Class-A, Class-B, Class -C, Class-D, and Class-E turn-off methods of SCR.*
- *Practical experiments on triggering and turn-off circuits.*

The concept of various triggering and turn-off methods of thyristors are discussed and analysed for generating further curiosity and creativity as well as improving problem-solving capacity with some numerical problems.

Besides giving a large number of multiple-choice questions as well as questions of short and long answer types marked in two categories following the lower and higher order of Bloom's taxonomy, assignments through several numerical problems, a list of references, and suggested readings are given in the unit so that one can go through them for practice.

Some practical experiments related to the course covered in Unit III are also appended at the end of this unit to make the students aware of the hands-on on these topics.

After the related practical experiments on the topic, based on the content, there is a "Know More" section appended. This section has been designed to supplement additional information and higher learning skills on the topic.

RATIONALE

This fundamental unit on turn-on and turn-off methods helps students to get a primary idea about the basic methods of turn-on and turn-off and its applications in various power electronics circuits involving SCR.

The various triggering circuits and their components for SCR firing are presented in this unit. The analysis of these circuits is also depicted. The various Class of turn-off methods are also presented.

The physics behind various turn-on and turn-off methods are discussed at length to develop the basic idea about these methods.

Some related problems are pointed out after each section with their solutions which can help further for getting a clear idea of the concerned topics. The mathematics behind the triggering and turn-off methods will certainly help students with numerical problem-solving.

As a student in the field of electrical engineering, this unit on turn-on and turn-off methods help students to grasp the basic knowledge of various types of power electronic circuits used for turn-on and turn-off of thyristors.

PRE-REQUISITES

ESC101: Basic Electrical Engineering

UNIT OUTCOMES

After completion of Unit-3 students will be able to:

U3-O1: Identify the turn-on and turn-off methods.
U3-O2: Explain the working of various turn-on and turn-off circuits for SCR.
U3-O3: Analyse the various triggering circuits for firing SCR.
U3-O4: Analyse the turn-off methods of SCR.

Unit-3 Outcomes	EXPECTED MAPPING WITH COURSE OUTCOMES (1- Weak Correlation; 2- Medium correlation; 3- Strong Correlation)				
	CO-1	CO-2	CO-3	CO-4	CO-5
U3-O1	2	2	1	...	1
U3-O2	2	2	3	3	2
U3-O3	1	2	3	3	3
U3-O4	..	2	3	3	3

3.1 INTRODUCTION

Power electronics uses various types of power semiconductor devices (PSDs). Examples of PSDs are power diodes, various power transistors, different thyristors, etc. Except for power diodes, the other PSDs have two stable states: a forward-blocking state called off-state and a forward-conducting state known as on-state. Power transistors and thyristors need to be turned on from the off-state to the on-state for operation. This switching operation is called turn-on or triggering or firing. It is also required to switch off the device from on-state to off-state. This process is called the turn-off process. There are certain methods to do the same. In this Unit, the details about the turn-on and turn-off methods for thyristors will be presented.

3.2 SCR TURN-ON METHODS

There are several methods of turn-on of an SCR. They are listed as follows.

(a) High voltage triggering.
(b) Thermal triggering.
(c) Illumination triggering.
(d) *dv/dt* triggering.
(e) Gate triggering.

(a) **High voltage triggering:** Keeping the gate open and increasing the anode-to-cathode voltage beyond the breakdown voltage, V_{BD}, will increase the flow of leakage current through the device and avalanche breakdown of junction J_2 occurs at V_{BD}. The flow of leakage current initiates the regenerative turn-on.

(b) **Thermal triggering:** The width of the depletion layers of any semiconductor device reduces with an increase in junction temperature. This is true in the case of SCR also. With high temperatures, the number of electron-hole pairs increases which in turn increases the leakage current. An increase in leakage current again increases the forward current gains (α_1 and α_2). As a result, the sum of α_1 and α_2 tends to unity and SCR will ger turned-on.

(c) **Illumination triggering:** The energy is imparted by radiation. The energy particles like neutrons or photons is utilized to turn-on SCR. Bombardment of SCR with such energy particles generates electron-hole pairs in the device resulting in increased carrier density. An increase in carrier leads to the flow of current through the SCR. Thus, the thyristor is turned on.

(d) ***dv/dt* triggering:** When the anode is positive with respect to (w.r.t) to cathode the SCR is forward biased. There are three junctions (J_1, J_2, and J_3) in the SCR as shown in Figure.3.1(a). At this forward-biased condition, J_1 and J_3 are forward-biased and J_2 is reverse-biased. Thus J_2 behaves like a capacitor due to the charge existing across the junction. If the rate of rise of the anode-to-cathode voltage is high, the charging current of the capacitor is high enough to trigger the SCR. If the impressed voltage is V_{ak}, C_{J2} is the capacitance, Q_C is the charge of the capacitor, then the capacitor charging current i_c is given by equation (3.1).

$$i_c = \frac{dQ_C}{dt} = \frac{d}{dt}(C_{J2}V_{AK}) = C_{J2}\frac{dV_{AK}}{dt} + V_{AK}\frac{dC_{J2}}{dt} \tag{3.1}$$

Since junction capacitance C_{J2} is almost constant, the term $\frac{dC_{J2}}{dt}$ can be neglected. Thus, the capacitor charging current can be written as (3.2).

$$i_c = C_{J2}\frac{dV_{AK}}{dt} \tag{3.2}$$

Hence, if the rate of change of voltage across the anode to the cathode is large, the SCR may get turned on.

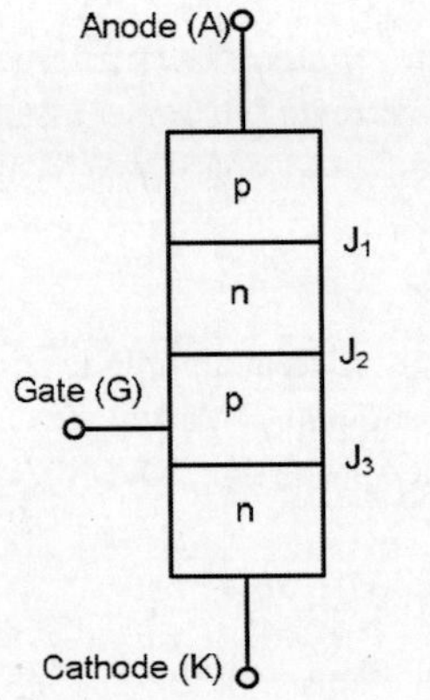

Figure.3.1 SCR junctions

(e) **Gate triggering:** This method of triggering is commonly used in SCR firing. When SCR is forward biased, the application of a positive gate voltage with respect to (w.r.t) cathode turns on the SCR. In this method, the conduction period of the SCR can be controlled by the application of the gate signal within the specified maximum and minimum values of gate current. Three types of gate signals namely D.C. signals, pulse signals, or ac signals are used for gate triggering. Either of these three signals can be used. The basic gate-triggering methods are shown in Figure.3.2

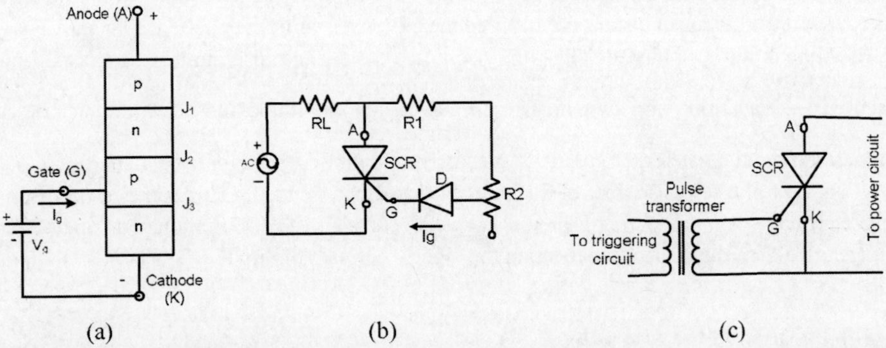

(a) (b) (c)

Figure.3.2 Basic gate triggering (a) DC gate triggering, (b) AC gate triggering, (c) Pulse triggering

In DC gate triggering method, a DC voltage of proper magnitude and polarity is applied between the gate and the cathode. The gate terminal is connected to the positive terminal and the cathode is connected to the negative terminal of the supply. When this voltage is sufficient to produce the required gate current, the SCR starts conducting. In this method, there is no isolation between power and the source.

The AC signal triggering is used for the devices applied for AC applications. Proper isolation of the power circuit and control circuit is provided in this method which is one of the advantages.

Pulse signal triggering is popularly used. In this method, the gate drive supplies a single pulse of high frequency on a periodic basis or on a sequential basis. A pulse transformer is used here for isolation.

3.3 BASIC REQUIREMENT OF THE FIRING OF AN SCR

The SCR is the most commonly used power semiconductor device because the device responds to small signals for turning off. The source of the gate triggering signal should be such that it can supply suitable voltage and current at the gate terminal. The performance of an SCR depends on the magnitude as well as the wave shape of the gate current. The proper instant of firing depends on the magnitude and wave shape of the gate current. The type of firing system is also a deciding factor for the generation of harmonics in the device operation. Based on these factors, the requirements of the proper triggering of an SCR are summarised as follows.

(a) The gate current should have sufficient amplitude and of short rise time.
(b) The duration of gate current should be of adequate duration.
(c) The current should be supplied at the time when the main circuit is favourable for conduction.

(a) Gate current magnitude and rise time

Though, the minimum gate current that is required to trigger a given type of thyristor is decided at a standard temperature, practically a margin must be provided for errors and uncertainties, and to ensure the turn-on time. Usually, a gate current of 1.4 times that of the minimum gate current is considered if the main circuit conditions do not have any special requirements. Most thyristors require a considerably higher gate current to reduce the turn-on time for reducing switching losses and to increase the $\frac{di}{dt}$ rating of the thyristor. The gate voltage must also exceed the specified minimum gate triggering voltage. The gate drive must be such that it must trigger the thyristor under the most severe condition such as the lowest junction temperature that may likely occur. The rise time of the gate pulse is also important because an adequate gate current must be reached before the thyristor turn-on within the turn-on delay time. Even if the given conditions are fulfilled the effectiveness of the pulse still increases somewhat as the rise time is further reduced. Thus, the rise time of the gate pulse must be specified. Thus for any specified $\frac{di}{dt}$ capability and switching performance, the amplitude and rise time must be considered while designing the gate drive circuit. A gate drive circuit designed for a very short gate pulse has to allow for the finite response time of the gate-cathode junction itself. This makes it necessary to apply a higher voltage than is apparent from the static gate characteristics. Commonly, an open-circuit source voltage of 15-20 Volt is specified to attain the highest current amplitude with a corresponding short rise time.

(b) Duration of the gate pulse

In general, a thyristor may be triggered satisfactorily by a pulse on the gate of a duration nearly equal to the turn-on time. But practically, a longer pulse duration is necessary for the reasons mentioned below.

i) A comparatively long period may be required for the anode current, to increase to the latching current level.
ii) Factors like oscillations, reflections, or any other disturbances may lead to tum-off the thyristor shortly after it is first triggered.
iii) There may be uncertainty in favourable conditions at which the triggering pulse is initiated.

The determination of pulse duration is dependent on which these factors are modified. To ensure the adequate duration of pulse some ways are identified. In most cases the duration is extended to the whole period for which the thyristor is likely to conduct. Pulse to cover at least one period of uncertainty can also alternatively be considered. But to generate pulses of long duration, particularly high amplitude and short time are required which increases complexity and cost. Hence, if possible, pulses of short duration are generally preferred. If a short-duration pulse coincides with reverse anode voltage, it results in the reduction of reverse losses. This is one advantage of a short-duration pulse. Generally, a pulse duration < 10 ps needs extensive care in the anode circuit design. A duration of 30-60ps is usually sufficient to avoid problems till the anode circuit conditions are favourable for conduction.

In order to cover the region of uncertain triggering the following alternative are suggested.

i) A control system where the gate pulse timing and thyristor voltage and current to zero is synchronized.
ii) An extended train of short-duration pulses.

The first one is more appropriate for comparatively simple applications in which isolation between the control and the main circuit is not required. The second one is for the more common application and it gives the thyristor repeated opportunities to turn on. However, if the first pulse of the train is not effective, the actual instant of firing is indeterminate.

(c) Pulse waveforms

Though an anode circuit requires a gate pulse of high amplitude and rise time, it is not required to maintain it for the whole duration. The thyristor starts conducting at the beginning of the gate pulse and the high amplitude of the pulses disappears after a few microseconds. At the same time, the thyristor is forward-biased in the duration of the pulse, at $\frac{dv}{dt}$ capability. There is neither a $\frac{di}{dt}$ condition nor necessity for a very short switching time. A pulse of continuous high amplitude is disadvantageous because it results in the demand of the firing circuit of increasing mean gate dissipation and latching current. Thus, it is a good practice to use a pulse waveform having a leading edge of the required amplitude and rise time followed by a tail of not more than the practical minimum amplitude required for triggering. Figure.3.3 provides a such waveform. A negative overshoot at the end of the gate pulse is normally avoided because in some cases this may lead to turn-off the anode current.

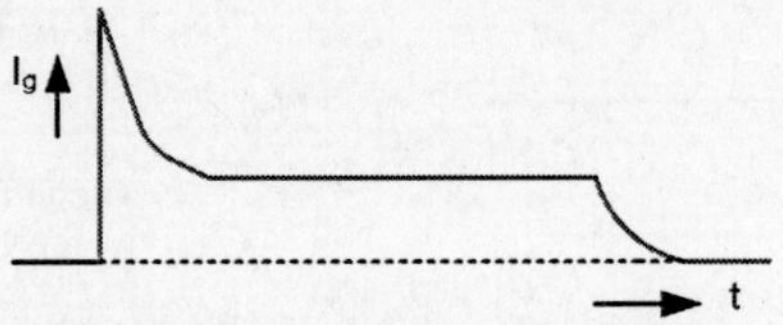

Figure.3.3 A gate firing pulse

(d) Spurious triggering

Spurious triggering may be due to the excessive anode-to-cathode voltage or $\frac{dv}{dt}$ in the thyristor circuit or stray pick-up of the gate triggering circuit. It may lead to direct damage to the thyristor because insufficient $\frac{di}{dt}$ capability to withstand the anode circuit conditions. Thus, the self-protecting circuits based on automatic breakover in the case of excessive voltage are to be avoided unless the thyristor is specifically rated for such duty.

(e) Determination of firing angle in naturally commutating converters

The main function of a triggering angle control circuit is generally to find the firing angle, concerning the A.C. supply voltage, so that the desired voltage is applied to the load in a naturally commutating converter. The following systems are mainly employed to determine the delay or firing angle.

a) Variable phase shift circuits.
b) Magnetic amplifiers.
c) Circuits that utilize the intersection of a repetitive waveform.
d) Digital timing circuits.

a) Variable Phase Shift circuits produce desired voltage and current based on the phase relationship between the supply voltage and triggering pulses. These circuits do not readily provide the full 180^0 range.

b) Magnetic amplifiers can be made to produce output voltage waveforms closely resembling those of a half-controlled bridge rectifier. The ideal linear relationship between the average output voltage and average control voltage suggests the exact variation of the triggering angle that is required to obtain similar characteristics in a controlled rectifier. This makes magnetic amplifiers attractive for thyristor-triggering circuits. Because of its poor rise time, and variation of its initial amplitude with firing angle, the output from a conventional magnetic amplifier is not suitable for firing a thyristor. This problem can be overcome by using a pulse-forming network. Still, the disadvantage remains because of the existence of serious non-linearity in the region of minimum triggering angle.

c) In the case of waveform intersections Systems, the triggering point is determined by the coincidence of the instantaneous level of a ramp or timing waveform and a triggering level. This can be done in two ways. They are by varying the slope of the timing waveform and altering the triggering level. Figure.3.4 shows such waveform intersection methods.

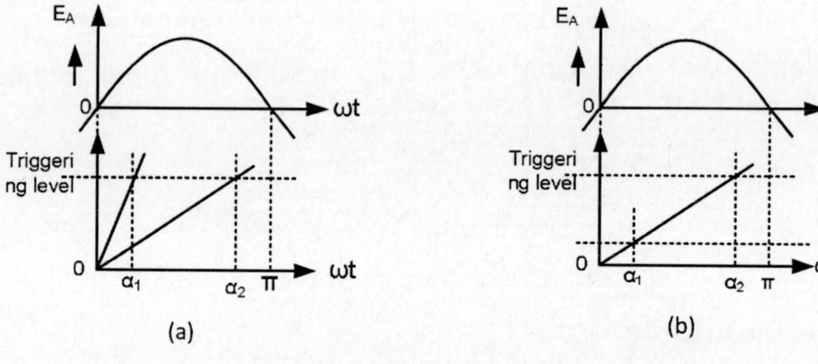

Figure.3.4 (a) ramp of variable slop, (b) variable triggering level

d) Like the waveform intersection method, digital systems are usually dependent on measuring off the required time interval from the supply voltage zero and are considered simply as an alternative to the analog approach to this particular requirement they do not appear to have any great advantage. They do, however, have a significant advantage as sections of wholly or mainly digital control systems, and lend themselves particularly well to microelectronic techniques.

(f) Additional features

Apart from the basic function like the determination of the triggering angle of the triggering circuit, some additional features need to be considered. They are limitations of firing angle, pulse distribution or gating, and pulse duplication. The limitation value of the firing angle should be such that it does not lead to loss of commutation. In the case of gating, it is essential to generate direct pulses to each thyristor of a converter in a proper region of the supply cycle and prevent in the region where the pulses may lead to malfunction or other desirable effects. In three-phase fully controlled converters, there are two thyristors in series and each need two triggering pulses in each cycle. This can be done by supplying the pulses from two appropriate phases. This is called pulse duplication.

3.4 PULSE TRANSFORMER

A pulse transformer (PTR) is like an ordinary transformer designed for the transmission of pulses between two windings. These transformers can handle pulses. It has one primary and one or more secondary windings. A PTR having multiple secondary winding is required for most of the power converters to provide the gating signal to individual switches. It is mainly used to isolate the control signal from the power circuit. In power electronics, PTRs are mainly used to trigger the thyristors.

Figure.3.5 Pulse transformers

Some pictures of pulse transformers are shown in Figure.3.5. A pulse transformer used in the SCR circuit is shown in Figure.3.6. In this figure the complete circuit is shown. Here diode is used to prevent reverse gate current flow. The resistance, R connected in series reduces the holding current. Here a 1:1 pulse transformer is shown helps to provide isolation between the gate-triggering circuit and the power circuit.

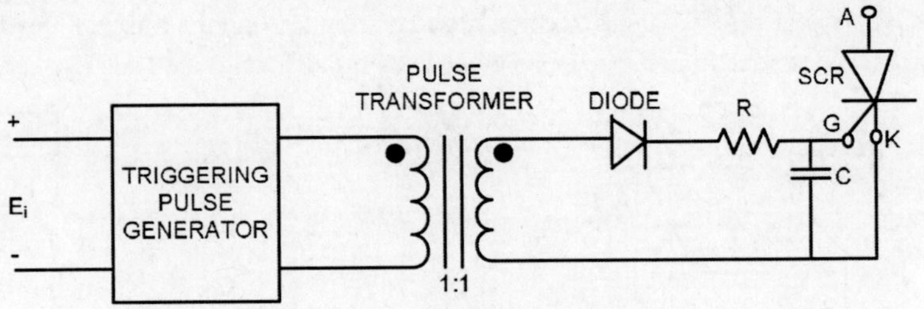

Figure.3.6 A pulse transformer in SCR triggering circuit

There are many other configurations of pulse transformers. It may be a 1:1:1 three-winding transformer (Figure.3.7(a) that provides pulses of anti-parallel SCR separately and also provides complete isolation. If isolation of the triggering circuit from the power circuit is not so important, a 1:1 pulse transformer (shown in Figure.3.7(b) in series mode or parallel mode (Figure.3.7(c) is used for anti-parallel SCRs.

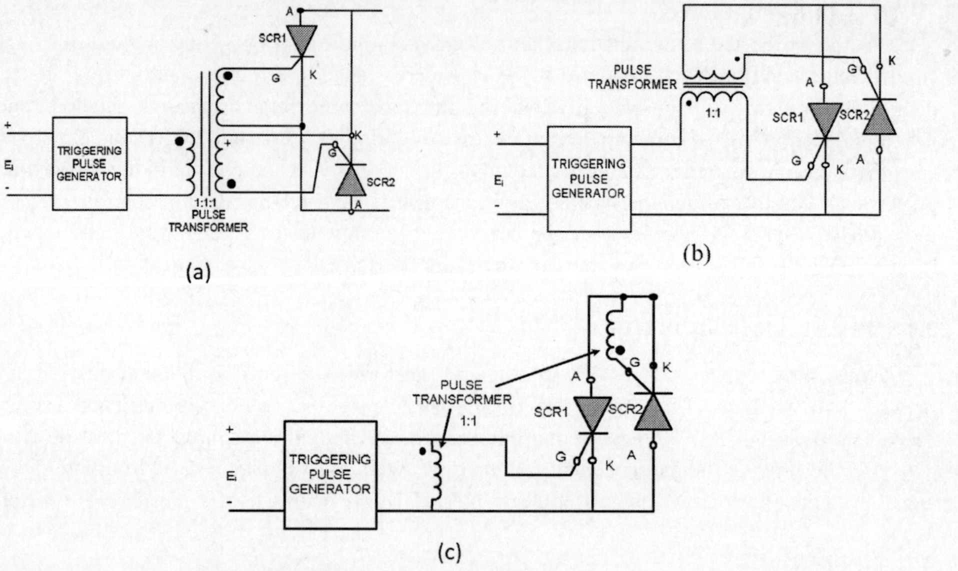

Figure.3.7 (a) A 1:1:1 Pulse transformer for two anti-parallel SCRs and complete isolation, (b) A 1:1 Pulse transformer in series mode for two anti-parallel SCRs, and (c) A 1:1 Pulse transformer in parallel mode for two anti-parallel SCRs.

3.5 OPTICAL ISOLATORS

An optical isolator or optoisolator is a device that is used to isolate the low voltage low current from the load circuit. It uses light energy. It comprises of a light source, a light-sensitive device, and a switching device. The light source may be an infrared emitter diode (IRED). The photosensitive device may be a phototransistor, LASCR, TRIAC, etc. Most of cases light sources and switching devices are the same.

In most situations, the low-power logic circuit normally an integrated circuit (IC) is used to trigger the thyristors which handle high voltage and high current load as shown in Figure.3.8(a). If there is a fault in the load circuit, it may damage the SCR as well as destroy the low-power control circuit. In this case, both low-power and high-power circuits need to be isolated. The relays (Figure.3.8(b) can be used, but the relays have several disadvantages such as high cost, bulky, shorter life than devices, producing inductive kicks, and production of sparks. To avoid such disadvantages optical isolators are used.

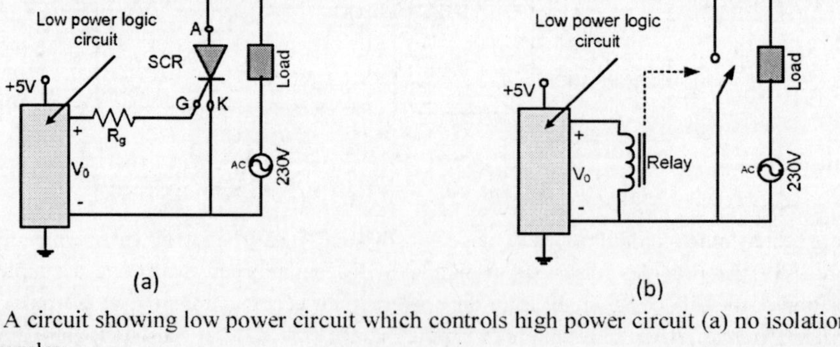

Figure.3.8 A circuit showing low power circuit which controls high power circuit (a) no isolation, (b) isolation by relay

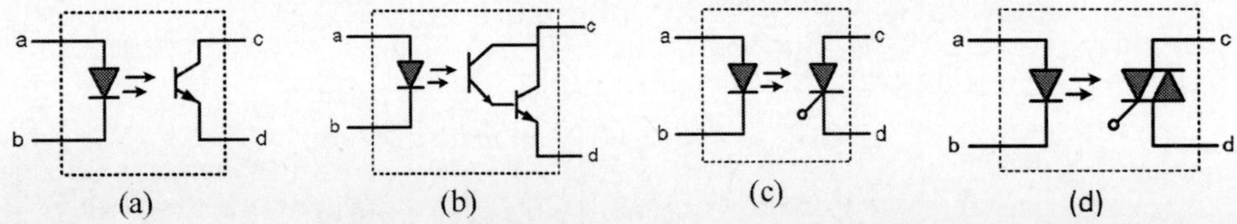

Figure.3.9 Various optical isolators (a) IRED and Phototransistor, (b) IRED and photodarlington, (c) IRED and LASAR, (d) IRED and TRIAC

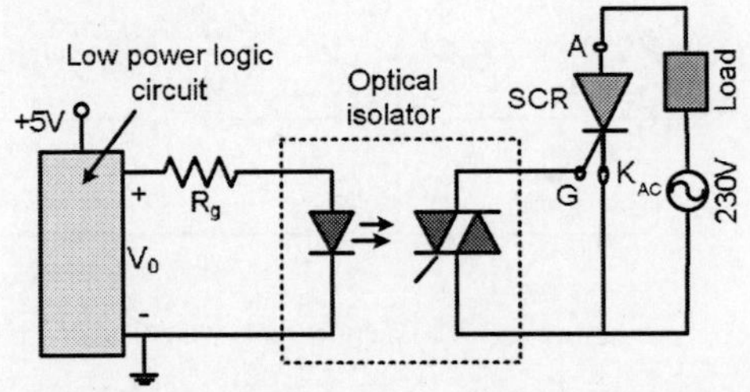

Figure.3.10 Circuit showing isolation provided by optical isolator

Figure.3.9 shows various types of optical isolators. In this figure, the terminal a-b is the input terminal and c-d is the output terminal. The optical isolator is shown inside the dotted box. A complete circuit in which an optical isolator is used to isolate high-power and low-power logic circuits is shown in Figure.3.10. In this circuit when the output of the logic circuit $V_0 = 0$, the IRED is inactive and the TRIAC is in off-state and hence load do not receive A.C. On the other hand, if $V_0 = 5V$, it activates the IRED and the radiant energy turns the TRIAC on, and the load receives the A.C.

3.6 GATE TRIGGERING CIRCUITS

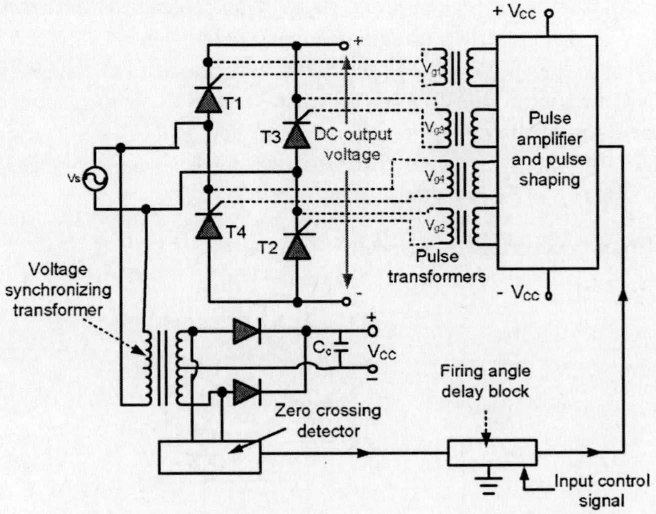

Figure.3.11 (a) Block diagram of a common firing circuit

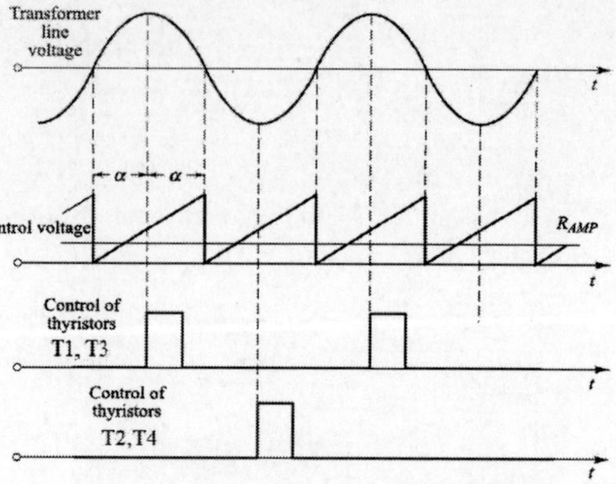

Figure.3.11(b) Triggering waveform

The gate-triggering circuit is a vital part of any power converter comprising of power semiconductor devices. The output of a power converter depends on how the gate drive circuit drives the switching devices. The characteristics of the gating circuit are the key components for the proper output of a converter. The design of triggering circuits requires the knowledge of gate characteristics as well as the devices for which it is designed. The common block diagram of a gate-trigger circuit is shown in Figure.3.11(a).

The converter is an SCR-based single-phase converter. The SCRs are at line potential. The triggering circuit is referred w.r.t a logic gate associated with the control input. With the help of a transformer called a voltage synchronizing transformer (VST), the gate drive circuit is isolated from the line voltage. The gate drive circuit requires a D.C supply. This is obtained from a rectifier circuit feeding from the secondary of the VST. The firing angle delay block generates a ramp signal (voltage) of constant peak-to-peak voltage is generated from the AC synchronization voltage. The ramp voltage is compared with that of a control signal. When the ramp voltage equals the control voltage in alternate half cycles, a pulse signal of controllable duration is generated. In this way, the triggering angle can be varied over 0-180°, and the delay angle is proportional to the control voltage. The waveform for triggering the thyristors is shown in Figure.3.11(b).

3.6.1. Resistance gate triggering circuit

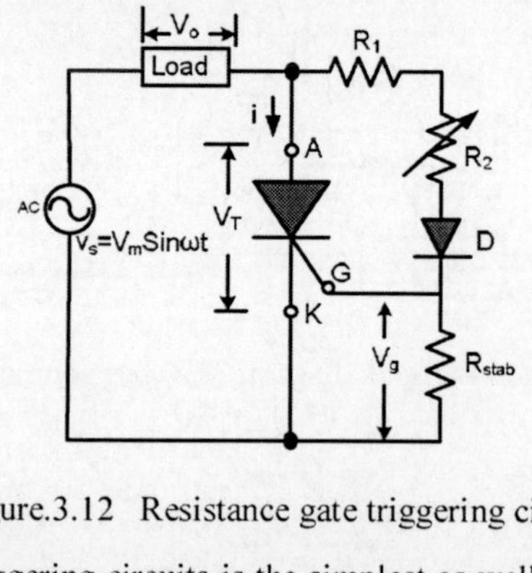

Figure.3.12 Resistance gate triggering circuit

The resistance gate triggering circuits is the simplest as well as most economical. Nowadays these circuits are not commercially used. However, *R* gate triggering can be limited in the firing angle range 0^0-90^0. The circuit diagram is shown in the Figure.3.12. In this circuit the gate current is provided from an A.C source via resistance, R_1, R_2, diode (D). R_2 is a variable resistance. A stabilizing resistance is connected (R_{stab}) is connected across the gate and the cathode. The flow of gate current is from source v_s, via Load, R_1, R_2, diode D to gate. The function of diode D is to prevent the gate to cathode from junction reverse biased. It allows the current flow during the positive half cycle only.

The operation of the resistance triggering circuit is as follows. When the voltage, v_s goes on positive the SCR is forward biased and it will conduct only when the gate current exceeds the minimum value of gate current (I_{gmin}). When gate current is less than I_{gmin} the output voltage across the load, $V_0 = 0$. Also with positive v_s, D and gate to cathode junction become forward biased. Gate current increases with increase in v_s towards maximum voltage V_m. When $I_g = I_{gmin}$, the SCR will turn on. The output voltage is now approximately equal to v_s. The SCR is in on state until v_s decreases from V_m to 0. At this instant, the load current is below the holding current. Now, the v_s decreases from zero to negative and the SCR becomes reverse-biased and remains in the off state. The sequence of the operation is repeated in the next cycle and so on.

The gate current should not exceed the maximum allowable value (I_{gmax}). The value of R_1 which limit the maximum value of gate current can be obtained from the peak value of supply voltage V_m and I_{gmax} as follows.

$$\frac{V_m}{R_1} \le I_{g\max} \text{ or } R_1 \ge \frac{V_m}{I_{g\max}}$$

Thus, R_1 limits the gate current to a safe value while R_2 is varied. The stabilizing resistance R_{stab} is such that it does not allow the gate to cathode voltage to exceed its maximum allowable voltage (V_{gm}). This situation will occur only when $R_2 = 0$. Thus,

$$\frac{V_m}{R_1 + R_{stab}}.R_{stab} \le V_{gm} \text{ or } R_{stab} \le \frac{V_{gm}.R1}{V_m - V_{gm}}$$

The values of R_1, R_2 are large and hence the firing circuits draw a small amount of current. The amplitude of DC pulse voltage is controlled by varying R_2.

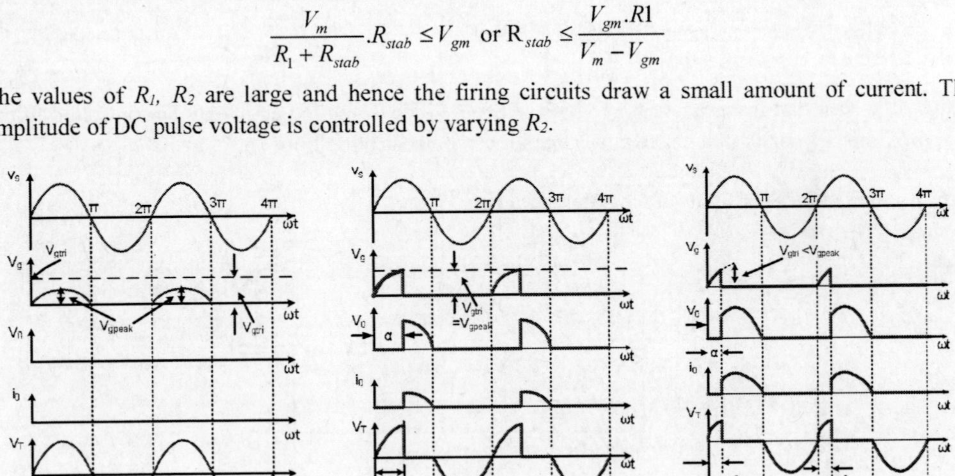

Figure.3.13 The waveform of resistance triggering (a) without triggering, (b) SCR triggering at firing angle = 90^0, (c) SCR triggering at firing angle < 90^0

Thus, the amplitude of the gate voltage dependent on the value of R_2. If R_2 is large i is small and hence $V_g = iR_{stab}$ is also small. Let, the gate triggering voltage V_{gtri} is given by

$$V_{gtri} = V_{gpeak} Sin\alpha \tag{3.3}$$

When $V_{gpeak} < V_{gtri}$, the SCR will not conduct, both the output voltage (V_O) and output current (i_O) are zero. The voltage across the SCR is same as the supply voltage v_s. The trigger circuit comprises of resistance only and hence v_g is in same phase with the v_s. The waveform of the V_g, i_g, V_O, i_O, and V_T are shown in Figure.3.13(a).

Now, the value of R_2 so adjusted that $V_{gpeak} = V_{gtri}$, and there will give the value of the firing angle (α) as 90^0. In this case, the waveform of the V_g, i_g, V_O, i_O, and V_T are shown in Figure.3.13(b).

Again, by adjusting R_2 in such a way that $V_{gpeak} > V_{gtri}$. In this case, $\alpha < 90^0$, and the values of V_g, i_g, V_O, i_O, and V_T are shown in Figure.3.13(c). When the $V_g = V_{gtri}$, the SCR will turn on for the first time. Increasing the value of V_g beyond V_{gtri} turns on the SCR at a firing angle less than 90^0. When $V_g = V_{gtri}$ for the first time, SCR fires, the gate losses its control, and V_g reduces to zero. In resistance firing the firing angle cannot go beyond 90^0.

The firing angle cannot be zero. A large value of V_{gpaek} may bring the SCR firing angle nearer 2^0 to 4^0 firing angles. The relationship between V_{gpeak} and V_{gtri} can be written as follows.

$$\text{From (3.3)},\quad \alpha = \sin^{-1}\left(\frac{V_{gtri}}{V_{gpeak}}\right) \tag{3.4}$$

$$V_{gpeak} = \frac{V_m R_{gstab}}{R_1 + R_2 + R_{gstab}} \tag{3.5}$$

$$\alpha = \sin^{-1}\left[\frac{V_{gtri}(R_1 + R_2 + R_{stab}}{V_m R_{stab}}\right] \tag{3.6}$$

Since, R_1, R_2, R_{stab}, and V_m are fixed, hence

$$\alpha \propto \sin^{-1}(R_2) \text{ or } \alpha \propto R_2 \tag{3.7}$$

Thus, it is seen that firing angle is proportional to R_2. The triggering angle can be controlled from 0^0 (Approximately) to 90^0 and hence power output can be controlled from 100% to 50%.

3.6.2. Resistance – Capacitance gate triggering circuit

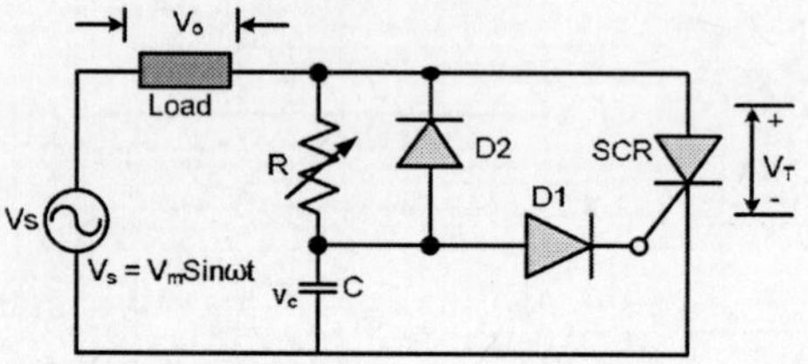

Figure.3.14(a) R-C half wave triggering circuits

In case of resistance triggering, the firing angle can be controlled is limited within 0^0 to 90^0. This limitation is overcome by using resistance (R) - capacitance (C) triggering circuit simply R-C triggering circuit. There are several such R-C circuits, two of them named as R-C half wave triggering and R-C full wave triggering circuits are shown in Figure.3.14(a) and Figure.3.15(b).

3.6.2.1. R-C half wave firing circuit

Let us consider the Figure.3.14(a) meant for R-C half wave firing circuit. In the negative half cycle of the AC waveform, the capacitor C charges through the diode $D2$ to peak of the supply voltage $-V_m$ (at the instant when $\omega t = -90^0$). The lower plate of C is positive and upper plate is negative. After reaching $-V_m$, the source voltage decreases to zero (at $\omega t = 0^0$). The capacitor voltage during -90^0 to 0^0 decreases to a value say - *oa* as shown in the Figure.3.14(b). After that the anode voltage of the SCR passing 0 and then it becomes positive. The capacitor charges from its initial voltage i.e - *oa* through *R*. When the capacitor voltage is equal to V_{gtri}, the SCR will be triggered. The diode *D1* is used to prevent from breakdown of gate to cathode junction during the negative half cycle. By varying the value of *R*, the firing angle can be varied from 0 to 180^0. The empirical formula for the value of RC is given by

$$RC \geq \frac{1.3T}{2} \cong \frac{4}{\omega} \tag{3.8}$$

Where, T= Time period =$1/f$, f is the supply frequency, $\omega=2\pi f$

When the $v_c = V_{gtri} + v_d$, the SCR will fire. Here v_d is the voltage across the diode *D1*. If at the time of triggering, capacitor voltage v_c is constant, the gate triggering current (I_{gt}) is governed by voltage source (v_s) through the circuit comprising of *R* and *D1*. The maximum value of *R* is given by

$$R \leq \frac{v_s - V_{gtri} - v_d}{I_{gtri}} \tag{3.9}$$

Thus, the value of *R* and *C* can be calculated from equation (3.8) and (3.9).

There is voltage drop across the SCR while it is triggered and normally it is falls from 1 to 1.5 V. This falls in voltage also lowers the voltage across the *R* and *C* from 1 to 1.5 V. The capacitor *C* is discharged during conduction period i.e in the positive half cycle till negative half cycle of the supply voltage appear across the SCR. With higher value of *R*, the time taken for charging *C* from $-oa$ to $\left(V_{gtri} + v_d\right) \approx V_{gtri}$ is more and hence, bigger the firing angle and lower is the output voltage. With lower value of *R*, the time taken for charging *C* from $-oa$ to $\left(V_{gtri} + v_d\right) \approx V_{gtri}$ is less and hence, the smaller the firing angle and the more is the output voltage. The corresponding waveforms for high *R* and low *R* are shown in Figure.3.14(b)

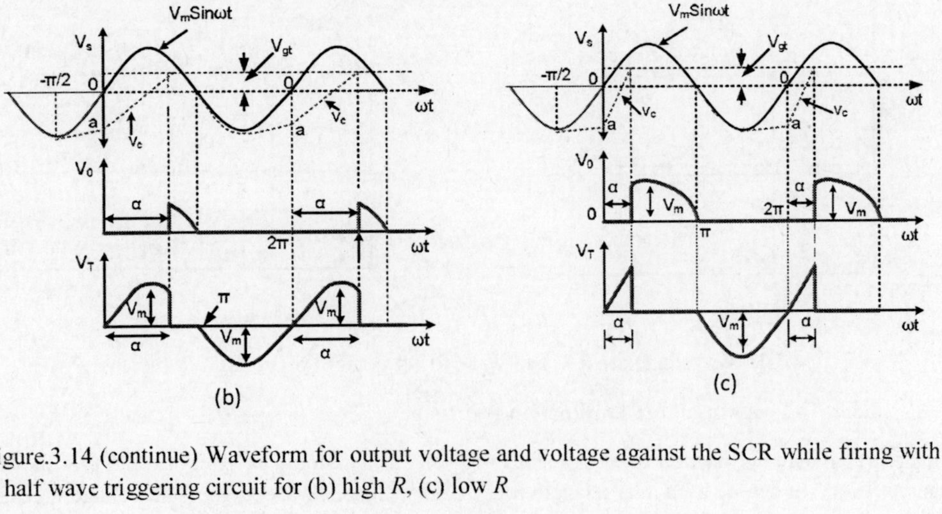

Figure.3.14 (continue) Waveform for output voltage and voltage against the SCR while firing with R-C half wave triggering circuit for (b) high *R*, (c) low *R*

3.6.2.2. R-C full wave firing circuit

Let us consider the Figure.3.15(a) meant for R-C full wave firing circuit. Here, there is a full wave bridge having diodes *D1, D2, D3, D4*. The initial voltage required for charging the *C* is almost zero. By clamping action, the capacitor *C* is fixed at this low voltage with upper plate positive. When the charging voltage reached to V_{gtri}, the SCR fires and a DC voltage v_d appear across the load as V_O. The empirical formulae using R and C is given by

$$RC \geq 50\frac{T}{2} \cong \frac{157}{\omega} \tag{3.10}$$

From equation (3.9), the formulae for R is given by

$$R \leq \frac{v_s - V_{gtri}}{I_{gtri}} \tag{3.11}$$

In (3.11), v_s is the source voltage at which the SCR is turn on. The waveform for triggering using this R-C full wave triggering circuit is given in Figure.3.15(b) and(c).

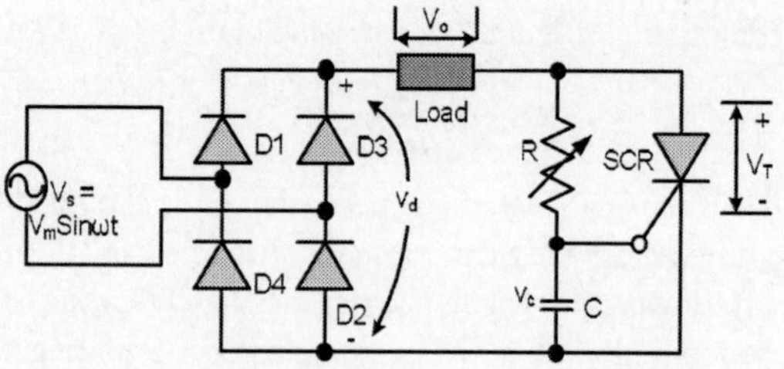

Figure.3.15(a) R-C full wave firing circuit

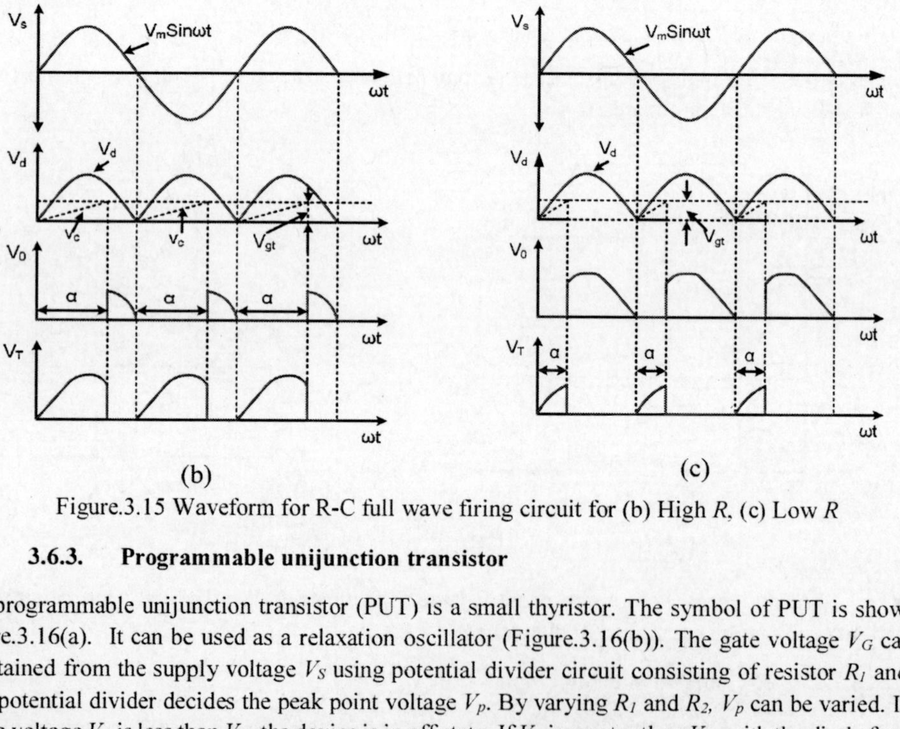

Figure.3.15 Waveform for R-C full wave firing circuit for (b) High R, (c) Low R

3.6.3. Programmable unijunction transistor

The programmable unijunction transistor (PUT) is a small thyristor. The symbol of PUT is shown in Figure.3.16(a). It can be used as a relaxation oscillator (Figure.3.16(b)). The gate voltage V_G can be maintained from the supply voltage V_S using potential divider circuit consisting of resistor R_1 and R_2. This potential divider decides the peak point voltage V_p. By varying R_1 and R_2, V_p can be varied. If the anode voltage V_A is less than V_G, the device is in off state. If V_A is greater than V_G, with the diode forward voltage V_D, the peak point is reached and PUT is turn on. The peak current Ip, and valley point current

I_p are depend on the equivalent impedance (R_G) of parallel resistor R_1 and R_2, and supply voltage V_S. The value of R_K as shown in Figure is limited to a value below 100Ω.
The V_P is given by

$$V_p = \frac{R_2}{R_1 + R_2} . V_S \tag{3.12}$$

This intrinsic ratio is given by

$$\eta = \frac{V_p}{V_S} = \frac{R_2}{R_1 + R_2} \tag{3.13}$$

The period of oscillation, T is approximately given by

$$T = \frac{1}{f} \approx RC \ln \frac{V_S}{V_S - V_p} = RC \ln\left(1 + \frac{R_2}{R_1}\right) \tag{3.14}$$

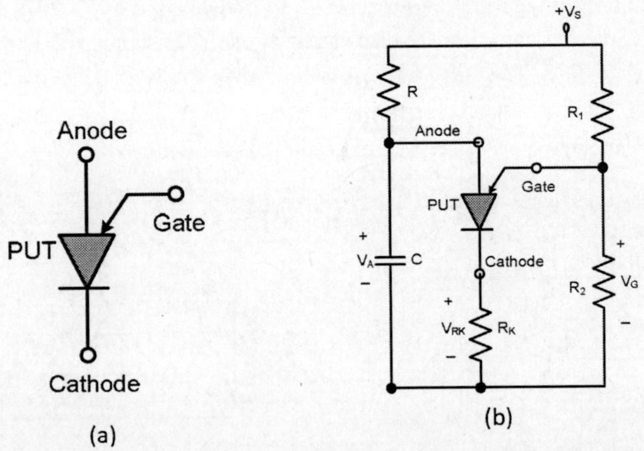

Figure.3.16 (a) Symbol of PUT, (b) Gate triggering circuit using PUT

The gate current at valley point is given by

$$I_G = (1 - \eta)\frac{V_S}{R_G} \tag{3.15}$$

In which
$$R_G = \frac{R_1 R_2}{R_1 + R_2} \tag{3.16}$$

The value of R_1 and R_2 can be found from the following equations.

$$R_1 = \frac{R_G}{\eta} \text{ and } R_2 = \frac{R_G}{1 - \eta} \tag{3.17}$$

3.6.4. UJT oscillator triggering

The basics of Unijunction transistor (UJT) have been presented in Section 2.7 of Unit-II. The UJT relaxation oscillator triggering will be presented here. The UJT is a very highly efficient switch whose switching time is in nanosecond range. UJT have negative resistance characteristics due to which it can be use as relaxation oscillator. This relaxation oscillator can be used for triggering SCR. The Figure.3.17 (a) shows the circuit diagram of UJT which works as an oscillator.

On application of the voltage V_{BB}, the capacitor, C charges to V_{BB} through the resistance R. The emitter circuit of the UJT is open circuit during capacitor charging. The voltage across the capacitor which is also equal to emitter voltage is given by

$$v_c = v_e = V_{BB}\left(1-e^{-t/RC}\right) = V_{BB}\left(1-e^{-t/\tau_1}\right) \tag{3.18}$$

Here, $\tau_1 = RC$ is the time constant of the charge circuit
The peak point voltage, V_p of the UJT oscillator circuit is given by

$$V_p = \eta V_{BB} + V_D \tag{3.19}$$

Where $\eta = \dfrac{R_{B1}}{R_{B1}+R_{B2}}$, V_D = threshold voltage, R_{B1} and R_{B2} are the resistance between emitter to base1 terminal, and resistor between emitter to base2 terminal.

When the UJT'S emitter voltage v_e (= v_c) reaches the V_p (expressed by (3.20)), *E-B1* junction breaks down and UJT turns on and capacitor, C discharges through resistance R_l. The time constant in the discharging case is $\tau_2 = R_1C$. The value of τ_2 is smaller than τ_1. With decaying of the emitter voltage to the value at valley point V_v, the UJT will turn off. The time required (T) to charge the capacitor C through R from the voltage V_v to V_p is calculated from the following

$$V_p = \eta V_{BB} + V_D = V_v + V_{BB}\left(1-e^{-\frac{T}{RC}}\right) \tag{3.20}$$

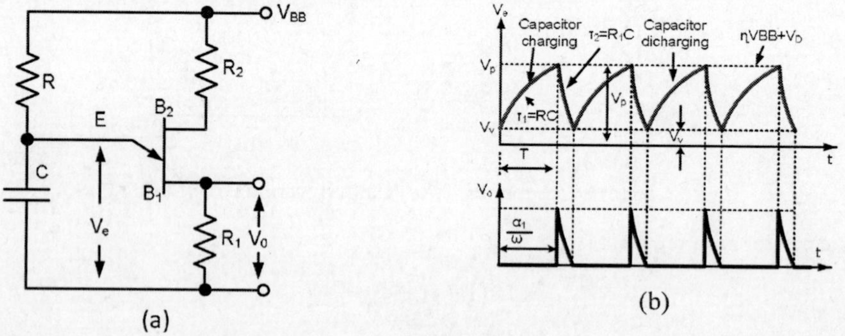

Figure.3.17 (a) Circuit diagram for UJT oscillator, (b) Voltage waveform for the oscillator

If it is assumed that the threshold voltage is equal to valley point voltage, i.e $V_D = V_v$. The time T is given by

$$T = \frac{1}{f} = RC \ln\left(\frac{1}{1-\eta}\right) \tag{3.21}$$

Neglecting the small discharge time, it can be assumed that the duration of the pulse is equal to T. In this case the firing angle (α) is given by

$$\alpha_1 = \omega T = \omega RC \ln\frac{1}{1-\eta}, \tag{3.22}$$

where ω is the angular frequency of the oscillator. The amplitude of the pulses is determined from the V-I characteristics of UJT by drawing a load line. When the output pulses are used to trigger SCR, the value of R_l should be small so that when UJT is off and leakage current flows through the R_l, the drop across it is normal which is not sufficient to trigger the SCR. Thus, we can write

$$\frac{V_{BB}.R_1}{(R_{B1}+R_{B2})+R_1+R_2}<V_{gtri} \tag{3.23}$$

The value of $(R_{B1}+R_{B2})=R_{BB}(\text{say})$ increases and threshold voltage, V_D decreases with increase in temperature. Thus, it effects the emitter diode forward characteristics. To provide compensation against the thermal effect, the resistance R_2 is used. The value of R_2 is attained from the following relation

$$R_2=\frac{10^4}{\eta.V_{BB}} \tag{3.24}$$

In some cases, the wide of the firing pulse is taken equal to R_1C.
The maximum value of R is obtained from peak point voltage (V_p) and current (I_p). The maximum value of R is given by

$$R_{max}=\frac{V_{BB}-V_P}{I_P}=\frac{V_{BB}-(\eta V_{BB}+V_D)}{I_P} \tag{3.25}$$

The valley point voltage (V_v) and current (I_v) is used to determine the minimum value of R and is given by

$$R_{min}=\frac{V_{BB}-V_v}{I_v} \tag{3.26}$$

3.6.4.1. Synchronized UJT circuit or Ramp triggering

This triggering circuit comprises of four diodes D_1, D_2, D_3, D_4, three resistors (R, R_1, R_2), a Zener diode Z, a capacitor, C, an UJT, and a Pulse transformer. The circuit is shown in Figure.3.18(a). The functions of diodes are to rectify the ac to DC (V_{dc}) and function R_1 is to lower the V_{dc} to a constant value (V_z) suitable for Z. The voltage V_z is applied to the R and C circuit. The charging current (i_1) of the C is passing through R; and R decides the magnitude of i_1. The rate of rise of capacitor voltage can be control by varying R. The voltage across C is denoted by v_c. When $v_c = \eta V_z$, the E-$B1$ junction breakdown and C discharges through the primary winding of pulse transformer and pass current i_2. Since, i_2 is in the form of pulse, the secondary output is also pulse voltage. The output secondary pulse is used to trigger the SCRs. In this method the firing angle can be control up to about 150^{0}C. This method of control of output power by varying R is called ramp control or open loop control or may be called as manual control. The necessary waveform is shown in Figure.3.18 (b). The V_z across Z goes to zero at the end of a half cycle, and hence synchronization of the trigger circuit with the supply voltage across the SCR is achieved. The time at which the firing pulse applied for the first time ($t = \alpha/\omega$) is remain constant for same value of R. However, slight variation of frequency and voltage of the supply will not affect the circuit operation. In case if R is reduced to such a value that $v_c = V_D$ (threshold voltage) twice in each half cycle. It is shown in Figure.3.18(c), then there will be two pulses in each half cycle. Since first pulse can trigger the SCR, the second pulse is useless.

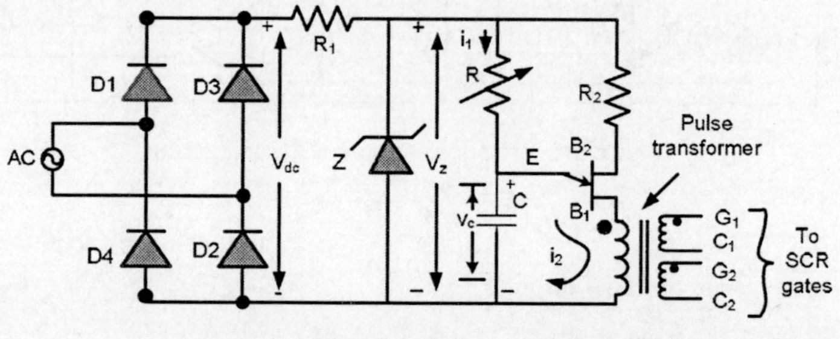

Figure.3.18 (a) Ramp triggering circuit

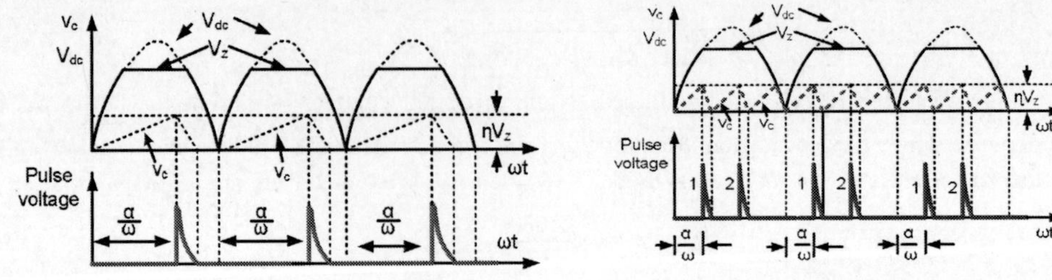

(b) (c)

Figure.3.18 (continue) Output pulse of circuit of Figure.3.18(a), (b) for high R, (c) for low R

3.6.4.2 Ramp and Pedestal triggering

An improved version of synchronized UJT triggering called ramp-and pedestal triggering is also there. It is shown in Figure.3.19(a). In this circuit two SCRs connected in antiparallel and triggered from the UJT circuit to control an AC load. The same circuit can also be used for single phase semi converter or full converter. The voltage across the Zener diode V_z is held constant at threshold voltage. The potential divider circuit formed by the variable resistance R_2. The pedestal voltage V_{pd} is controlled by varying R_2. The diode $D1$ in the circuit permits the capacitor to charge quickly to a value equal to V_{pd} via the upper portion of the wiper point of R_2. The set point of R_2 is done in such a way that V_{pd} is less than ηV_z i.e the UJT firing point voltage. The R_2 is varied in such a way that V_{pd} is small and V_z charges through the resistor R. This situation is shown in Figure.3.19(b). When the voltage v_c across the capacitor attains ηV_z, the UJT is trigger and gate voltage V_g need to fire the SCR1 and SCR2 is obtained across the pulse transformer secondary. The SCR1 is forward biased and turned on. After this instant, the capacitor voltage v_c reduces to V_{pd} and then reach 0 at $\omega t = \pi$. From 0 to π, SCR1 is forward biased and it is turned on. From π to 2π, the forward biased SCR2 is turned on. In this way, the AC load is controlled. Two sets of waveforms are shown in the Figure.3.19(b) and Figure.3.19(c). Figure.3.19(b) is corresponds to low pedestal voltage across the capacitor C, in which charging of C to ηV_z takes longer time and firing angle delay is more. Thus, the output voltage is low. Figure.3.19(c) corresponds to high pedestal voltage across the capacitor C, in which charging of C to ηV_z takes smaller time and firing angle delay is small. Thus, the output voltage is high.

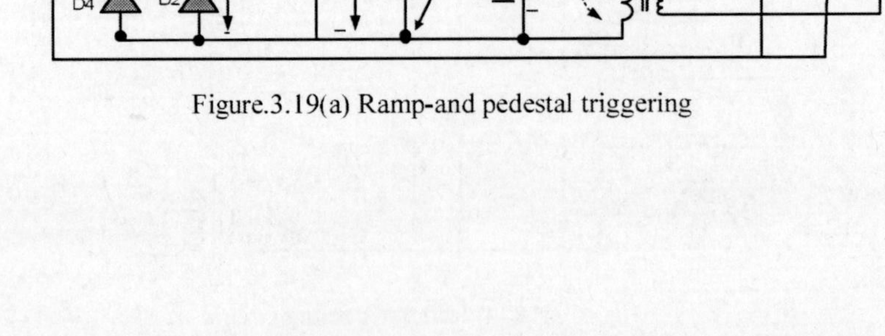

Figure.3.19(a) Ramp-and pedestal triggering

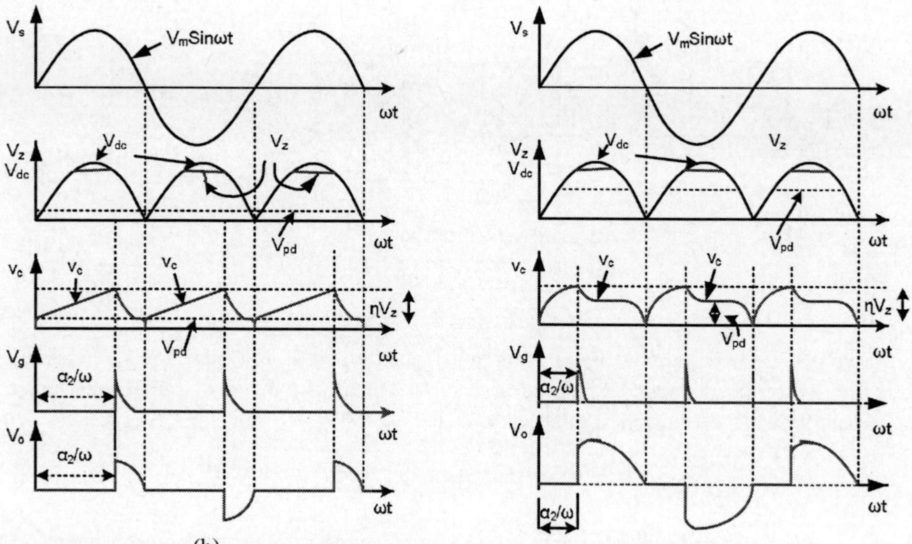

(b) (b)

Figure.3.19 Waveform for various voltages for circuit shown in Figure.3.19(a) (b) Low pedestal voltage, (c) high pedestal voltage

The time to charge the capacitor C from V_{pd} to ηV_z is given by

$$\eta V_z = V_{pd} + (V_z - V_{pd})(1 - e^{-T/RC}) \tag{3.27}$$

Here, $\left(V_z - V_{pd}\right)$ is the voltage that charges the capacitor C from pedestal voltage to ηV_z.

From (3.21), the time T to charge the capacitor C from V_{pd} to ηV_z is found as

$$T = RC \ln \frac{(V_z - V_{pd})}{V_z(1-\eta)} \tag{3.28}$$

The firing angle α_2 is given by

$$\alpha_2 = \omega RC \ln \frac{\left(V_z - V_{pd}\right)}{V_z\left(1-\eta\right)} \tag{3.29}$$

3.6.5. Pulse Transformer and opto-coupler based triggering

3.6.5.1. Pulse transformer triggering

The introduction of pulse transformer has been provided in Section 3.4. In this section the theory related to pulse transformer firing circuit is provided. A general layout of pulse transformer triggering circuit is shown in Figure.3.20 (a). The circuits have resistor R_L to limit the primary current of the pulse transformer, a transistor is used as switch and it turns on when pulse of high level is applied to its base, thus connect the primary of the pulse transformer to the bias voltage V_B. The diode D allows the flow of current after the pulse i.e when the transistor is off. The equivalent circuit of the Figure.3.20(a) is drawn as Figure.3.20(b).

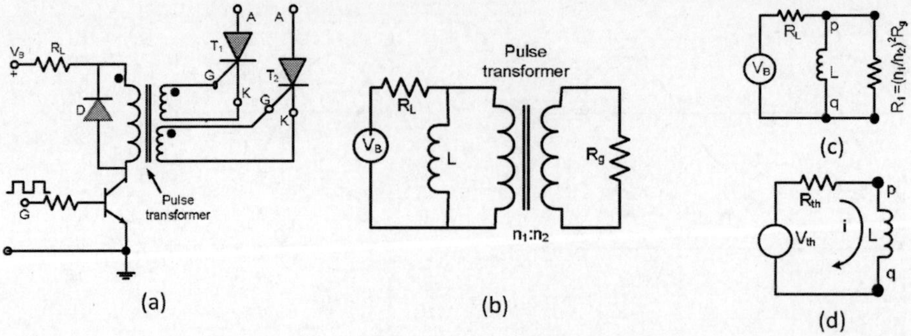

Figure.3.20 (a) Triggering circuit using pulse transformer, (b) to (d) are equivalent circuit

In this figure L represents magnetizing inductance, R_g represents the gate-to-cathode resistance of the SCR circuit. When R_g is transferred to primary, it is equal to R_1. The value of R_1 is given by

$$R_1 = \left(\frac{n_1}{n_2}\right)^2 R_g \tag{3.30}$$

Where N_1 and N_2 are the number of turns in primary and secondary of the pulse transformer respectively. Application of Thevenin theorem will give the circuit of Figure.3.20(c), in which V_{th} is the Thevenin voltage, R_{th} is the Thevenin resistance. The Thevenin equivalent circuit is shown in Figure.3.20(d). The expression for V_{th} and R_{th} are given by following equations.

$$V_{th} = \frac{R_1}{R_1 + R_L}.V_B \text{ and } R_{th} = \frac{R_1 R_L}{R_1 + R_L} \tag{3.31}$$

Application KVL across the point p-q will give the following equation.

$$V_{th} = R_{th}.i + L\frac{di}{dt} \tag{3.32}$$

Putting the values of V_{th} and R_{th} in the above equation, we will get

$$\frac{R_1}{R_1 + R_L}.V_B = \frac{R_1 R_L}{R_1 + R_L}.i + L\frac{di}{dt}$$

$$\text{Or } V_B = R_L.i + L.\left(\frac{R_1 + R_L}{R_1}\right)\frac{di}{dt} \tag{3.33}$$

Solving for I, we will get

$$i = \frac{V_B}{R_L}\left(1 - e^{-\frac{R_1 R_L}{L(R_1 + R_L)}.t}\right) \tag{3.34}$$

The voltage appearing across L is given by

$$e = L\frac{di}{dt} = \frac{V_B R_1}{R_1 + R_2} e^{-\left(\frac{R_{th}}{L}\right)t} \tag{3.35}$$

There may be two modes (Mode1 and Mode2) of operation of a pulse transformer based on the value of R_{th} and L. The modes are

Mode1: L is so large as compared to R_{th} so that $\frac{L}{R_{th}} > 10T$, in which T = pulse width of the signal applied to the gate G. In this case from the equation (3.35)

$$e = V_B \frac{R_1}{R_1 + R_L} . e^{-\left(\frac{t}{10T}\right)} \tag{3.36}$$

At $t = 0$, the value of voltage across L is given by

$$e_0 = V_B \frac{R_1}{R_1 + R_L} \tag{3.37}$$

At $t = T$, the voltage across L is given by

$$e_T \approx 0.904 V_B \frac{R_1}{R_1 + R_L} \approx 0.904 e_0 \tag{3.38}$$

It is observed that the fall in pulse level is very small at $t = T$ and thus when $\frac{L}{R_{th}} > 10T$, the input pulse is faithfully transmitted to the output of the pulse transformer (Figure.3.21 (a)).

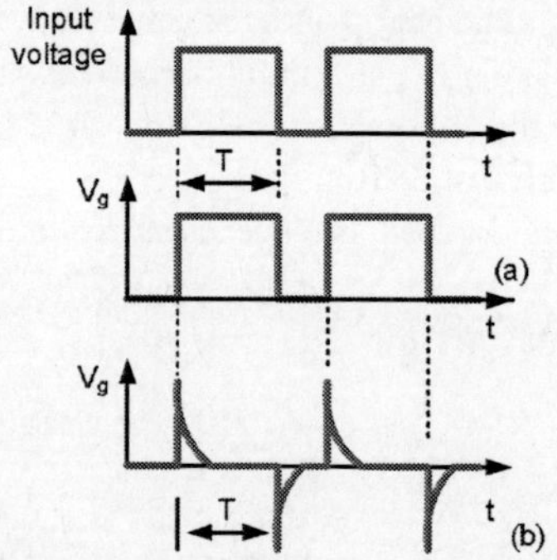

Figure.3.21 Output waveform of pulse transformer (a) $\frac{L}{R_{th}} > 10T$, (b) $\frac{L}{R_{th}} < \frac{T}{10}$

Mode2: R_{th} is very large as compared to L such that $\frac{L}{R_{th}} < \frac{T}{10}$. The operation of pulse transformer in this mode is achieved by using an air core pulse transformer. In this case, voltage across L becomes

$$e = V_B . \frac{R_1}{R_1 + R_L} e^{-(10/T)t} \tag{3.39}$$

$$\text{At } t = 0,\ e_0 = V_B . \frac{R_1}{R_1 + R_L} \tag{3.40}$$

At $t = T$,

$$e_T = V_B . \frac{R_1}{R_1 + R_L} e^{-10} = 4.53 \text{x} 10^{-5} e_0 \tag{3.41}$$

It is observed that the transmitted input pulse is decaying exponentially (Figure.3.21(b)). Thus, for a step rise in input voltage, the output is a positive pulse whereas for fall in input voltage, the output is a negative pulse. Thus, the value of the L decides the output of the pulse transformer. The negative pulse can be removed by using a clipper. The output voltage at the secondary of pulse transformer is given by

$$V_g = \frac{n_2}{n_1} . V_B . \frac{R_1}{R_1 + R_L} \tag{3.42}$$

To produce the triggering voltage V_{gtri}, V_g (i.e $\frac{n_2}{n_1}.V_B.\frac{R_1}{R_1+R_L}$) need to be greater or equal to V_{gtri}.

$$\text{Or } V_B \geq V_{gtri}\frac{n_1}{n_2}\left(1+\frac{R_L}{R_1}\right) \tag{3.43}$$

Since,

$$R_1 = \left(\frac{n_1}{n_2}\right)^2 R_g \tag{3.44}$$

$$V_B \geq V_{gtri}\frac{n_1}{n_2}\left[1+\left(\frac{n_2}{n_1}\right)^2\frac{R_L}{R_g}\right] \tag{3.45}$$

Normally, exponentially decaying type of triggering pulses are favoured because of the following facts.

(a) It is suitable for injecting large charge into the gate. Thus, turn on is reliable.
(b) There is no significant heating of gate circuit because of small duration of the pulse.
(c) With the same gate to cathode power, it the permissible to increase the V_B to a suitably high value to achieve a hard drive of SCR.
(d) The size of pulse transformer is reduced because it requires smaller value of L.

3.6.5.2. Opto-coupler based triggering

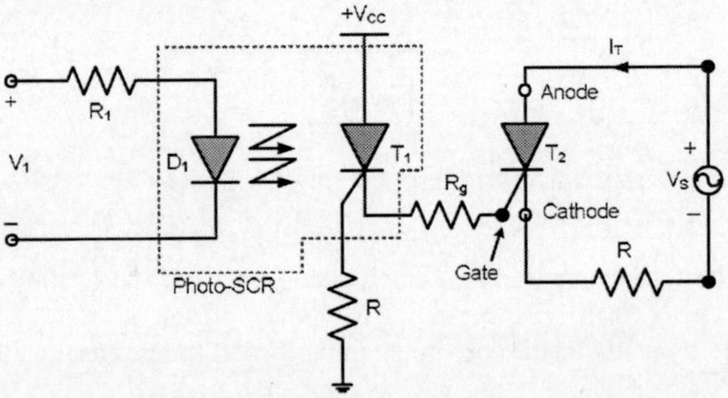

Figure.3.22 Opto-coupler based Triggering

The details about the optical isolator or optoisolator is depicted in section 3.5. In opto-coupler based triggering optoisolator is used. A circuit using optical isolator to trigger a thyristor is shown in Figure.3.22. When a short pulse is applied to infrared light emitting diode (ILED), D_1 turns on the photo SCR, T_1 and the power thyristor T_2 is triggered. The optocoupler provides isolation between power and the control circuit. However, this type of triggering circuit requires a separate power supply V_{CC}. The cost and weight of the triggering circuit is increases.

3.7 SCR TURN-OFF METHODS

An SCR can be turned on by the application of a gate signal to its gate terminal. It is required to turn the SCR off for power control purpose. Already we know that the SCR can be turned off by bringing the SCR from conduction state to forward blocking state. It is done by either bring the anode current below the holding current or by application of reverse voltage to such a value that the SCR will be brought to the reverse blocking state. The commutation of thyristor or SCR is the process by which the device is turned off. It is also known that it is impossible to make the SCR off while it is in conducting state. The commutation is the process by which transfer of current takes place from one path to another path or from one thyristor to another thyristor. To turned off the current, the thyristor current must be brought to zero. Basically, there are two methods of turning off of thyristors. They are natural commutation and forced commutation.

Natural commutation

It is a simple and widely used method. It uses the alternating or reversing nature of alternating voltage for transferring current from one thyristor to another thyristor. In every half cycle alternating current reaches zero or current zero. When alternating current passes through the natural zero a reverse voltage is simultaneously appeared across the device and the thyristor is immediately turned off. This process of commutation is called natural commutation. No external circuit is required in natural commutation. Natural commutation make use of A.C supply voltage or generated A.C voltage by the generators or by resonant circuits to turn off the SCR. Line commutated converters use this method of commutation. The line and natural commutation circuits are also called as Class F commutation circuits.

Forced commutation

As we know, to turned off thyristor it is required to bring the thyristor current to zero for sufficient time for removal of charged carriers. In case of using SCRs in DC circuits, to turn off a thyristor it is required to use an external circuit. When some external circuit is used to turned off the SCR, the process is called forced commutation. The external circuit used for commutation are called as commutation circuits. The components that are used in the external circuits are called commutation components. With the help of the commutation circuit, a reverse voltage is developed across the device and this voltage help in bring the SCR forward current to zero. In this way the SCR is turned off. In forced commutation normally inductance (*L*) and capacitance (*C*) or underdamped resistance - inductance - capacitance (*RLC*) components are used to force current or voltage to zero to turned off the thyristor. Under forced commutation, there are various techniques available. They are as follows.

(a) Class A Commutation circuit
(b) Class B Commutation circuit
(c) Class C Commutation circuit
(d) Class D Commutation circuit
(e) Class E Commutation circuit

3.7.1 Class A Series resonant commutation circuit

This type of commutation circuit is also called resonant commutation circuit or load commutation circuit. Here *LC* commutation components are used in series with the load. The circuit comprises of resistance R_L as load. The R_L may be used in series with *L* and *C* or R_L may be used in parallel with the *C*. The circuits are shown in Figure.3.23(a) and Figure.3.23(b).

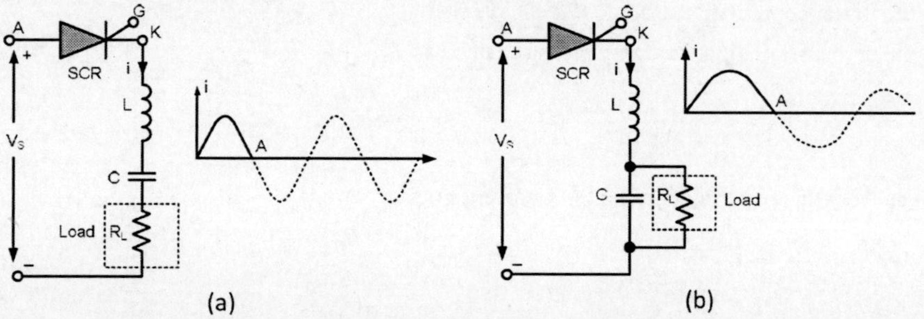

Figure.3.23 Class A Commutation circuit (a) Load and *C* are in series, (b) Load and *C* are in parallel

The voltage and currents under Class A triggering circuit is shown in Figure.3.23(c)

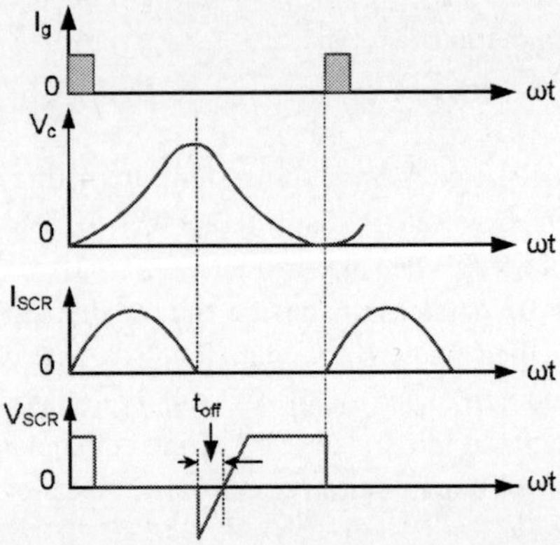

Figure.3.23(c) Voltage and current waveform for Class A triggering circuit

A requirement of both the circuits is that, both circuits should be essentially underdamped. When energized with DC, circuit current is waveform the corresponding to the respective is also shown in figures. The circuit current rises up to maximum and then decays to zero when it reaches the point *A* at which the SCR is turned off. This type of commutation is possible in DC circuit only.

3.7.1.1 Design consideration when load is in series with capacitance

In the circuit of Figure.3.23(a), the governing equation is

$$V_S = iR_L + L\frac{di}{dt} + \frac{1}{C}\int idt \qquad (3.46)$$

Differentiating above equation, the obtained equation is

$$\frac{1}{L}\frac{d}{dt}V_S = \frac{R_L}{L}\frac{di}{dt} + \frac{d^2i}{dt^2} + \frac{i}{LC} \qquad (3.47)$$

The second order homogeneous equation is

$$\frac{d^2i}{dt^2} + \frac{R_L}{L}\frac{di}{dt} + \frac{i}{LC} = 0 \qquad (3.48)$$

Solving this equation for underdamped situation is obtained as

$$i = e^{-\varepsilon t}\left[A_1 \cos\omega t + A_2 Sin\omega t\right] \qquad (3.49)$$

In this equation, $\varepsilon = \frac{R_L}{2L}, \omega_0 = \frac{1}{\sqrt{LC}}, \omega = \omega_0\sqrt{1-\varepsilon^2} = \sqrt{\frac{1}{LC} - \frac{R_L^2}{4L^2}}$

From the initial condition, $i(0^+) = i(0^-) = 0$, we will get $A_1=0$, $A_2=V_S/L$

The time domain equation for current, i is obtained as

$$i(t) = e^{-\left(\frac{R_L}{2L}\right)t}\left[\frac{V_S}{\omega L} Sin\omega t\right] \qquad (3.50)$$

It is seen from this equation that the SCR current goes to zero at $\omega t = \pi$ or $t = \frac{\pi}{\sqrt{\frac{1}{LC} - \frac{R_L^2}{4L^2}}}$

Also, $$\frac{di}{dt} = -e^{-\left(\frac{\pi R_L}{2\omega L}\right)}\left(\frac{V_S}{L}\right) \tag{3.51}$$

It is seen that the capacitor voltage at the end of conduction period is $V_c = V_S - V_L$ in which $V_L = L\frac{di}{dt}$

Thus, $$V_c = V_s\left[1 + e^{-\frac{\pi R_L}{2\omega L}}\right] \tag{3.52}$$

The current equation and voltage V_c is given by

$$i(t) = e^{-\left(\frac{R_L}{2L}\right)t}\left[\frac{V_S - V_0}{\omega L}\sin\omega t\right] \tag{3.53}$$

$$V_c = V_S + e^{-\left(\frac{\pi R_L}{2\omega}\right)t}\left(V_S - V_0\right) \tag{3.54}$$

In which V_o is the initial capacitor voltage.
The underdamped condition for ω>0 is obtained as follows

$$\left(\frac{1}{LC} - \frac{R_L{}^2}{4L^2}\right) > 0 \text{ or } R < \sqrt{\frac{4L}{C}} \tag{3.55}$$

3.7.1.2 Design considerations when load is in parallel with capacitance

The circuit shown in Figure.3.23(b) is considered here. Let us consider load voltage is V_0. The governing voltage and current equation for the circuit are

$$V_S = L\frac{di}{dt} + V_0 \tag{3.56}$$

$$i = C\frac{dV}{dt} + \frac{V_0}{R_L} \tag{3.57}$$

Taking Laplace transform of these equations

$$V_S(s) - \mathrm{V}_0(s) = sLi(s) \tag{3.58}$$

$$i(s) = \frac{V_o(s)}{R_L} + sCV_o(s) \tag{3.59}$$

From (1), $$V_S(s) = \mathrm{V}_0(s) - sLi(s) \tag{3.60}$$

Again, $$V_S(s) = \frac{V_S}{s} \tag{3.61}$$

Putting the values $V_S(s)$ in (3.59), we have, $$i(s) = \frac{V_S}{R_L.s} - \frac{sLi(s)}{R_L} + sC\left[\frac{V_S}{s} - sLi(s)\right] \tag{3.62}$$

Simplifying (3.62) and after taking inverse Laplace transformation, we will get

$$i(t) = \frac{V_S}{R_L}\left[1 + \frac{1}{\sqrt{1-\varepsilon^2}}.\frac{\omega_n}{\varepsilon}.e^{-(t/R_LC)}\sin(\omega t + \phi)\right] \tag{3.63}$$

In which ,

$$\varepsilon = \text{Damping ratio} = \frac{1}{2R_L}\sqrt{\frac{L}{C}}, \tag{3.64}$$

$$\omega_n = \text{ Undamped natural frequency} = \frac{1}{\sqrt{LC}} \tag{3.65}$$

$$\omega = \omega_n \sqrt{1-\varepsilon^2} \tag{3.66}$$

Thus,

$$\omega = \sqrt{\frac{1}{LC} - \frac{1}{4R_L{}^2C^2}} \tag{3.67}$$

The expression for angle Φ is given by $\phi = \tan^{-1}\dfrac{2R_L C\omega}{-\varepsilon} - \tan^{-1}\dfrac{\sqrt{1-\varepsilon^2}}{-\varepsilon} = \tan^{-1} 2R_L C\omega$ (3.68)

The expression for load voltage V_O is obtained from (3.58) and (3.59) can be obtained as

$$V_0(s) = \frac{V_S}{LC\left(s^2 + \frac{1}{R_L C}s + \frac{1}{LC}\right)} \tag{3.69}$$

Taking inverse Laplace transform, we will get

$$V_0(t) = V_S \frac{\omega_n}{\sqrt{1-\varepsilon^2}} e^{-\left(\frac{1}{2R_L C}\right)} \sin \omega t + V_S \tag{3.70}$$

The triggering frequency of the particular SCR should be less than ω_n to complete the conduction cycle.

3.7.2 Class B-Shunt Resonant commutation circuit

The circuit diagram of this is shown in Figure.3.24 (a). The LC resonating circuit is connected across the SCR not in series with the load. There are two thyristors (SCRs). They are main SCR (MT_1) and auxiliary SCR (AT). The polarity of source voltage, V_S is shown in the figure. At this voltage the capacitor is charged to V_S with polarity shown near the left- and right-hand side plates. Now, MT_1 is forward biased and turned on at $t = 0$. It is assumed that output current (I_o) is constant and is passed through the load. Up to time t_1 the voltage across the capacitor, $v_c = V_S$, capacitor current, $i_c = 0$, $i_O = I_o$, and main thyristor current, $i_{MT1} = I_o$. These are shown in Figure.3.24(b).

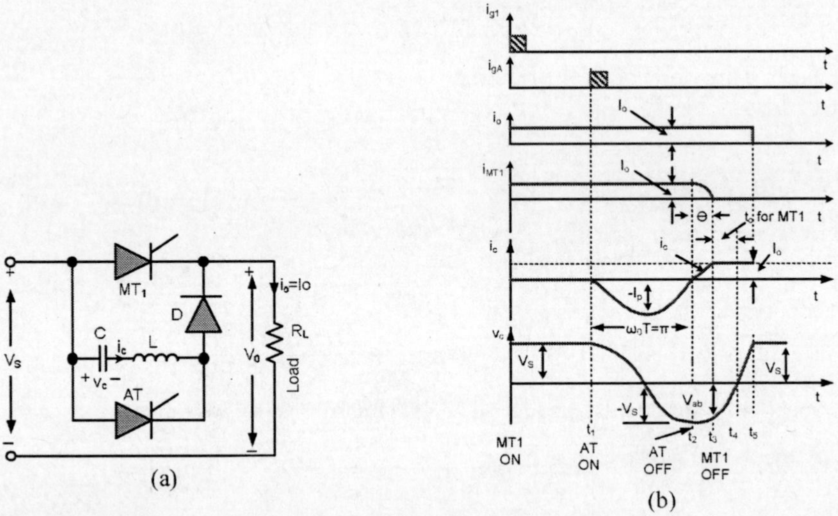

Figure.3.24 Class B resonant commutation circuit and related waveform of voltage and current (a) Circuit, (b) The waveform

For start the commutation of MT_1, AT is triggered at $t = t_1$. When AT is turned on, a resonant current i_c start flowing from C through AT, through L and finally to C.

This resonant current is given by, $$i_c = -V_S\sqrt{\frac{C}{L}}\sin\omega_0 t = -I_p \sin\omega_0 t \tag{3.71}$$

In this equation, the negative sign is inserted because, the current i_c flows just opposite to that of positive direction as shown in the Figure.3.24(a). The capacitor voltage v_c is given by

$$v_c(t) = \frac{1}{C}\int i_c dt = V_S \cos\omega_0 t \tag{3.72}$$

After half cycle from t_1; $i_c = 0$, $v_c = -V_S$, and $i_{MT1} = I_o$. After angular distance π from t_1 i.e from t_1 to t_2, i_c tend to flows in the reverse direction and AT is turned off at t_2. With $v_c = -V_S$, the right hand plat of the capacitor has positive polarity and the resonant current i_c is taking the path through C, L, D, MT_1. Thus, i_c flows just opposite to forward SCR current of MT_1. The net forward current through the MT_1 is I_0-i_c and will start decreasing. Finally, when i_c reverses and becomes equal to I_o, the forward current reaches zero value and MT_1 is turned off. It is to be noted that peak value of reverse current (I_p) must be greater than I_o. This method is also called current commutation, Class B commutation or resonant commutation.

When MT_1 is turned off at instant t_3, constant current I_0 is flowing from source V_S to load via C, L, and diode D. The capacitor now starts charging from -V_{ab} to 0 at instant t_4 and then to V_S at instant t_5. At t_5, when $v_c = V_S$, $i_0 = i_c = I_o$ reduces to zero.
Thus, it is seen from the Figure.3.24(b) that the MT_1 is turned off when,

$$V_S\sqrt{\frac{C}{L}}\sin\omega_0(t_3 - t_2) = I_0 \tag{3.73}$$

Or $$\omega_0(t_3 - t_2) = \sin^{-1}\left(\frac{I_o}{I_p}\right) \tag{3.74}$$

where, $$I_p = V_S\sqrt{\frac{C}{L}} = \text{ peak resonant current} \tag{3.75}$$

MT_1 is commutated at t_3. The capacitor C charges from the constant load current I_0 linearly from -V_{ab} at instant t_3 to zero at instant t_4. MT_1 is now reversed biased by the capacitor voltage v_c for the period from t_4-$t_3 = t_c$.

Thus the turn-off time t_c is given by $$t_c = t_4 - t_3 = C\frac{V_{ab}}{I_0} \tag{3.76}$$

Thus, t_c is dependent on load current. The magnitude of reverse voltage V_{ab} is given by

$$V_{ab} = V_S \cos\omega_0(t_3 - t_2) \tag{3.77}$$

3.7.3 Class C-Complimentary Symmetry Commutation Circuit

The circuit diagram for Class-C commutation circuit is shown in Figure.3.25(a). In this circuit there are two SCRs. One is main SCR, *MT1* which is to be turned off and other is auxiliary SCR, *AT*. The *MT1* is connected in series with the load (R_1). The *AT* is connected in parallel with the *MT1*. The operation of the circuit is given below.

The operation of the commutation circuit is described as follows. The polarity of voltages and the currents are shown in the Figure.3.25(a). It is assumed that the capacitor is initially uncharged.
At time $t = 0$, *MT1* is turned on. The current through R_1 and R_2 are given by equations (3.78) and (3.79) respectively.

$$i_1 = \frac{V_S}{R_1} \tag{3.78}$$

$$i_c = \frac{V_S}{R_2} \tag{3.79}$$

The total current flows through the *MT1* is given by

$$i_{MT1} = i_1 + i_c = V_S\left(\frac{1}{R_1} + \frac{1}{R_2}\right) \tag{3.80}$$

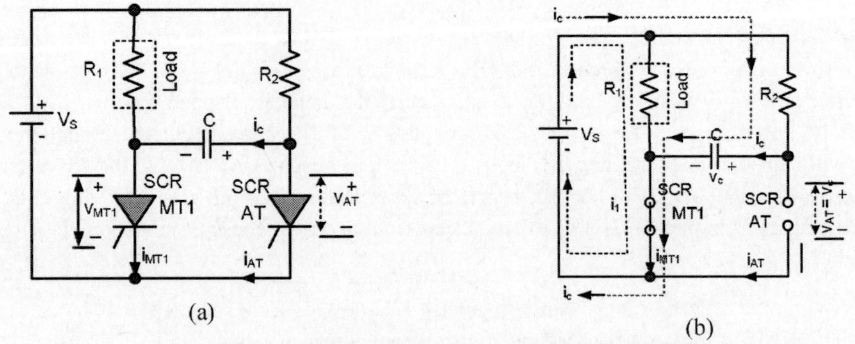

Figure.3.25(a) Class-C commutation Circuit

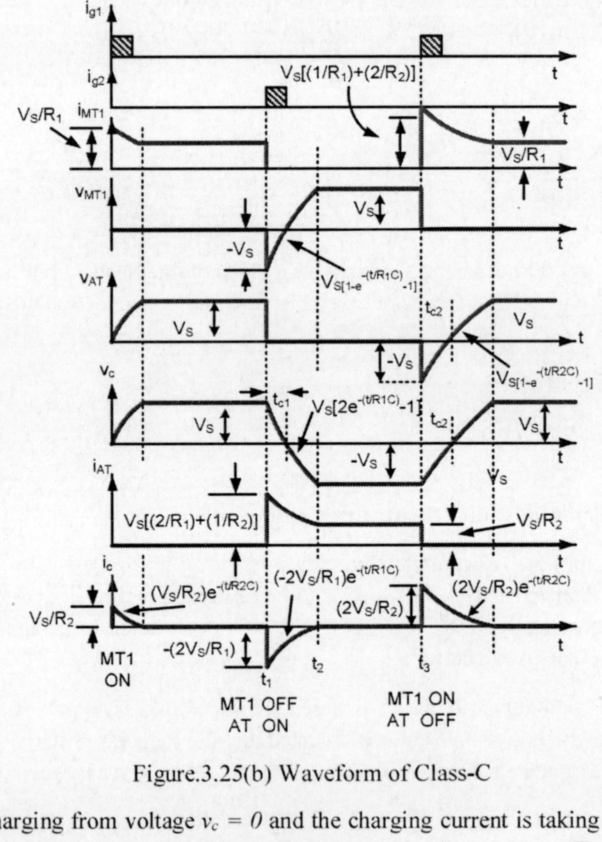

Figure.3.25(b) Waveform of Class-C

The capacitor is charging from voltage $v_c = 0$ and the charging current is taking the path from V_S, R_2, and C. The capacitor current and voltage across the same, and voltage across *AT* are given by equations (3.81), (3.82), and (3.83) respectively are given by

$$i_c(t)=\frac{V_S}{R_2}e^{-\left(t/R_2C\right)} \tag{3.81}$$

$$v_c(t)=V_S\left(1-e^{-\left(t/R_2C\right)}\right) \tag{3.82}$$

$$v_{AT}(t)=v_c(t) \tag{3.83}$$

After the transient is over, the voltage across *AT* is equal to v_c which is equal to V_S, i.e $v_{AT}=v_c=V_S$. The current i_c decays after the transient is over. The waveforms are shown in Figure.3.25(c).
When *MT1* is turned off, *AT* is fired at instant t_1, voltage v_c applies a reverse voltage V_S across *MT1* and it gets turned off. Thus, at t_1, $V_{AT}=0$, $v_{MT1}=-V_S$. The capacitor current and voltage across *MT1* are given by equations (3.84) and (3.85) respectively.

$$i_c=-\frac{2V_S}{R_1} \tag{3.84}$$

$$i_{AT}=V_S\left(\frac{2}{R_1}+\frac{1}{R_2}\right) \tag{3.85}$$

Applying KVL, in the path V_S, R_1, C, and *AT* in which the capacitor is charging from V_S to $-V_S$, the governing equation is

$$R_1 i_c+\frac{1}{C}\int i_c dt=V_S \tag{3.86}$$

Taking Laplace transform of this equation and solving for i_c, we will get

$$i_c(t)=\frac{2V_S}{R_1}e^{-\left(t/R_1C\right)} \tag{3.87}$$

$i_c(t)$ is flowing in the opposite direction to that shown in Figure.3.25(a), hence the equation for capacitor current is given by

$$i_c(t)=-\frac{2V_S}{R_1}e^{-\left(t/R_1C\right)} \tag{3.88}$$

Now, the voltage across C is given by

$$v_c(t)=\left[\frac{1}{C}\int_0^t i_c(t)dt+V_S\right]=\left[\frac{1}{C}\int_0^t -\frac{2V_S}{R_1}e^{-\left(t/R_1C\right)}dt+V_S\right]=V_S\left[2e^{-\left(t/R_1C\right)}-1\right] \tag{3.89}$$

In (3.88) and (3.89), the time t is measured from t_1. The current i_{AT} falls from $V_S\left[\frac{2}{R_1}+\frac{1}{R_2}\right]$ to $\frac{V_S}{R_2}$. The time constant in this case is R_1C. The plots are shown in Figure.3.25(c).
When the transient dies out after the instant t_1, the various values of voltage and currents are as follows.

$$v_{MT1}=V_S;\ v_c=-V_S;\ i_c=0;\ v_{AT}={V_S}/{R_2};\ \text{and } i_{MT1}=0$$

When SCR, *MT1* is turned on to commutate SCR "*AT*" at time t_3, the various voltage and currents are as follows.

$$v_{MT1}=0;\ i_c=\frac{2V_S}{R_2};\ v_{AT}=-V_S;\ i_{AT}=0,\ i_{MT1}=V_S\left[\frac{2}{R_2}+\frac{1}{R_1}\right]$$

It is seen that at t_1, SCR *AT* is turned on, voltage across C equal to V_S suddenly appears as reverse bias voltage across SCR *MT1*. Likewise, at instant t_3, voltage across C equal to V_S applies a reverse bias across SCR *AT* and turn it off. Hence class C commutation is also called complementary commutation.

From the waveform shown in Figure.3.25(c) it is clear that the voltage across *MT1* indicate a reverse voltage equal to $-V_S$ to zero for a particular period. This period is called circuit turn off time represented by t_{c1} for SCR *MT1* and t_{c2} for SCR *AT* and are given by (3.90) and (3.91) respectively.

$$t_{c1} = R_1 C \ln(2) \tag{3.90}$$

$$t_{c2} = R_2 C \ln(2) \tag{3.91}$$

3.7.4 Class D - Auxiliary commutation

The class D commutation is also called auxiliary commutation or Impulse commutation. In this method to commutate a thyristor or SCR another thyristor or SCR is used. The SCR which is to be commutate is called main SCR (*MT1*) and the other thyristor is called auxiliary thyristor (*AT*). The circuit diagram is shown in the Figure.3.26(a). A capacitor (*C*), inductor (*L*) and diode (*D*) are there in the circuit. The connection of these components are as shown in Figure.3.26(a).
Initially, both *MT1* and *AT* are at off condition and it is also assumed that *C* is charged up to the voltage V_S and the polarity of capacitor plats are as shown in the Figure.3.26(a).

At instant $t = 0$, the *MT1* is turned on and V_S is applied across the load. The load current, I_0 begin to flow. It is assumed that I_0 is constant. When *MT1* is on, an oscillator circuit is formed by the circuit components comprising of *C, MT1, L* and *D*. The capacitor current is given by

$$i_c = V_S\sqrt{\frac{C}{L}}\sin\omega_0 t = I_p \sin\omega_0 t \tag{3.92}$$

In this equation, if $\omega_0 t = \pi$, $i_c = 0$

Thus, in the time range $0 < t < \dfrac{\pi}{\omega_0}$, the current through the *MT1* is given by

$$i_{MT1} = I_0 + I_p Sin\omega_0 t \tag{3.93}$$

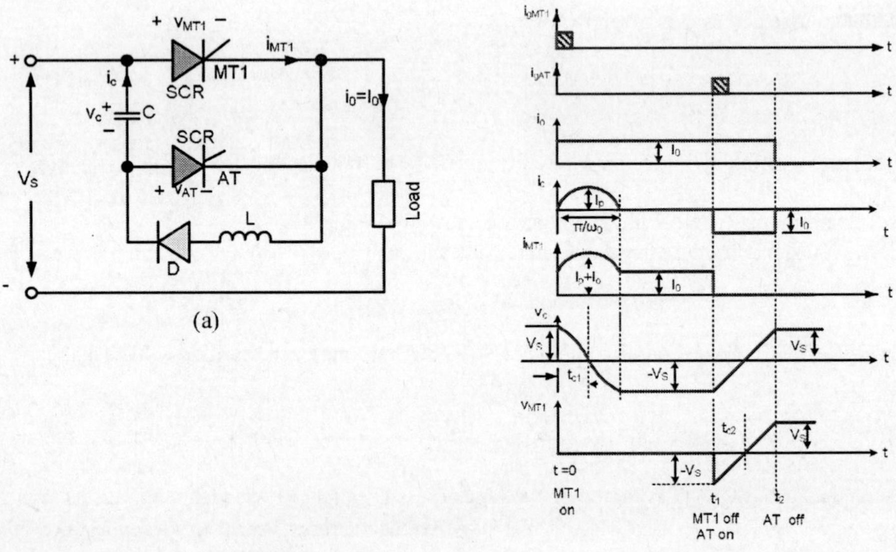

(b)

Figure.3.26(a) Circuit diagram of Class D commutation, (b) Waveforms for class D commutation

Now, the voltage across C changes from V_S to $-V_S$. The lower plat becomes positive. At $\omega_0 t = \pi$, $i_c = 0$, $i_{MT1} = I_0$, $v_c = -V_S$. The corresponding waveforms are presented in Figure.3.26(b).

At $t = t_1$, SCR "*AT*" is turned on. Just after, SCR "*AT*" is turned on, $-V_S$ is applied across *MT1* from the capacitor voltage and reverse biases the *MT1,* thus turned off the same. Accordingly, $i_{MT1} = 0$. Thus, load current is passing through the C and SCR "*AT*" and C charges from $-V_S$ to V_S and current is I_0. This is a linear change. At $t = t_2$, capacitor voltage v_c is equal to V_S and capacitor current $i_c = 0$. SCR "*AT*" is turned off. During t_1 to t_2, $v_c = V_{MT1}$, and SCR "*AT*" is turned on. The turn of time t_c of *MT1* is shown in the Figure.3.26(b).

Thus, in this method when SCR "*AT*" is fired, a reverse voltage is applied across *MT1* which turned off the SCR *MT1*. Hence, this method of commutation is also sometimes called voltage commutation. The sudden application of this reverse voltage, reduces the current though the *MT1* to zero.

3.7.5 Class E - External pulse commutation

This is a method of commutation in which a reverse voltage is applied from an external source of voltage pulse to commutate an SCR or a thyristor. The external voltage is called auxiliary supply. A typical circuit for external pulse commutation is shown in the Figure.2.27(a). The commutating pulse is applied from the auxiliary supply via the pulse transformer. In this case it is the pulse generator. The pulse transformer is specially designed so that there should be tight coupling between primary and secondary. For commutation of the SCR *T1*, a pulse having duration equal or greater than the turn off time of the SCR.

On Triggering SCR *T1*, the later is in conducting mode and the current is flowing through the load as well as the secondary of the pulse transformer. After application of voltage V_P of negative polarity i.e $-V_p$, this negative voltage appears across the SCR *T1* and turn off the same. The induced pulse is of high frequency and hence the C offers almost zero impedance. After SCR *T1* is turned off, load current decreases to zero. The various voltage and current waveform of the Class-E commutation is shown in Figure.3.27(b).

This commutation technique is efficient. However, industry do not normally adopt this technique for SCR commutation.

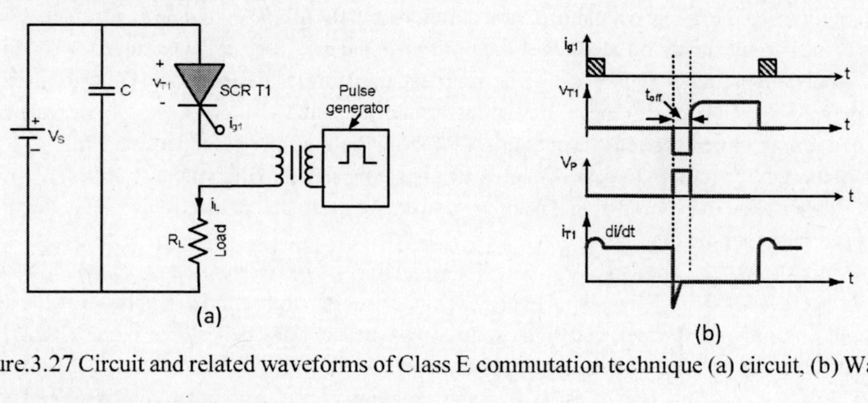

Figure.3.27 Circuit and related waveforms of Class E commutation technique (a) circuit, (b) Waveforms

3.7.6 Class F- Line or Natural commutation

This technique is also called natural commutation. If the supply voltage is alternating, the load current will flow during positive half cycle only. The device is turned off during the negative half of the alternating voltage. The time duration of a half cycle of the applied alternating voltage should be greater than turn off time of the SCR. The maximum frequency of operation of this circuit is decided by turn off time of the SCR. The circuit for line commutation of natural commutation or Class F commutation is shown in Figure.3.28(a). The corresponding voltage and current waveforms are shown in Figure.3.28(b).

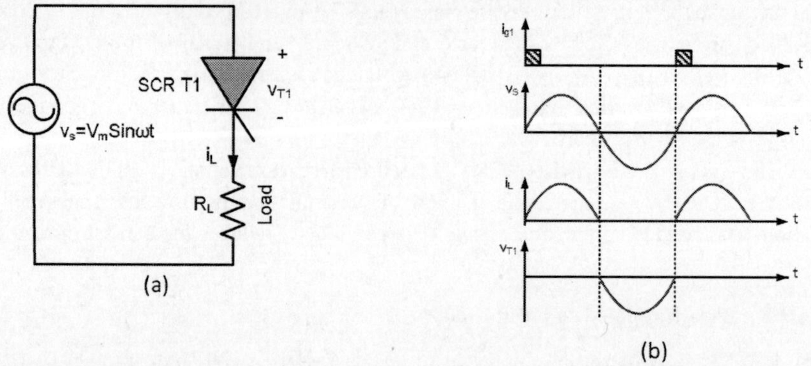

Figure.3.28 Circuit and related waveform of Class F commutation technique (a) circuit, (b) Waveform

Unit Summary

This Unit explores the various turn on and turn off methods of SCRs. Various circuits under each method are presented with detail operating procedure. Following is the summary of this unit.

1. There are several methods of turn-on of an SCR. They are high voltage triggering. thermal triggering, illumination triggering, *dv/dt* triggering, and gate triggering.
2. Gate triggering method is commonly used. At the forward biased of the SCR, the application of a positive gate voltage with respect to (w.r.t) cathode turns on the SCR. In this method, the conduction period of the SCR can be controlled by the application of the gate signal within the specified maximum and minimum values of gate current.
3. Three types of gate signals namely D.C. signals, pulse signals, or AC signals are used for gate triggering. Either of these three signals can be used.
4. The basic requirement for triggering an SCR are (a) The gate current should have sufficient amplitude and of short rise time, (b) the duration of gate current should be of adequate duration, (c) the current should be supplied at the time when the main circuit is favorable for conduction.
5. A pulse transformer (PTR) is like an ordinary transformer designed for the transmission of pulses between two windings. These transformers are able to handle pulses. It has one primary and one or more secondary windings. A PTR having multiple secondary winding is required for most of the power converters to provide the gating signal to individual switches. It is mainly used to isolate the control signal from the power circuit. In power electronics, PTRs are mainly used to trigger the thyristors.
6. Optical isolator or optoisolator is a device which is used to isolate the low voltage low current from the load circuit. It uses light energy. It comprises of a light source, a light-sensitive device, and a switching device. The light source may be an infrared emitter diode (IRED). The photosensitive device may be a phototransistor, LASCR, TRIAC, etc.
7. The gate-triggering circuit is a vital part of any power converter comprising of power semiconductor devices. The output of a power converter depends on how the gate drive circuit drives the switching devices.
8. There are many types of triggering circuit. They are (a) Resistance triggering circuit, (b) Resistance capacitance (R-C) triggering circuit, (c) Programmable unijunction transistor (PUT) triggering, (d) Pulse transformer Triggering, (e) Opto-coupler based Triggering.

9. An SCR can be turned on by application of gate signal to the gate. An SCR can be turned off by bringing the SCR from conduction state to forward blocking state. It is done by either bring the anode current below the holding current or by application of reverse voltage to such a value that the SCR will bring the reverse blocking state.
10. The commutation of thyristor or SCR is the process by which the device is turned off. Basically, there are two methods of turning of thyristors. They are natural commutation and forced commutation.
11. Natural commutation is a simple and widely used method. It uses the alternating or reversing nature of alternating voltage for transferring current from one thyristor to another thyristor. In every half cycle alternating current reaches zero or current zero. When alternating current passes through the natural zero a reverse voltage is simultaneously appeared across the device and the thyristor is immediately turned off. No external circuit is required in natural commutation. In natural commutation make use of A.C supply voltage or generated A.C voltage by the generators or by resonant circuits. Line commutated converters uses this method of commutation.
12. In forced Commutation, some external circuit is used to turned off the SCR. The external circuit used for commutation are called commutation circuits. The components that are used in the external circuits are called commutation components. With the help of the commutation circuit, a reverse voltage is developed across the device and this voltage help in bring the SCR forward current to zero. In this way the SCR is turned off.
13. In forced commutation normally inductance (*L*) and capacitance (*C*) or underdamped resistance -inductance - capacitance (*RLC*) components are used to forced current or voltage to zero to turned off the thyristor.
14. The commutation circuits are classified into the following. (a) Class A Commutation circuit, (b) Class B Commutation circuit, (c) Class C Commutation circuit, (d) Class D Commutation circuit, (e) Class E Commutation circuit, (f) Class F Commutation circuit

Exercises

Example.3.1

Consider the circuit shown in Figure.3.29. The SCR has the following data. I_{gmin}= 0.1mA, V_{gmin} = 0.5V. The diode is a silicon diode. The peak value of the input voltage is 24V. Given that R_2=100kΩ and R_1=10kΩ. Find the Firing angle, α

Solution:

Given values are I_{gmin}= 0.1mA, V_{gmin} = 0.5V, R_1=10kΩ, R_2=100kΩ, Peak amplitude of input voltage = 24V

Applying KVL in the triggering circuit, we get

$$e_s = I_g(R_1 + R_2) + V_D + V_g$$

For the Silicon diode, V_D = 0.7V

Using minimum value gate voltage and current,

$$V_g = V_{gmin} = 0.5V,\ I_g = I_{gmin} = 0.1mA$$

$e_{s(triggering)} = 0.1 \times 10^{-3}(100+10) \times 10^3 + 0.7 + 0.5 = 12.2V$

e_s is a sine wave, hence we can write

$$e_s = V_{max}\text{Sin}\omega t$$

In this case, ωt is the phase angle at which SCR is fired.

Thus, $e_s = 24\text{Sin}\alpha = 12.2$,

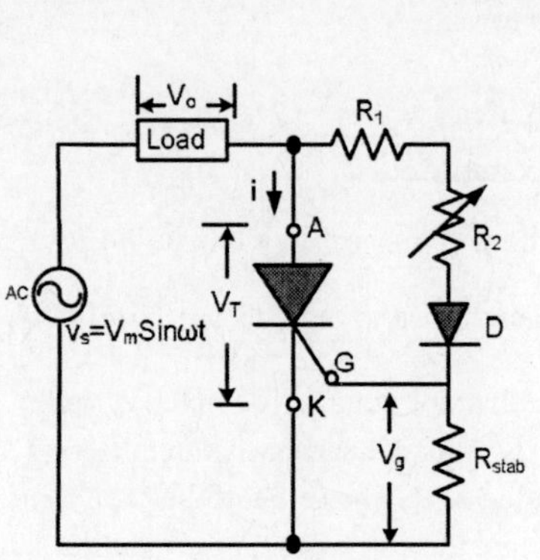

Figure.3.29 Circuit related to example3.1

Solving for α, we will get

$$\alpha = \sin^{-1}\left(\frac{12.2}{25}\right) = 30.6^0$$

Example.3.2

The UJT is used in a relaxation oscillator to trigger an SCR. The circuit is shown in Figure.3.30. The data related to the UJT are as follows. The ratio, $\eta = \frac{R_{B1}}{R_{B1}+R_{B2}} = 0.72$, peak point current, $I_p = 0.6$mA, peak point voltage, V_p =18.0V, valley point voltage, V_v =1.0V, valley point current, I_v = 2.5mA, R_{BB} = 5kΩ, leakage current with emitter open = 4.2mA. The switching frequency is 2kHz. The value of the capacitor, C = 0.04μF. Find the value of R_1, R_2, and R.

Solution:

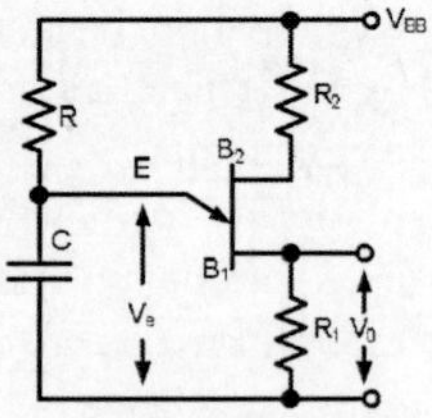

Figure.3.30 Circuit related to example3.2

From $T = \frac{1}{T} = RC\ln\left(\frac{1}{1-\eta}\right)$, the value of R can be found as

$$R = \frac{T}{C\ln\left(\frac{1}{1-\eta}\right)} = \frac{1}{fC\ln\left(\frac{1}{1-\eta}\right)} = \frac{10^6}{2000\text{x}0.04\ln\left(\frac{1}{1-0.72}\right)}$$

$$\therefore R = 9.82k\Omega$$

The valley point voltage is given by

$$V_p = \eta V_{BB} + V_D$$

Since, VD is not given, we can assume $V_p = \eta V_{BB}$ from which we can find the value of VBB as follows.

$$V_{BB} = \frac{V_p}{\eta} = \frac{18}{0.72} = 25V$$

Thus from $$R_2 = \frac{10^4}{\eta . V_{BB}},\ R_2 = \frac{10^4}{0.72\text{x}25} = 555.55\Omega$$

The relation between the V_{BB} and emitter leakage current with emitter open is given by

$$V_{BB} = \text{ leakage current x } (R_1 + R_2 + R_{BB})$$

Here, R_{BB} =5kΩ, leakage current = 4.2mA, and R_2 = 555.55Ω, and V_{BB} = 25V, we can find R_1 as follows.

$$R_1 = \frac{25}{4.2\text{x}10^{-3}} - 5000 - 555.55 = 396.83\Omega$$

Example.3.3

The UJT is used in a relaxation oscillator to trigger an SCR. The circuit is shown in Figure.3.30. The data related to the UJT are as follows. The ratio, $\eta = \frac{R_{B1}}{R_{B1}+R_{B2}} = 0.72$, peak point current, $I_p = 0.6$mA, peak point voltage, V_p =18.0V, valley point voltage, V_v =1.0V, valley point current, I_v = 2.5mA, R_{BB} = 5kΩ, leakage current with emitter open = 4.2mA. The switching frequency is changed by changing the value of R. The value of the capacitor, C = 0.04μF. Find (1) V_{BB}, (2) the maximum and minimum value R and the corresponding switching frequency.

Solution:

The valley point voltage is given by

$$V_p = \eta V_{BB} + V_D$$

Since, VD is not given, we can assume $V_p = \eta V_{BB}$ from which we can find the value of VBB as follows.

$$V_{BB} = \frac{V_p}{\eta} = \frac{18}{0.72} = 25V$$

The maximum value of R is obtained from

$$R_{max} = \frac{V_{BB} - V_p}{I_p} = \frac{V_{BB} - (\eta V_{BB} + V_D)}{I_p}$$

Since, V_D not supplied, hence we can write the expression for R_{max} as follows

$$R_{max} = \frac{V_{BB} - V_p}{I_p} = \frac{V_{BB} - (\eta V_{BB} + V_D)}{I_p}$$

$$= \frac{V_{BB} - \eta V_{BB}}{I_p} = \frac{V_{BB}(1-\eta)}{I_p} = \frac{25(1-0.72)}{0.6x10^{-3}}$$

$$= 11.67k\Omega$$

The minimum value of R is obtained from

$$R_{min} = \frac{V_{BB} - V_v}{I_v}$$

Putting the values, of V_{BB}, V_v, and I_v

$$R_{min} = \frac{25.0 - 1.0}{2.5x10^{-3}} = 9.6k\Omega$$

The minimum value of switching frequency is given by

$$f_{min} = \frac{1}{R_{max} C \ln\left(\frac{1}{1-\eta}\right)} = \frac{1}{11.67x10^3 x0.4x10^{-6} \ln\left(\frac{1}{1-0.72}\right)} = 1.683kHz$$

The maximum value of switching frequency is given by

$$f_{max} = \frac{1}{R_{min} C \ln\left(\frac{1}{1-\eta}\right)} = \frac{1}{9.6x10^3 x0.4x10^{-6} \ln\left(\frac{1}{1-0.72}\right)} = 2.05kHz$$

Example.3.4

If the emitter resistance R_E of an UJT is 1kΩ, and valley point current, Iv is 5mA, find the supply voltage to emitter circuit, V_{EE}. The UJT equivalent circuit is shown in Figure.3.31. The valley point voltage V_v is 2V

Solution:

At the valley point, $V_E = V_v = 2V$ and $I_E = I_v = 5mA$

Applying KVL,

$V_{EE} = I_E R_E + V_E = 5x10^{-3} x 1x10^3 + 2 = 7V$

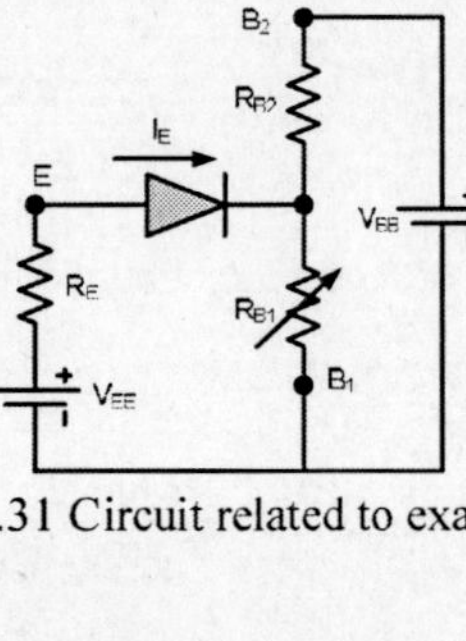

Figure.3.31 Circuit related to example 3.4

Example.3.5

The data of an UJT are given below.

The ratio, $\eta = \frac{R_{B1}}{R_{B1}+R_{B2}} = 0.7$, peak point current, I_p = 50μA, alley point voltage, V_v =2.0V, valley point current, I_v = 6mA, R_{BB} = 7kΩ, Emitter leakage current, I_{EO} = 2mA. The value of the capacitor, C = 0.1μF. The V_{BB} is equal to 20V. The minimum value of gate voltage (I_{gmin}) required to trigger the SCR is 0.2V. Design the relaxation oscillator circuit and find the maximum and minimum value of switching frequency.

Solution:

Given data,

$\eta = \frac{R_{B1}}{R_{B1}+R_{B2}} = 0.7$, I_p = 50μA, V_v =2.0V, I_v = 6mA, R_{BB} = 7kΩ, I_{EO} = 2mA, C = 0.1μF. V_{BB} = 20V,

I_{gmin} = 0.2V

The maximum value of R is obtained from

$$R_{max} = \frac{V_{BB}-V_p}{I_p} = \frac{V_{BB}-(\eta V_{BB}+V_D)}{I_p}$$

Since, V_D not supplied, hence we can write the expression for R_{max} as follows

$$R_{max} = \frac{V_{BB}-V_p}{I_p} = \frac{V_{BB}-(\eta V_{BB}+V_D)}{I_p}$$

$$= \frac{V_{BB}-\eta V_{BB}}{I_p} = \frac{V_{BB}(1-\eta)}{I_p} = \frac{20(1-0.7)}{50\text{x}10^{-6}} = 120k\Omega$$

The minimum value of R is obtained from

$$R_{min} = \frac{V_{BB}-V_v}{I_v}$$

Putting the values, of V_{BB}, V_v, and I_v

$$R_{min} = \frac{20.0-2.0}{6\text{x}10^{-3}} = 3.0k\Omega$$

The value of R_2 is approximately given by

$$R_2 = \frac{10^4}{\eta.V_{BB}}$$

$$\therefore R_2 = \frac{10^4}{0.7\text{x}20} = 714.29\Omega$$

The value of R_1 can be obtained from

$$R_1 = \frac{V_{g\min}}{I_{EO}} = \frac{0.2}{2\text{x}10^{-3}} = 100\Omega$$

The limit of switching frequency is obtained as follows.

$$T_{max} = R_{max}C\ln\frac{1}{1-\eta} = 120x10^3x0.1x10^{-6}\ln\frac{1}{1-0.7} = 14.44ms$$

The maximum value of the switching frequency is

$$f_{max} = \frac{1}{R_{min}C\ln\left(\frac{1}{1-\eta}\right)} = \frac{1}{3x10^3x0.1x10^{-6}\ln\left(\frac{1}{1-0.7}\right)} = 2.78kHz$$

The minimum value of switching frequency is

$$f_{min} = \frac{1}{T_{max}} = \frac{1}{14.44x10^{-3}}$$

Example.3.6

A class C commutation circuit is shown in Figure.3.32. The dc supply voltage (E_{DC}) is 120V. The current through R1 and R2 is equal to 20A. The turn off time of SCR MT1 and SCR AT is equal to 60µs. Find the value of C for complete commutation.

Solution:

The value of resistances R1 and R2 are obtained as

$$R_1 = R_2 = \frac{E_{dc}}{I} = \frac{120}{20} = 6\Omega$$

For complete commutation, the value of C is calculated as follows.

$$t_{off} = R_1C\ln(2)$$

$$\text{or } t_{off} = 0.6931R_1C$$

$$C = 1.44\frac{t_{off}}{R_1} = 1.44x\frac{60x10^{-6}}{6} = 14.4\mu F$$

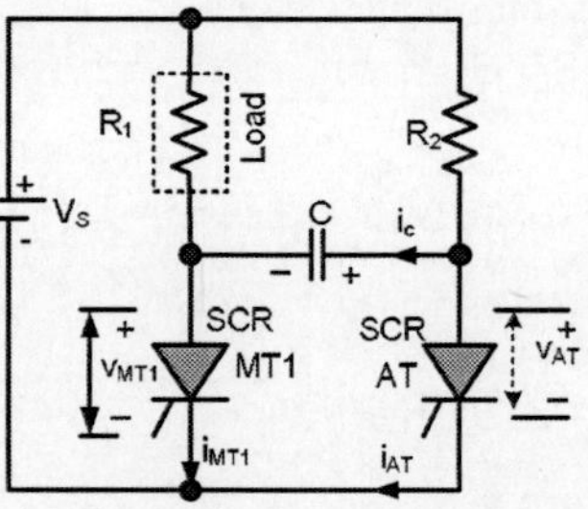

Figure.3.32 Circuit related to example 3.6

Example.3.7

A Class D commutation circuit is shown in Figure.3.33. Following data are provided. The dc supply voltage (E_{DC}) is 50V. Maximum value of inductor current, I_{Lmax} = 50A. Turn of time for SCR MT1 is 30µs. The chopping frequency, f_s = 500Hz. Variation of load is from 10 %-100%. Find the value of C and L

Solution:

Given data

E_{DC} = 50V, I_{Lmax} = 50A, t_{off} = 30µs for MT1, f_s = 500Hz, Load variation = 10 %-100%.

Considering 50% tolerance on turn off time, the turn off time of MT1 is

$$t_{off} = \left(30 + \frac{50}{100}x30\right) = 45\mu s$$

The value of C is obtained from

$$CE_{dc} = I_Lt_{off}$$

$$C = \frac{I_Lt_{off}}{E_{dc}} = \frac{50x45x10^{-6}}{50} = 45\mu F$$

The minimum value of load voltage is given by the relation

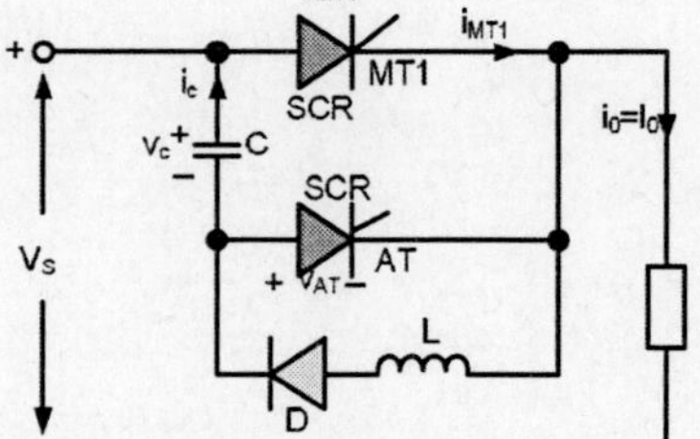

Figure.3.33 Circuit related to example 3.7

$$V_{o(\min)} = \frac{t_1 - t_2}{T} E_{dc} = \frac{\pi\sqrt{LC}}{T} E_{dc}$$

From this equation, we can confirm that the value of L will be

$$L \leq \left(\frac{V_{0(\min)}}{E_{dc}}\right)^2 \text{x} \frac{T^2}{\pi^2 C}$$

The $V_{0(min)}$ is given by

$$V_{0(\min)} = 10\% \text{ of } 50 = 5\text{V}$$

The time period is given by

$$T = 1/f_s = 2\text{x}10^{-3}\text{ s}$$

Thus,

$$L \leq \left(\frac{5}{50}\right)^2 \text{x} \frac{(2\text{x}10^{-3})^2}{\pi^2 \text{x} 45\text{x}10^{-6}} \text{ or } L \leq 90\mu\text{H}$$

From the following relation

$$L \geq 45\text{x}10\text{x}\left(\frac{E_{dc}}{I_{L\max}}\right)^2$$

$$L \geq 45\mu H$$

Thus, range for L is $45\mu H < L < 90\mu H$

Example.3.8

A relaxation oscillator circuit is shown in the Figure.3.34. Find the value of frequency of the output voltage spikes for two values of resistances (R) (a) 5Ω, (b) 10kΩ. Following data are provided.
C = 0.2μF, η = 0.64, R_1=10kΩ, R_2 = 100Ω

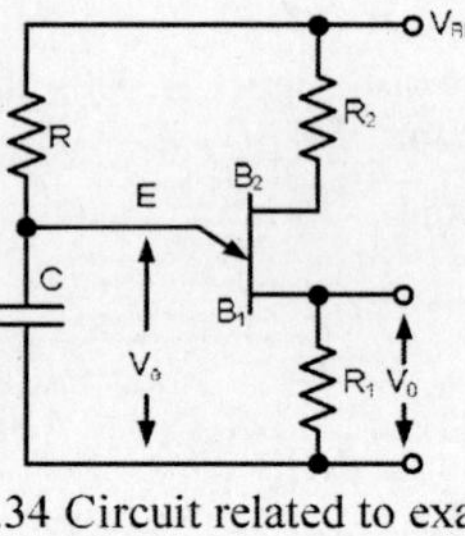

Figure.3.34 Circuit related to example 3.8

Solution:

(a) When R=5kΩ,

$$T = RC \ln \frac{1}{1-\eta} = 5\text{x}10^3 \text{x} 0.2\text{x}10^{-6} \ln \frac{1}{1-0.64} = 1.02ms$$

$$\therefore\ f = \frac{1}{T} = \frac{1}{1.02\text{x}10^{-3}} = 980Hz$$

(b) When R= 10kΩ,

$$T = RC\ln\frac{1}{1-\eta} = 10x10^3x0.2x10^{-6}\ln\frac{1}{1-0.64} = 2.043ms$$

$$\therefore\ f = \frac{1}{T} = \frac{1}{2.043x10^{-3}} = 490Hz$$

Multiple Choice Questions

1. Which of the following is the best-suited triggering method for turning ON the SCR?
   (a) Forward voltage triggering
   (b) dv/dt triggering
   (c) Thermal triggering
   (d) Gate triggering
2. The SCR triggering method in which the junction temperature can be maintained at low value is?
   (a) Thermal triggering
   (b) di/dt triggering
   (c) Gate triggering
   (d) dv/dt triggering
3. SCR can be turned on by
   (a) Applying anode voltage at a sufficiently fast rate
   (b) Applying sufficiently large anode voltage
   (c) decreasing the temperature of SCR
   (d) Applying sufficiently large gate current
4. Which of the above statements are correct?
   (a) 1, 2 and 3
   (b) 1, 3 and 4
   (c) 1, 2 and 4
   (d) 2, 3 and 4
5. In case of class A type commutation or load commutation with low value of R load the
   (a) L is connected across R
   (b) L-C is connected across R
   (c) L is connected in series with R
   (d) L-C is connected in series with R
6. The class A commutation or load commutation is possible in case of
   (a) dc circuits only
   (b) ac circuits only
   (c) both DC and AC circuits
   (d) none of the above mentioned
7. In case of class B commutation or resonant-pulse commutation with L = 5 μH and C = 20 μC with initial voltage across the capacitor (Vs) = 230 V. Find the peak value of resonant current.
   (a) 560 A
   (b) 460 A
   (c) 360 A
   (d) 260 A
8. In case of class B commutation or resonant-pulse commutation with L = 5 μH and C = 20 μC with the initial voltage across the capacitor (Vs) = 230 V. Find the conduction time for auxiliary thyristor.

(a) 0.23 μs
(b) 6.57 μ
(c) 31.41 μs
(d) 56 μs

9. The type of commutation when the load is commutated by transferring its load current to another incoming thyristor is
(a) class A or load commutation
(b) class B or resonant commutation
(c) class C or complementary commutation
(d) class D or impulse commutation

10. The type of commutation in which the pulse to turn off the SCR is obtained by separate voltage source is
(a) class B commutation
(b) class C commutation
(c) class D commutation
(d) class E commutation

11. The natural reversal of ac supply voltage commutates the SCR in case of
(a) forced commutation
(b) only line commutation
(c) only natural commutation
(d) both line & natural commutation

12. Parallel-capacitor commutation is
(a) line commutation
(b) load commutation
(c) forced commutation
(d) external-pulse commutation

13. Class E commutation is a/an
(a) line commutation technique
(b) load commutation technique
(c) forced commutation technique
(d) external-pulse commutation technique

14. Class A commutation is often referred to as :
(a) Load commutation
(b) Forced commutation
(c) Natural commutation
(d) External pulse commutation

15. Which of the following statements is true pertaining to the commutation of SCR?
(a) Class-B commutation is also known as complementary commutation.
(b) Class-D commutation is also known as load commutation.
(c) Class-C commutation is also known as complementary auxiliary commutation.
(d) Class-A commutation is also known as load commutation.

16. Which commutation technique is also called line commutation?
(a) Class F
(b) Class D
(c) Class E
(d) Class A

17. In the case of DC circuits, since the supply voltage does not go through a zero value, some external source is required to commutate the device. This process with the thyristor inverter is known as the ______ commutation process.
    (a) Natural
    (b) Self
    (c) Forced
    (d) Load
18. A thyristor circuit having $V_s = 230$ V, L = 10 μH and C = 40 μF is commutated using Class-D commutation. The peak value of current through the capacitor is given by:
    (a) 460 A
    (b) 230 A
    (c) 23 A
    (d) 46 A
19. In which of the following commutation technique, the triggering of one SCR commutates the already conducting SCR and vice versa?
    (a) Class A commutation
    (b) Class C commutation
    (c) Class E commutation
    (d) Class D commutation
20. Which one of the following is also known as resonant commutation?
    (a) Class A commutation
    (b) Class C commutation
    (c) Class D commutation
    (d) Class E commutation
21. Light triggering mainly used in
    (a) low - voltage direct current transmission.
    (b) medium voltage direct current transmission.
    (c) high voltage direct current transmission.
    (d) all of these.
22. Gate circuit or triggering circuit of a thyristor is
    (a) lower power circuit.
    (b) high power circuit.
    (c) magnetic circuit.
    (d) may be low power or high-power circuit.
23. Which of the following is the function of an R-C snubber circuit connected in parallel to an SCR?
    (a) Limiting the di/dt through the SCR
    (b) Triggering the SCR
    (c) Preventing over voltages across the SCR
    (d) Forced commutation of the SCR
24. Which of the following is referred as TRIGGERING a thyristor?
    (a) Turning the SCR from reverse blocking state to forward blocking state
    (b) Turning the SCR from forward conduction state to forward blocking state
    (c) Turning the SCR from forward blocking state to reverse blocking state
    (d) Turning the SCR from forward blocking state to forward conduction state
25. False turn-on of SCR by large dv/dt can be prevented by using a ______ with the SCR.
    (a) Snubber circuit in parallel
    (b) Snubber circuit in series

(c) Filter circuit in parallel
(d) Filter circuit in series

26. The snubber circuit used to shape the turn-on switching trajectory of thyristor and/or to limit di/dt during turn on is
 (a) L - R snubber polarized
 (b) R - C snubber polarized
 (c) R - C snubber unpolarized
 (d) L - R snubber unpolarized
27. The basic requirement of proper firing of an SCR should have
 (a) The gate current should have sufficient amplitude and of short rise time.
 (b) The duration of gate current should be of adequate duration.
 (c) The current should be supplied at the time when the main circuit is favourable for conduction
 (d) All of these
28. After proper turn on of thyristor
 (a) gate signal is always present.
 (b) gate signal must be removed.
 (c) gate signal should present but can be removed.
 (d) none of the above.
29. Pulse gate triggering is achieved by means of
 (a) An LC circuit
 (b) A UJT relaxation oscillator circuit
 (c) A diac-triac circuit
 (d) A rheostatic arrangement
30. RC triggering is preferred over resistance triggering because
 (a) Provides a large value of the triggering angle
 (b) Provides accurate triggering
 (c) Causes quick triggering
 (d) Protects the device from getting damaged
31. The duration of the pulse in a pulse triggering system for scrs should be at least
 (a) 60 us
 (b) 40 us
 (c) 20 us
 (d) 10 us
32. For thyristors, pulse triggering is preferred over dc triggering because
 (a) The gate dissipation is low
 (b) The pulse system is simpler
 (c) The triggering signal is required for a short duration
 (d) all of the above
33. At a room temperature of 30o C, minimum voltage and current required to fire a SCR is
 (a) 3V,40 mA
 (b) 0.6V,40 mA
 (c) No limit
 (d) 3V,100 mA
34. Turn on time of an SCR can be reduce by using a
 (a) Rectangular pulse of high amplitude and narrow width
 (b) Rectangular pulse of low amplitude and narrow width
 (c) Triangular pulse

(d) Trapezoidal pulse

35. UJT oscillator are used for gate triggering of thyristor for
    (a) Better phase control
    (b) Snap action
    (c) Being cheap simple
    (d) None of these
36. A PUT relaxation oscillator has values V_{BB} =15 V, R = 22 kΩ, R_2 = 6 kΩ, I_P=100 μA, V_V = 1 V, I_V = 7 mA, C =1μF, R_K = 100 kΩ, R_3 = 12 kΩ. The value of V_P will be
    (a) 0.7 V
    (b) 10 V
    (c) 10.7 V
    (d) 15 V
37. When a UJT is used for triggering of an SCR, the waveshape of the voltage is a
    (a) Sine wave
    (b) Saw-tooth wave
    (c) Trapezoidal wave
    (d) Square wave
38. Optocouplers combine
    (a) SIT and BJT
    (b) IGBTs and MOSFETs
    (c) Power transistor and silicon transistor
    (d) Infrared light emitting diode and silicon phototransistor
39. In a UJT, maximum value of charging resistance is associate with
    (a) Peak Point
    (b) Valley point
    (c) Any point between peak and valley point
    (d) After the valley point
40. External circuit that is used to turn off the in SCR
    (a) Communication Circuit
    (b) Commutation Circuit
    (c) Conducting Circuit
    (d) Conversion Circuit

Answers to multiple-Choice Questions

1. (d)	**9. (c)**	**17. (c)**	**25. (a)**	**33. (a)**
2. (c)	**10. (d)**	**18. (a)**	**26. (a)**	**34. (a)**
3. (b)	**11. (d)**	**19. (b)**	**27. (d)**	**35. (a)**
4. (c)	**12. (c)**	**20. (a)**	**28. (b)**	**36. (c)**
5. (d)	**13. (d)**	**21. (c)**	**29. (b)**	**37. (b)**
6. (a)	**14. (a)**	**22. (a)**	**30. (a)**	**38. (d)**
7. (b)	**15. (d)**	**23. (c)**	**31. (c)**	**39. (a)**
8. (c)	**16. (a)**	**24. (d)**	**32. (a)**	**40. (b)**

Short and Long Answer Type Questions

1. What are the different methods of SCR turn on? Explain briefly.
2. Discuss the essential features that triggering circuits for SCRs should possess.
3. Explain the functions of all the components of a general triggering circuit. Give necessary figure.

4. Draw the circuit and the waveform of a resistance triggering circuit. Describe the features of resistance triggering circuit.
5. Is it possible to trigger an SCR with a delay angle greater than 90^0 using resistance triggering circuit? Illustrate your comments with the help of waveforms.
6. Show how firing delay angle (α) in a resistance firing circuit proportional to the variable resistance.
7. Draw RC half wave triggering circuit and the corresponding waveforms. Discuss the function of each component of the circuit.
8. Draw RC full wave triggering circuit and the corresponding waveforms. Discuss the function of each component of the circuit.
9. Describe RC full wave triggering circuit when the load is (i) AC, (ii) DC
10. Discuss the synchronized UJT triggering circuit using a Zener diode. Also explain the relevant voltage and current waveform.
11. Describe the pulse transformer used in triggering the circuit.
12. What is the main requirements of pulse transformer use in triggering circuit,
13. What is optoisolator? Why it is necessary to consider optoisolator?
14. Innumerate the advantages of optoisolator in comparison to electromechanical relays.
15. How the operation of PUT is different from UJT? Explain.
16. Draw the block diagram of an SCR firing circuit.
17. Discuss the working of an oscillator circuit that employs UJT.
18. Derive the expression for switching frequency in an oscillator circuit that utilize the UJT.
19. Explain V-I characteristics of UJT. Provide necessary diagram.
20. Explain the turn off methods of thyristors.
21. What do you mean by commutation of SCR? At what conditions, the thyristors are turn off?
22. What do you mean by natural commutation and forced commutation of thyristors?
23. What are different types of forced commutation circuit?
24. Explain the operation of Class-A forced commutation circuit. Provide necessary circuit and waveform diagram for the same.
25. Explain the operation of the Class -B forced commutation circuit. Provide the necessary circuit and waveform diagram for the same.
26. Explain the operation of the Class -C forced commutation circuit. Provide the necessary circuit and waveform diagram for the same.
27. Explain the operation of the Class -D forced commutation circuit. Provide the necessary circuit and waveform diagram for the same.
28. Explain the operation of the Class -E forced commutation circuit. Provide the necessary circuit and waveform diagram for the same.
29. Explain the operation of the Class -F forced commutation circuit. Provide the necessary circuit and waveform diagram for the same.

Numerical problems

1. Consider the circuit shown in Figure.3.29. The SCR has the following data. I_{gmin}= 0.15mA, V_{gmin} = 0.6V. The diode is a silicon diode. The peak value of the input voltage is 25V. Given that R_2=110kΩ and R_1= 5kΩ. Find the Firing angle, α
2. The UJT is used in a relaxation oscillator to trigger an SCR. The circuit is shown in Figure.3.30. The data related to the UJT are as follows. The ratio, $\eta = \frac{R_{B1}}{R_{B1} + R_{B2}} = 0.71$, peak point current, I_p = 0.6mA, peak point voltage, V_p=18.0V, valley point voltage, V_v=1.0V, valley point current,

I_v = 3.0mA, R_{BB} = 6kΩ, leakage current with emitter open = 4.0mA. The switching frequency is 2kHz. The V_D = 0.7V, The value of the capacitor, C = 0.06μF. Find the value of R_1, R_2, and R.

3. The UJT is used in a relaxation oscillator to trigger an SCR. The circuit is shown in Figure.3.30. The data related to the UJT are as follows. The ratio, $\eta = \frac{R_{B1}}{R_{B1}+R_{B2}} = 0.72$, peak point current, I_p = 0.6mA, peak point voltage, V_p=18.0V, valley point voltage, V_v=1.2V, valley point current, I_v = 2.0mA, R_{BB} = 6kΩ, leakage current with emitter open = 4.1mA. The V_D = 0.7V, The switching frequency is changed by changing the value of R. The value of the capacitor, C = 0.05μF. Find (1) V_{BB}, (2) the maximum and minimum value R and the corresponding switching frequency.
4. If the emitter resistance R_E of an UJT is 1.5kΩ, and valley point current, Iv is 6mA, find the supply voltage to emitter circuit, V_{EE}. The UJT equivalent circuit is shown in Figure.3.31. The valley point voltage V_v is 3V
5. The data of an UJT are given below.

   The ratio, $\eta = \frac{R_{B1}}{R_{B1}+R_{B2}} = 0.71$, peak point current, I_p = 60μA, alley point voltage, V_v=1.5V, valley point current, I_v = 6mA, R_{BB} = 5kΩ, Emitter leakage current, I_{EO}= 1mA. The value of the capacitor, C = 0.2μF. The V_{BB} is equal to 20V. The minimum value of gate voltage (I_{gmin}) required to trigger the SCR is 0.2V. Design the relaxation oscillator circuit and find the maximum and minimum value of switching frequency.
6. A class C commutation circuit is shown in Figure.3.32. The dc supply voltage (E_{DC}) is 100V. The current through R1 and R2 is equal to 18A. The turn off time of SCR MT1 and SCR AT is equal to 50μs. Find the value of C for complete commutation.
7. A Class D commutation circuit is shown in Figure.3.33. Following data are provided. The dc supply voltage (E_{DC}) is 60V. Maximum value of inductor current, I_{Lmax} = 60A. Turn of time for SCR MT1 is 20μs. The chopping frequency, f_s = 600Hz. Variation of load is from 20 %-100%. Find the value of C and L

Practical Experiments

Experiment No.3.1

Title: Test the effect of variation in R and C in R and RC triggering circuits on the firing angle of SCR

Objectives:

(a) To test the effect of variation of R in the R triggering circuit of SCR
(b) To test the effect of variation of R and C in RC triggering circuit of SCR

Resources required:

Table.3.1 Apparatus/ component required

Sl No.	Apparatus/ Component	Specification Range/ Rating	Nos/Quantity
1.	Circuit Kit for SCR triggering		1
2.	Transformer	230V/12 V	1
3.	Load resistor	50Ω, 2A	1
4.	Unearthed CRO	20MHZ	1
5.	Connecting probes		1 set

Sl No.	Apparatus/ Component	Specification Range/ Rating	Nos/Quantity
6.	Capacitor	0.47µF or available along with the kit	1

Theory:

(a) Resistance (R) Gate Triggering Circuit

The resistance gate triggering circuits is the simplest as well as most economical. The *R* gate triggering can be limited in the firing angle range 0^0-90^0. The circuit diagram is shown in Figure.3.35 (a). In this circuit, the gate current is provided from an AC source via resistance, R_1, R_2, and diode (*D*). R_2 is a variable resistance. A stabilizing resistance is connected (R_{stab}) is connected across the gate and the cathode. The flow of gate current is from source v_s, via Load, R_1, R_2, diode *D* to gate. The function of diode *D* is to prevent the gate to cathode from junction reverse biased. It allows the current flow during the positive half cycle only.

When the voltage, v_s goes on positive the SCR is forward biased and it will conduct only when the gate current exceeds the minimum value of gate current (I_{gmin}). When gate current is less than I_{gmin} the output voltage across the load, $V_O = 0$. Also with positive v_s, *D* and gate to cathode junction become forward biased. Gate current increases with an increase in v_s towards maximum voltage V_m. When $I_g = I_{gmin}$, the SCR will turn on. The output voltage is now approximately equal to v_s. The SCR is in on state until v_s decreases from V_m to 0. At this instant, the load current is below the holding current. Now, the v_s decreases from zero to negative and the SCR becomes reverse-biased and remains in the off state. The sequence of the operation is repeated in the next cycle and so on.

(b) Resistance – Capacitance (RC) Gate Triggering circuit

In case of resistance triggering, the firing angle can be controlled is limited to 0^0 to 90^0. This limitation is overcome by using resistance (*R*)- capacitance (*C*) triggering circuit simply R-C triggering circuit. There are several such R-C circuits, two of them named as R-C half wave triggering and R-C full wave triggering circuits. Here will go for RC half wave firing circuit as shown in Figure.3.35(b).

The details of the operation of the resistance and resistance-capacitance triggering circuits are provided in the respective sections of this Unit.

Circuit diagram:

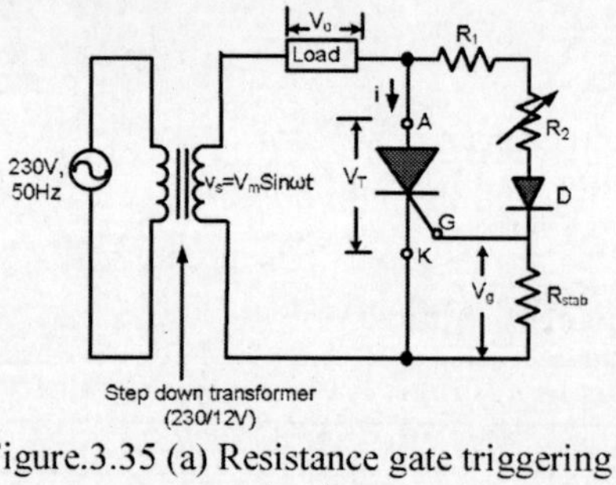

Figure.3.35 (a) Resistance gate triggering circuit

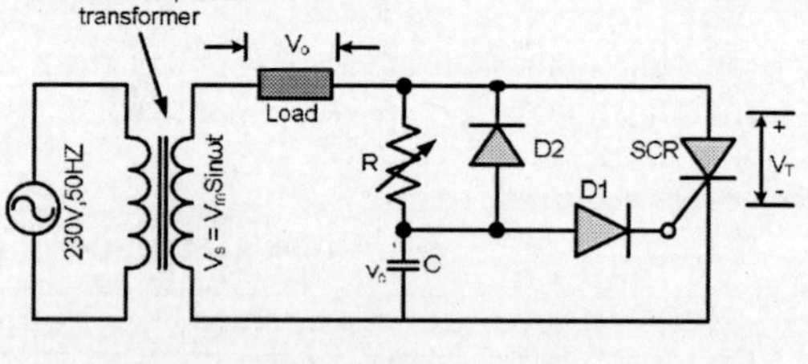

Figure.3.35(b) R-C half wave triggering circuits

Procedure:

R-Triggering

1. Connect all the components as per the circuit diagram shown in Figure.3.35(a).

2. Set the value of resistances at the maximum position.
3. Connect the probe of the oscilloscope across the load resistance (Load).
4. Switch on the power supply.
5. Vary the resistance R_2 to vary the triggering angle. Observe the output voltage and the triggering pulse on the CRO. The readings are tabulated in Table.3.2
6. Draw the respective waveforms.

RC-Triggering

1. Connect all the components as per the circuit diagram provided in Figure.3.35 (b).
2. Set the value of resistances at the maximum position.
3. Connect the probe of the oscilloscope across the load resistance (Load).
4. Switch on the power supply.
5. Vary the resistance R_2 to vary the triggering angle (0-180^0). Observe the output voltage and the triggering pulse on the CRO. Tabulate the data as shown in Table.3.3
6. Vary the resistance C to vary the triggering angle (0-180^0). Observe the output voltage and the triggering pulse on the CRO. Tabulate the data as shown in Table.3.4
7. Draw the respective waveforms.

Result and Discussion:
Using the procedure followed for the R triggering circuit, the various values are noted in Table.3.2

Table.3.2 Values of T_{ON}, T_{OFF}, V_{in}, V_O for various values of α in R-firing circuit

Sl No.	Firing angle (α)	Time in (ms)		Input voltage (V_{in})	Output voltage (V_O)
		T_{ON}	T_{OFF}		

Table.3.3 Values of T_{ON}, T_{OFF}, V_{in}, V_O for various values of α in RC-firing circuit (by varying R)

Sl No.	Firing angle (α)	Time in (ms)		Input voltage (V_{in})	Output voltage (V_O)
		T_{ON}	T_{OFF}		

Table.3.4 Values of T_{ON}, T_{OFF}, V_{in}, V_O for various values of α in RC-firing circuit (by varying C)

Sl No.	Firing angle (α)	Time in (ms)		Input voltage (V_{in})	Output voltage (V_O)
		T_{ON}	T_{OFF}		

The plots for the R firing circuit look like Figure.3.36(a). Similarly, the floats for the RC firing circuit look like in Figure.3.36 (b)

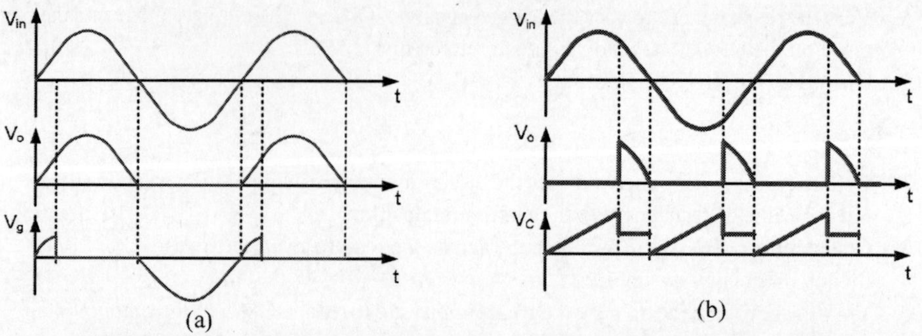

Figure.3.36 Waveform for (a) R-triggering circuit, (b) RC triggering circuit

Discussion: From this experiment, we can differentiate between R and R-C triggering.

Conclusion: From this experiment, we have learned how SCR can be turned on using resistance and RC triggering methods.

Experiment No.3.2

Title: Test the effect of variation of *R, C* in UJT triggering technique.

Objectives:

(a) To study the effect of variation in *R* and *C* in UJT triggering for SCR

Theory:

The Unijunction transistor (UJT) is a very highly efficient switch whose switching time is in the nanosecond range. UJT has negative resistance characteristics due to which it can be used as a relaxation oscillator. This relaxation oscillator can be used for triggering SCR. Figure.3.37 (a) shows the circuit diagram of UJT which works as an oscillator.

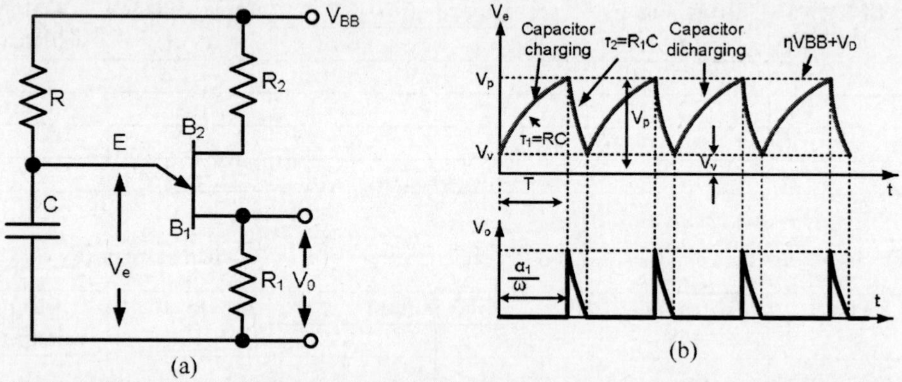

Figure.3.37 (a) Circuit diagram for UJT oscillator, (b) Voltage waveform for the oscillator

The details theory of the UJT triggering circuit is already provided in this Unit. Students are advised to follow the respective section and write a brief theory on it.

Circuit diagram:

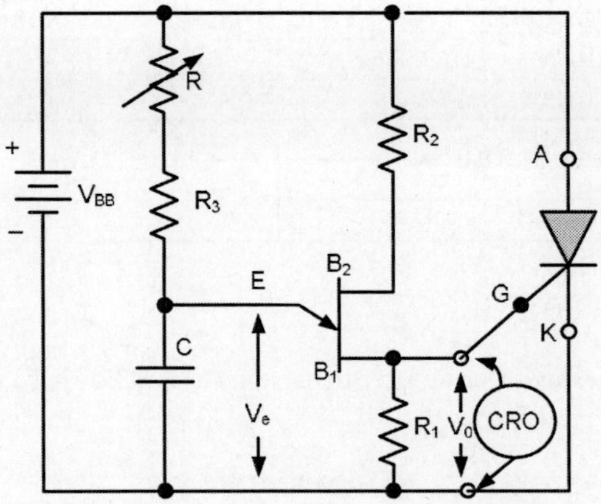

Figure.3.38 Resistance gate triggering circuit

Procedure:

UJT-Triggering

By varying R

1. Connect all the components as per the circuit diagram provided in Figure. 3.38
2. Set the value of resistances at the maximum position.
3. Connect the probe of the oscilloscope across the load resistance (Load).
4. Switch on the power supply.
5. Vary the resistance *R* to vary the triggering angle ($0\text{-}180^0$). Observe the output voltage and the triggering pulse on the CRO. Tabulate the data as shown in Table.3.5.
6. Draw the respective waveforms.
7. Switch off the power supply.

By varying C

1. Set the value of resistances at the maximum position.
2. Connect the probe of the oscilloscope across the load resistance (Load).
3. Switch on the power supply.
4. Vary the resistance *C* to vary the triggering angle ($0\text{-}180^0$). Observe the output voltage and the triggering pulse on the CRO. Tabulate the data as shown in Table.3.6
5. Draw the respective waveforms.

Result and Discussion:

Table.3.5 Values of T_{ON}, T_{OFF}, V_{in}, V_O for various values of α in UJT-firing circuit (by varying R)

Sl No.	**Firing angle (α)**	**Time in (ms)**		**Input voltage (V_{in})**	**Output voltage (V_O)**
		T_{ON}	**T_{OFF}**		

Table.3.6 Values of T_{ON}, T_{OFF}, V_{in}, V_O for various values of α in UJT-firing circuit (by varying C)

Sl No.	Firing angle (α)	Time in (ms)		Input voltage (V_{in})	Output voltage (V_O)
		T_{ON}	T_{OFF}		

Discussion: ....

Conclusion: From this experiment we have understood how SCR can be turn on using UJT triggering technique.

Experiment No.3.3

Title: Study the operation of Class - A, B, and C turn-off circuits.

Objectives:

To study the operation of Class A, B, and C commutation circuits

Resources required:

Table.3.7 Apparatus/ component required

Sl No.	Apparatus/ Component	Specification Range/ Rating	Nos/Quantity
1.	Circuit Kit for SCR forced commutation kits		1
2.	DC Supply	12V (Or as available that matches with the SCR)	
3.	Load resistor	50Ω, 2A (or available along with the kit)	1
4.	Unearthed CRO	20MHZ	1
5.	Connecting probes		1 set
6.	Capacitor	0.47μF (or available along with the kit)	1

Theory:

An SCR can be turned on by the application of a gate signal to the gate. It is required to turn the SCR off for power control purposes. Already we know that the SCR can be turned off by bringing the SCR from the conduction state to the forward blocking state. It is done by either bringing the anode current below the holding current or by application of reverse voltage to such a value that the SCR will bring the reverse blocking state. The commutation of thyristor or SCR is the process by which the device is turned off. It is also known that it is impossible to make the SCR off itself while it is in conducting state. The commutation is the process by which the transfer of current from one path to another path i.e one thyristor to another thyristor. To turn off the current, the thyristor current must be brought to zero. Basically, there are two methods of turning off thyristors. They are natural commutation and forced commutation.

(a) Natural Commutation
(b) Forced commutation

Natural commutation is a simple and widely used method. It uses the alternating or reversing nature of alternating voltage for transferring current from one thyristor to another thyristor. In every half cycle alternating current reaches zero or current zero. When alternating current passes through the natural zero a reverse voltage is simultaneously appeared across the device and the thyristor is immediately turned off. The natural commutation is also called Class F commutation.

In forced Commutation, an external circuit is required. It is mainly required in DC circuits. The external circuit used for commutation are called commutation circuits. The components that are used in the external circuits are called commutation components. With the help of the commutation circuit, a reverse voltage is developed across the device, and this voltage help in bringing the SCR forward current to zero. Under forced commutation, there are various techniques available. They are as follows.

(a) Class A Commutation circuit
(b) Class B Commutation circuit
(c) Class C Commutation circuit
(d) Class D Commutation circuit
(e) Class E Commutation circuit

Class A to E are used in DC and Class F is used in AC. Here, we will study the operation of Class A to C circuits.

(a) Class A Series Resonant Commutation Circuit

This type of commutation circuit is also called a resonant commutation circuit or load commutation circuit. The *LC* commutation components is used in series with the load. The circuit comprises of resistance R_L as load. The R_L may be used in series with *L* and *C* or R_L may be used in parallel with the *C*. Here we have presented only the circuit when R_L is in series with *C*. The circuit is shown in Figure.3.39(a). The requirement of the circuit is that it should essentially underdamped. When energized with DC, the circuit current is waveform corresponding to the respective is also shown in figures. The circuit current rises up to maximum and decays to zero when it reaches point A at which the SCR is turned off. This type of commutation is possible in DC circuits only.

The output voltage and current are expressed by (3.94) and (3.95) respectively, in which V_o is the initial capacitor voltage

$$V_c = V_S + e^{-\left(\frac{\pi R_L}{2\omega}\right)t}\left(V_S - V_0\right) \tag{3.94}$$

$$i(t) = e^{-\left(\frac{R_L}{2L}\right)t}\left[\frac{V_S - V_0}{\omega L}\sin\omega t\right] \tag{3.95}$$

(b) Class B-Shunt Resonant commutation circuit

The circuit diagram of this is shown in Figure.3.33 (b). The *LC* resonating circuit is connected across the SCR, not in series with the load. There are two thyristors (SCRs). They are main SCR (MT_1) and auxiliary SCR (*AT*). The polarity of the source voltage, V_S is shown in the figure. At this voltage, the capacitor is charged to V_S with the polarity shown near the left- and right-hand side plates. Now, MT_1 is forward-biased and turned on at t = 0. It is assumed that the output current (I_o) is constant and is passed through the load. Up to time t_1 the voltage across the capacitor, $v_c = V_S$, capacitor current, $i_c = 0$, $i_0 = I_o$, and main thyristor current, $i_{MT1} = I_o$. These are shown in Figure.3.34(b). To start the commutation of MT_1, the AT is triggered at $t = t_1$. When AT is turned on, a resonant current i_c start flowing from C through AT, through L, and finally to C.

This resonant current is given by, $$i_c = -V_S\sqrt{\frac{C}{L}}\sin\omega_0 t = -I_p \sin\omega_0 t \tag{3.71}$$

In this equation, the negative sign is inserted because the current i_c flows just opposite to that of the positive direction as shown in Figure.3.33(b). The capacitor voltage v_c is given by

$$v_c(t) = \frac{1}{C}\int i_c dt = V_S \cos\omega_0 t$$

After half cycle from t_1; $i_c = 0$, $v_c = -V_S$, and $i_{MT1} = I_o$. After angular distance π from t_1 i.e from t_1 to t_2, i_c tend to flow in the reverse direction and AT is turned off at t_2. With $v_c = -V_S$, the right-hand plate of the capacitor has a positive polarity and the resonant current i_c is taking the path through C, L, D, MT_1. Thus, i_c flows just opposite direction to the forward SCR current of MT_1. The net forward current through the MT_1 is $I_0 - i_c$ and it starts decreasing. Finally, when i_c reverses and becomes equal to I_o, the forward current reaches zero value and MT_1 will get turned off. It is to be noted that the peak value of reverse current (I_p) must be greater than I_o. This method is also called current commutation, Class B commutation or resonant commutation.

(c) Class C-Complimentary Symmetry Commutation Circuit

The circuit diagram for the Class-C commutation circuit is shown in Figure.3.39(c). In this circuit, there are two SCRs. One is the main SCR, *MT1* which is to be turned off, and the other is the auxiliary SCR, *AT*. *MT1* is connected in series with the load (R_1). *AT* is connected in parallel with the *MT1*. The polarity of voltages and the currents are shown in Figure.3.39(c). It is assumed that the capacitor is initially uncharged. The details of the circuit operation of this circuit is given in section 3.7

Circuit diagram: The circuit diagrams for Class A to C are shown in Figure.3.39

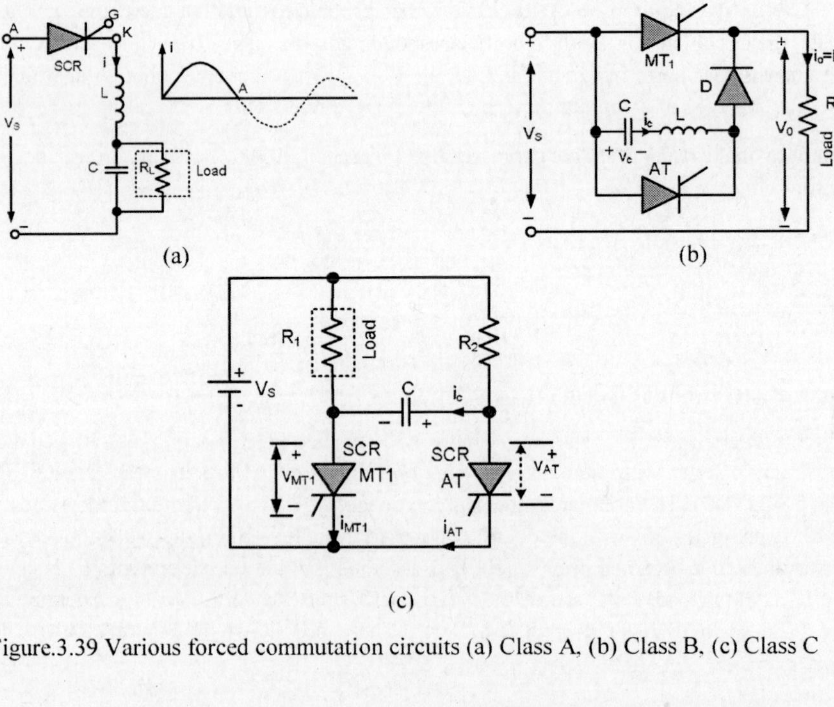

Figure.3.39 Various forced commutation circuits (a) Class A, (b) Class B, (c) Class C

Procedures:

Procedure for Class A forced commutation circuit

1. Connect the commutation components and other appliances as shown in Figure.3.39(a)
2. Connect the triggering circuit terminal to the gate (*G*) and cathode (*K*) terminals of the SCR.
3. Switch on the supply Vs and observe the waveform for voltage at the load, SCR and *C* by varying the frequency potentiometer
4. Change the value of *R, L,* and *C* and observe the waveform for voltage at the load, SCR and *C* by varying the frequency potentiometer
5. Repeat the step4
6. Plot the voltage waveforms

Procedure for Class B forced commutation circuit

1. Connect the commutation components and other appliances as shown in Figure.3.39(b).
2. Switch on the supply V_s and observe the output voltage and current across load resistance R_L
3. Connect the triggering circuit terminal to the gate (*G*) and cathode (*K*) terminals of the *MT* and observe the output voltage and current, and voltage across the *MT*.
4. Connect the triggering circuit terminal to the gate (*G*) and cathode (*K*) terminals of the *AT* and observe the output voltage and current, and the voltage across *MT* and *AT*.
5. Change the value of *R, L,* and *C* and repeat the step3 and step4. Observe the waveform for voltage at the load, SCR and *C* by varying the frequency potentiometer
6. Plot the voltage waveforms

Procedure for Class C forced commutation circuit

1. Connect the commutation components and other appliances as shown in Figure.3.39(c).
2. Switch on the supply *Vs* and observe the waveform for voltage at the load, SCR *MT1, AT,* R_2, and *C*.
3. Connect the triggering circuit terminal to the gate (*G*) and cathode (*K*) terminals of the SCR *MT1*. observe the waveform for voltage at the load, SCR *MT1, AT,* R_2, and *C*.
4. Connect the triggering circuit terminal to the gate (*G*) and cathode (*K*) terminals of the SCR *AT*. observe the waveform for voltage at the load, SCR *MT1, AT,* R_2, and *C*.
5. Change the value of *R, L,* and *C* and observe the waveform for voltage at the load, SCR and *C* by varying the frequency potentiometer
6. Plot the voltage waveforms

Result and Discussion:

Following the procedures mentioned above for different forced commutation circuits, the graphs are plotted and the plots are looked like Figure.3.40

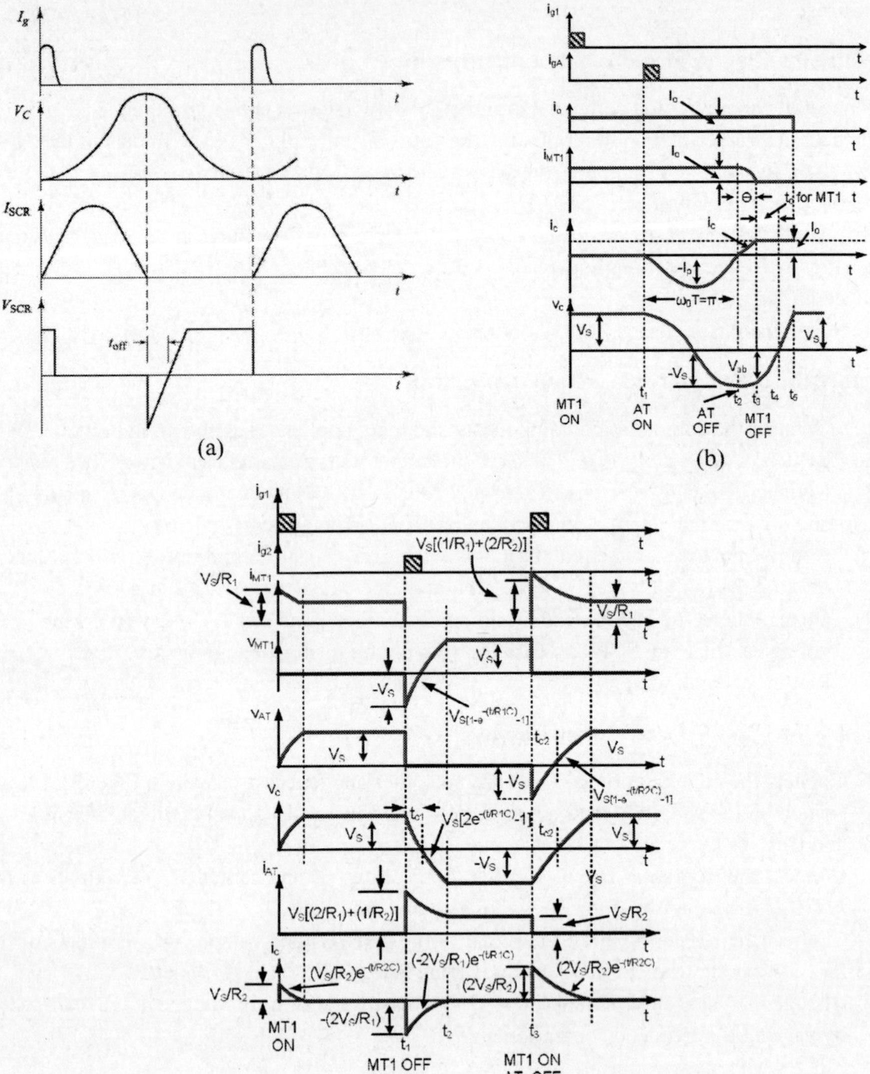

Figure.3.40 Voltage and current waveforms of various forced commutation circuits (a) waveform for Class A, (b) waveform for Class B, (c) waveform for Class C

Results: The operation of Class A, Class B, and Class C turn-off circuits have been performed.

Discussion:

1. From this experiment, we can differentiate between natural and forced commutation.
2. We have learned how to analyze the various waveform under the Class A, Class B, and Class C turn-off circuits

Experiment No.3.4

Title: Perform the operation of Class D, E, and F turn-off circuits.

Objectives:

To study the operation of Class D, E, and F commutation circuits

Resources required:

Table.3.8 Apparatus/ component required

Sl No.	Apparatus/ Component	Specification Range/ Rating	Nos/Quantity
7.	Circuit Kit for SCR forced commutation kits		1
8.	Transformer	230V/12 V (Or as available that match with the SCR)	1
9.	DC Supply	12V Or as available that match with the SCR)	
10.	Load resistor	50Ω, 2A (or available along with the kit)	1
11.	Unearthed CRO	20MHZ	1
12.	Connecting probes		1 set
13.	Capacitor	0.47µF (or available along with the kit)	1
14.	Pulse generator		1

Theory:

An SCR can be turned on by the application of a gate signal to the gate. It is required to turn the SCR off for power control purposes. Already we know that the SCR can be turned off by bringing the SCR from the conduction state to the forward blocking state. It is done by either bringing the anode current below the holding current or by application of reverse voltage to such a value that the SCR will bring the reverse blocking state. The commutation of thyristor or SCR is the process by which the device is turned off. It is also known that it is impossible to make the SCR off itself while it is in conducting state. The commutation is the process by which the transfer of current from one path to another path i.e one thyristor to another thyristor. To turn off the current, the thyristor current must be brought to zero. Basically, there are two methods of turning off thyristors. They are natural commutation and forced commutation.

Natural commutation is a simple and widely used method. It uses the alternating or reversing nature of alternating voltage for transferring current from one thyristor to another thyristor. In every half cycle alternating current reaches zero or current zero. When alternating current passes through the natural zero a reverse voltage is simultaneously appeared across the device and the thyristor is immediately turned off. The natural commutation is also called Class F commutation.

In forced Commutation, an external circuit is required. It is mainly required in DC circuits. The external circuit used for commutation are called commutation circuits. The components that are used in the external circuits are called commutation components. With the help of the commutation circuit, a reverse voltage is developed across the device, and this voltage help in bringing the SCR forward current to zero. Under forced commutation, there are various techniques available. They are as follows.

(a) Class A Commutation circuit
(b) Class B Commutation circuit
(c) Class C Commutation circuit
(d) Class D Commutation circuit
(e) Class E Commutation circuit

lass A to E are used in DC and Class F is used in AC. Here, we will study the operation of Class D to F circuits.

(a) Class D Commutation circuit

The class D commutation is also called auxiliary commutation or Impulse commutation. In this method to commutate a thyristor or SCR another thyristor or SCR is used. The SCR which is to be commutate is called the main SCR (*MT1*) and the other thyristor is called the auxiliary thyristor (*AT*). The circuit diagram is shown in Figure.3.41(a). A capacitor (*C*), inductor (*L*), and diode (*D*) are there in the circuit. The connection of these components is shown in Figure.3.41(a).

(b) Class E Commutation circuit

This is a method of commutation in which a reverse voltage is applied from an external source of voltage pulse to commutate an SCR or a thyristor. The external voltage is called the auxiliary supply. A typical circuit for external pulse commutation is shown in Figure.2.41(b). The commutating pulse is applied from the auxiliary supply via the pulse transformer. In this case, it is the pulse generator. The pulse transformer is specially designed so that there should be tight coupling between the primary and secondary. For commutation of the SCR *T1,* a pulse having a duration equal to or greater than the turn-off time of the SCR.

On Triggering SCR *T1,* the latter is in conducting mode and the current is flowing the load as well as the secondary of the pulse transformer. After the application of voltage V_P of negative polarity i.e - V_p, this negative voltage is appeared across the SCR *T1* and is turned off the same. The induced pulse is of high frequency and hence the *C* offers almost zero impedance. After SCR *T1* is turned off, the load current decreases to zero. The various voltage and current waveform of the Class-E commutation is shown in Figure.3.41(b).

(a) Class F- Line or Natural commutation

This technique is also called natural commutation. If the supply voltage is alternating, the load current will flow during the positive half-cycle only. The device is turned off during the negative half of the alternating voltage. The time duration of a half cycle of the applied alternating voltage should be greater than the turn-off time of the SCR. The maximum frequency of operation of this circuit is decided by the turn-off time of the SCR. The circuit for line commutation of natural commutation or Class F commutation is shown in Figure.3.41(c). The corresponding voltage and current waveforms are shown in Figure.3.41(c).

Circuit diagram:

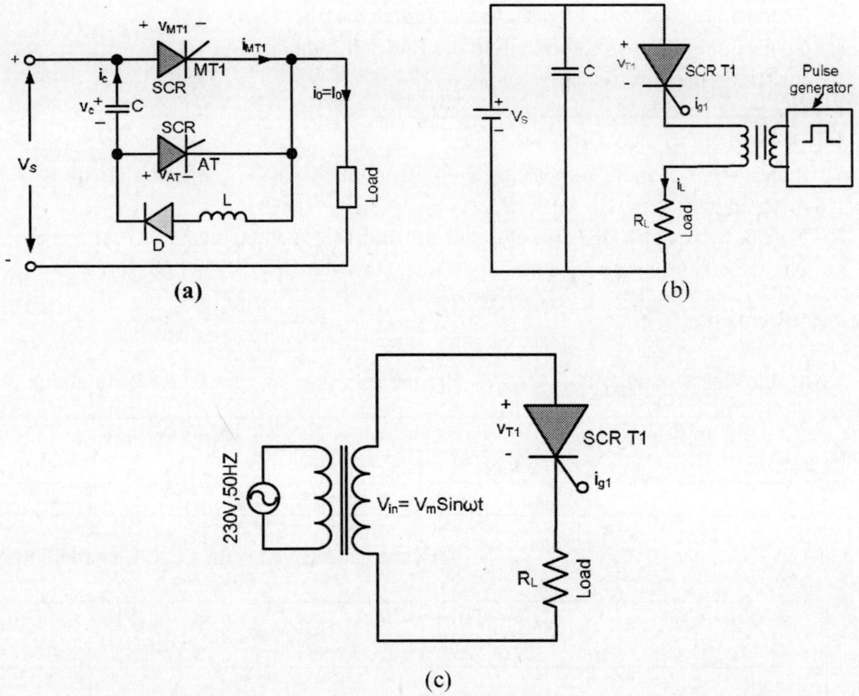

Figure.3.41 Various commutation circuits (a) Class D (b) Class E, (c) Class F

Procedure:

Procedure for Class D forced commutation circuit

1. Connect the components shown in **Figure.3.41(a).**
2. Assume that the load current is constant and the capacitors is charged to V_S.
3. Connect the Firing circuit of the SCRs.
4. Turn on *AT*.
5. After some time, the *AT* will turn off. This is observed by connecting CRO across the *AT*.
6. Turn on the SCR *MT1*. Now observe the SCR voltage and current.
7. Switch on the triggering circuit and note down the voltage waveform for various values of duty cycles.
8. Draw the output waveforms (It will be like Figure.3.42(a))

Procedure for Class E forced commutation circuit

1. Connect the components as per the circuit diagram shown in Figure.3.41(b).
2. Connect the firing circuit output to the gate and cathode of SCR.
3. Switch on the power supply.
4. Switch on the pulse generator
5. Observe the voltage waveform across the load, SCR, and capacitor by varying the potentiometer.
6. Draw the waveforms as shown. (It will be like Figure.3.42(b))

Procedure for Class F forced commutation circuit

1. Connect the components as shown in **Figure.3.41(c).**
2. Connect the firing circuit output to the gate and cathode of SCR.
3. Switch on the power supply.
4. Observe the voltage and output current using CRO.
5. Observe the voltage across the SCR.
6. Repeat steps 3 and 4 by varying the potentiometer of the triggering circuit to change the firing angle.
7. Note down the turn-on (T_{ON}) and turn-off time (T_{OFF}) and tabulate them in Table.
8. Plot the output voltage and output current. (It will be like Figure.3.42(c))

Result and Discussion:

Table.3.9 Values of T_{ON}, T_{OFF}, V_{in}, V_O for various values of α for Class D commutation

Sl No.	Firing angle (α)	Time in (ms)		Input voltage (V_{in})	Output voltage (V_O)
		T_{ON}	T_{OFF}		

Table.3.10 Values of T_{ON}, T_{OFF}, V_{in}, V_O for various values of α for Class E commutation

Sl No.	Firing angle (α)	Time in (ms)		Input voltage (V_{in})	Output voltage (V_O)
		T_{ON}	T_{OFF}		

Table.3.11 Values of T_{ON}, T_{OFF}, V_{in}, V_O for various values of α for Class F commutation

Sl No.	Firing angle (α)	Time in (ms)		Input voltage (V_{in})	Output voltage (V_O)
		T_{ON}	T_{OFF}		

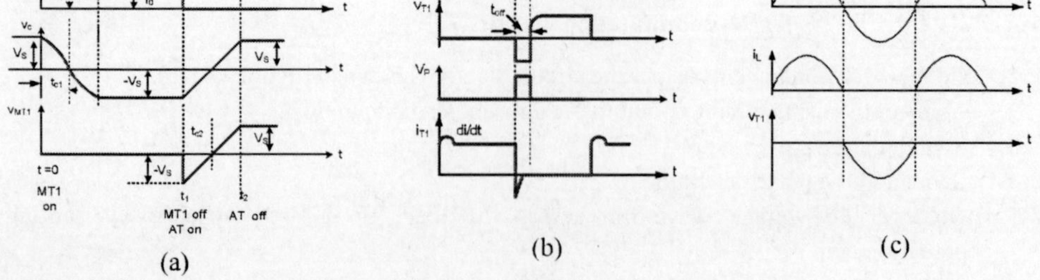

Figure.3.42 Voltage and current waveforms of various forced commutation circuits (a) waveform for Class D, (b) waveform for Class E, (c) waveform for Class F

Results: The operation of Class D, Class E, and Class F turn-off circuits has been performed.

Discussion:

1. From this experiment, we can differentiate between natural and forced commutation.
2. We can analyse the various waveform under the Class D, Class E, and Class F turn-off circuits

Know More

(a) The DIAC is used for triggering TRIC. A triggering circuit for the same is shown in Figure.3.43

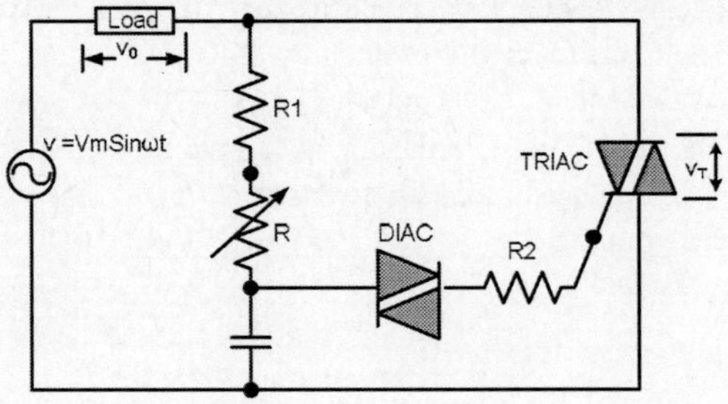

Figure.3.43 Triggering circuit for TRIAC using DIAC

(b) The generation of gating signals for thyristors of DC to AC converters have some requirement like (i) zero crossing detector, (ii) phase shifting signals, (ii) pulse shaping, and (iii) pulse isolation. An integrated circuit (IC) gate drive integrate most of the functions required to drive one high side power and low side device in a compact high performance package with minimum power dissipation. These ICs must have some protection functions for overload and faulted condition. There are various drive IC for converters. Example is MOS gated driver, ICE2AS01, etc.

(c) High voltage ICs are also available for motor drive. These are known as power conversion processors (PCPs). The architecture for the IC family is divided into (i) two-level power conversion processing, (ii) single-level conversion processing, (iii) mixed-mode power conversion processing.

(d) Microprocessors/microcontrollers are used to control the firing angle.

References and suggested readings

1. Ramamoorty M., An Introduction to Thyristors and their Applications, East-West Press Pvt. Ltd., New Delhi, ISBN: 8185336679.
2. Sugandhi, Rajendra Kumar and Sugandhi, Krishna Kumar, Thyristors: Theory and Applications, New Age International (P) ltd. Publishers, New Delhi, ISBN: 978-0-85226-852-0.
3. Bhattacharya, S.K., Fundamentals of Power Electronics, Vikas Publishing House Pvt. Ltd. Noida. ISBN: 978-8125918530.
4. Jain & Alok, Power Electronics and its Applications, Penram International Publishing (India) Pvt. Ltd, Mumbai, ISBN: 978-8187972228.
5. Rashid, Muhammad, Power Electronics Circuits Devices and Applications, Pearson Education India, Noida, ISBN: 978-0133125900.
6. Singh, M. D. and Khanchandani, K.B., Power Electronics, Tata McGraw Hill Publishing Co. Ltd, New Delhi, 2008 ISBN: 9780070583894.
7. Zbar, Paul B., Industrial Electronics: A Text –Lab Manual, McGraw Hill Publishing Co. Ltd., New Delhi, ISBN: 978-0070728226.
8. Grafham D.R., SCR Manual, General Electric Co., ISBN: 978-0137967711.

9. Mohan Ned, Undeland Tore.M., Robins William P., Power Electronics – Converters, Applications and Design, John Wiley & Sons. Inc,
10. https://www.google.com/search?q=picture+of+heat+sink+for+power+device.
11. Rashid, M. H. (2014). Power Electronics: Devices, Circuits, and Applications: International Edition. United Kingdom: Pearson Education.
12. Undeland, T. M., Mohan, N., Robbins, W. P. (2003). Power Electronics: Converters, Applications, and Design. India: John Wiley & Sons.
13. Fewson, D. (1998). Introduction to Power Electronics. United Kingdom: Elsevier Science.
14. Williams, B. W. (1992). Power electronics: devices, drivers, applications, and passive components. United Kingdom: McGraw-Hill.
15. Robbins, W. P., Mohan, N., Undeland, T. M. (2003). Power Electronics: Converters, Applications, and Design. India: John Wiley & Sons.
16. https://www.youtube.com/watch?v=d4sbVc-r7I4
17. https://www.youtube.com/watch?v=ekFmK3xtbK8
18. https://www.youtube.com/watch?v=9i61RREAWjI

4 Phase Controlled Rectifiers

UNIT SPECIFICS

This unit covers the following aspects:

- *Introduction to Phase control and phase-controlled rectifiers.*
- *Definitions of a few basic terms like firing angle or delay angle, extinction angle, and conduction angles.*
- *Performance parameters related to phase-controlled rectifier.*
- *Single phase half wave-controlled rectifier with different types of loads.*
- *Single phase full wave (or two pulse) controlled rectifier with different types of loads.*
- *Single phase full wave bridge rectifiers with different types of loads.*
- *Single-phase half-controlled bridge rectifier with different types of loads.*
- *Common anode and common cathode for three-phase controlled rectifier.*
- *Three phase half wave-controlled rectifier with different types of loads.*
- *Basics of three-phase fully controlled bridge converter.*
- *Three-phase fully controlled bridge rectifiers with different types of loads.*
- *Three-phase half-controlled bridge converters with different types of loads.*
- *Phase-controlled rectifier with input impedance.*
- *Single phase fully controlled bridge rectifier considering source inductance.*
- *Three phase fully controlled bridge rectifier considering source inductance.*

The concept of phase control and phase control rectifiers of single-phase and three-phase with various types of loads with and with considering source impedance are presented and analyzed using SCR for generating further curiosity and creativity as well as improving problem-solving capacity with some numerical problems.

Besides giving a large number of multiple-choice questions as well as questions of short and long answer types marked in two categories following the lower and higher order of Bloom's taxonomy, assignments through several numerical problems, a list of references, and suggested readings are given in the unit so that one can go through them for practice.

After the detailed analysis, based on the content, there is a "Know More" section appended. This section has been designed to supplement additional information and higher learning skills on the topic.

RATIONALE

This fundamental unit on phase-controlled rectifiers using the basic device SCR helps students to get a primary idea about the phase-controlled rectifiers being used in various power electronics circuits involving SCR.

The various single-phase and three-phase phase-controlled rectifier circuits with different types of loads are presented in this unit. The analysis of these circuits is depicted.

The physics behind various circuits for phase-controlled rectifiers using SCR are discussed at length to develop the basic idea about these circuits.

Some related problems are pointed out after each section with their solutions which can help further for getting a clear idea of the concerned topics. The mathematics behind the phase-controlled rectifiers will certainly help students with numerical problem-solving.

As a student in the field of electrical engineering, this unit on phase-controlled rectifiers helps students to grasp the basic knowledge of controlled rectifiers.

PRE-REQUISITES

ESC101: Basic Electrical Engineering

UNIT OUTCOMES

After completion of Unit-4 students will be able to:

U4-O1: Identify the types of phase-controlled rectifier circuits.
U4-O2: Explain the working of various phase-controlled rectifier circuits using SCR.
U4-O3: Analyse the working of various phase-controlled rectifier circuits for different loads.
U4-O4: Identify the application of various phase-controlled rectifier circuits.

Unit-4 Outcomes	EXPECTED MAPPING WITH COURSE OUTCOMES (1- Weak Correlation; 2- Medium correlation; 3- Strong Correlation)				
	CO-1	CO-2	CO-3	CO-4	CO-5
U4-O1	...	2	1	1	2
U4-O2	1	3	3	3	1
U4-O3	...	3	3	3	3
U4-O4	2	...	1	2	3

4.1 INTRODUCTION

The process of conversion of alternating voltage or current to a direct voltage or current is called rectification. The circuits used for rectification are called rectifier circuits. Prior to now, either motor generator sets or mercury-arc rectifiers, or thyratrons were used to convert ac power to dc power. These circuits contain switching devices. The most common switching devices are diodes, thyristors, and power transistors. The diode is an uncontrolled device. The other power electronics switches are controlled devices. Accordingly, the rectifier circuits are also classified as uncontrolled, half-controlled, and fully controlled. In uncontrolled rectifier circuits, only the diode is used as a switching element. The amplitude of the DC output of the uncontrolled rectifier is fixed and it is decided by the amplitude of the AC voltage applied. In controlled rectifiers, only thyristors are used. The output DC voltage is decided by the amplitude of the AC input and the instant at which the thyristors are fired. Thus, the output voltage is controlled. The half-controlled rectifiers contain both diodes and thyristors. Since both diodes and thyristors are present in the circuit, there is limited control over the output DC.

In the case of uncontrolled and half-controlled rectifiers, the power flows from the AC system to the DC load only. Hence, these converters are unidirectional. But, in the case of a fully controlled converter, the power can flow from both sides. When power flows from the AC system to the DC load it is called a fully controlled rectifier. On the other hand, when power flows from the DC side to the AC side, it is called an inverter. Thus, fully controlled converters are bidirectional. In this unit, only the rectifier mode of operation of fully controlled converters will be discussed.

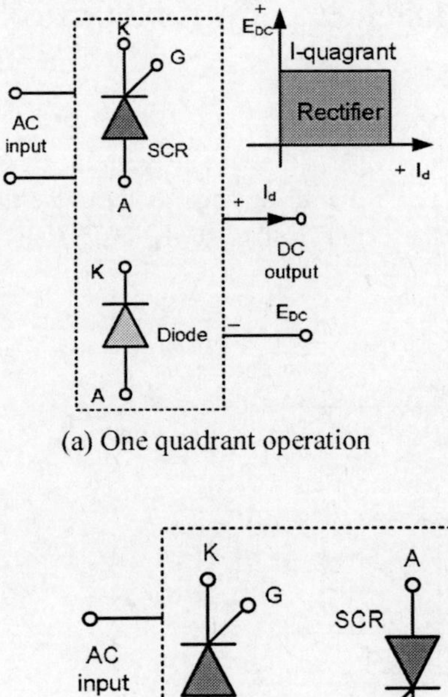

(a) One quadrant operation

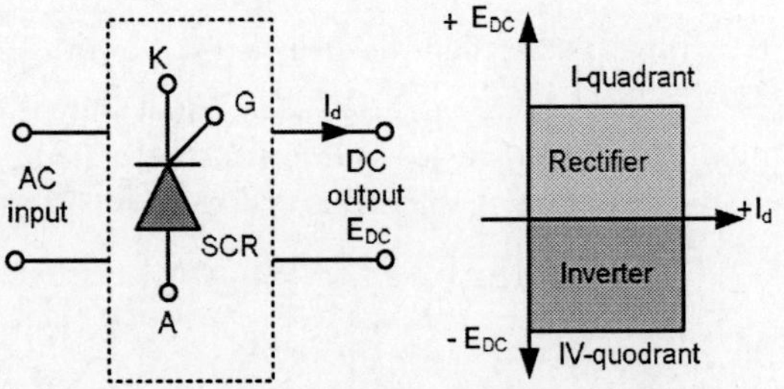

(b) Two-quadrant operation

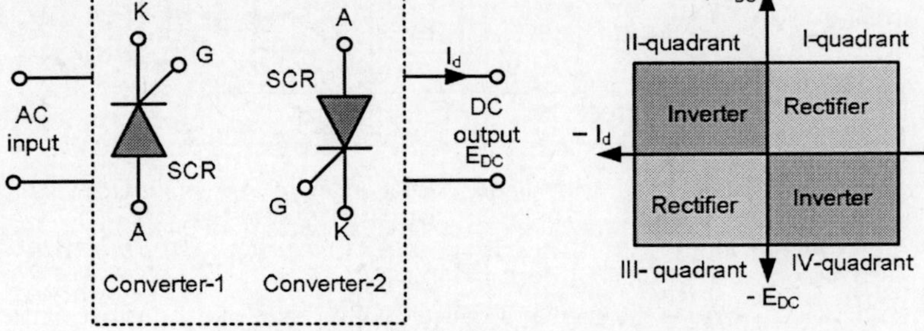

(c) Four quadrant converter operations

Figure.4.1 Various phase-controlled converters

The converters in which thyristors are present are called phase-controlled converters (PCC). Phase-controlled rectifiers (PCRs) are better than diode rectifiers because they can control the output voltage.

Uncontrolled rectifiers are diode rectifiers. These diodes transform into PCRs when they are switched using thyristors or silicon-controlled rectifiers (SCRs). Diodes do not offer any form of control over the voltage that is produced by the device, in contrast to thyristors, which can have their output voltage adjusted by changing the firing angle. A phase control thyristor is turned on by sending a pulse to its gate terminal, and it is turned off by line communication. In the event of a heavy inductive load, the rectifier deactivates it by firing another thyristor during the negative half-cycle of input voltage. As PCRs do not require a commutation circuit, they are simple, less expensive, and widely used in industries that require controlled dc power.

The PCC provides one-quadrant, two-quadrant, and four-quadrant operations on the DC side. Diagrammatically they are shown in Figure.4.1(a)-(c).

Thus, rectifiers are of two types. They are uncontrolled and controlled (phase-controlled) converters. Diodes are used in uncontrolled rectifiers. On the other hand, thyristors or SCRs and in some cases combinations of thyristors and diodes are used in phase-controlled rectifiers. Depending on the type of input voltage phase controlled rectifies are divided into two. They are (i) single phase and (ii) three phase. Each type is again classified into (i) semiconverter, (ii) full converter, and (iii) dual converter. Semiconverters are one quadrant and it has one polarity of the output voltage and current. Full converters are two-quadrant converters. The output voltage may be either positive or negative polarity. Output current is of positive polarity. Dual converters are four-quadrant converters. Both output voltage and current are either positive or negative.

4.2 DEFINITIONS OF FIRING ANGLE, EXTINCTION ANGLE, AND CONDUCTION ANGLE

(a) Firing angle or delay angle

The firing angle or delay angle (α) of a thyristor is an angular difference between the time it is triggered and the time it would conduct as a diode. As a result, the firing angle is the angle that is measured from the instant that produces the highest average output voltage to the instant that it is triggered.

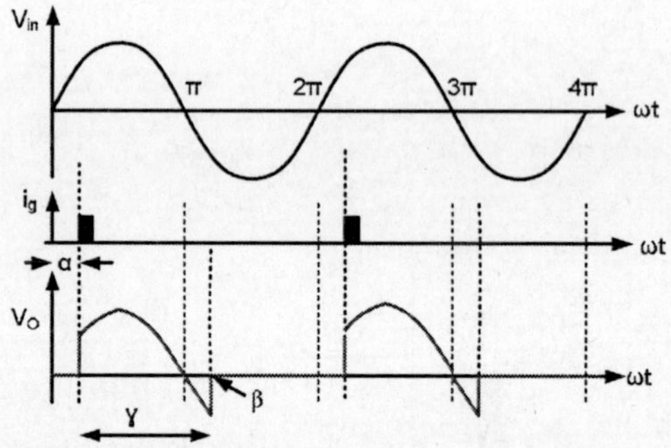

Figure.4.2 Figure showing firing angle, extinction angle, and conduction angle

(b) Extinction angle

The extinction angle (γ) is the angular distance from the start of the positive half cycle of the input supply to the instant the load current drops to zero. At $\omega t = \pi$, the supply voltage is negative, but the SCR continues to conduct because of the inductive load, the load current is greater than the holding current. The SCR will stop conducting when the load current is less than the holding current. So this angle depends on the load inductance value.

(c) Conduction angle

Conduction angle (β) is defined as the angle that measures the SCR period of conduction (how long it conducts in one cycle of time), from the waveforms (in the case of RL load) the conduction angle is given by

$$\beta = \gamma - \alpha \tag{4.1}$$

This angle depends on the α and the load impedance angle φ. For purely resistive load, the conduction angle $\beta = (\pi - \alpha)$.

All the above angles are shown in Figure.4.2.

4.3 PHASE CONTROL: FIRING ANGLE CONTROL

In the phase control rectifiers, unidirectional switches are used to convert AC into DC. The amplitude of output voltage depends on the time at which the switches are triggered. A simple circuit consisting of a switch, AC source, and load resistance is shown in Figure.4.3 (a). A source inductance L_S is also shown in the figure, which is generally neglected in most of the analysis. When the switch S is closed it conducts current in the direction as shown by the arrow. These switches are power electronic devices. They may be pulse triggered for which current pulses are required to turn on. Examples of pulsed-triggered devices are SCR, GTO, and MCTs. Some of the devices such as BJT, MOSFET, and IGBT are level-triggered for which voltage pulses are needed to turn on. The SCR is most commonly used for controlled rectifiers.

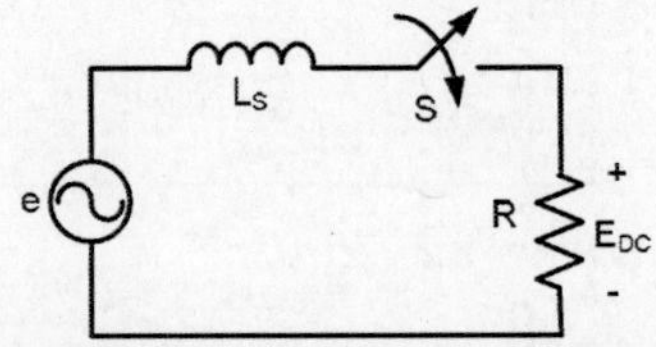

Figure.4.3 (a) Representation of a half-wave rectifier using a unidirectional switch

The phase control rectifier uses three methods of control to control the output. They are

(a) Firing angle control or phase angle control.
(b) Conduction angle control.
(c) Pulse wide modulation (PWM) control.

In firing angle control, the SCR is turned on by the application of current pulses (gate pulses) at a particular angle with respect to the input voltage. The firing angle α is measured from a reference point. Generally, the reference point is the zero-crossing point of the input AC waveform. The α is 0^0 means the zero-crossing point (reference point) at which anode voltage first reached positive in every cycle at which the firing pulse is applied. The α is 30^0 means the firing pulse is applied at 30^0 from the reference in every cycle. This method of controlling the output voltage is called phase angle control. Figure 4.3(b) shows the waveform of phase angle control.

In the conduction angle control or simple β control, the conduction period is controlled. The rising edge of the control pulse coincides with the beginning of the input AC waveform. The falling edge lies at an angle $\beta = (\pi - \alpha)$. In the triggering circuit, two short pulses are generated one is to turn on and the other is to turn off the SCR. The waveform for such control is shown in Figure.4.3(c).

In the PWM technique, the pulse is symmetrically positioned at the peak of the positive and negative half cycle. Here, pulse width (δ) is the control parameter. Like the conduction angle, the control pulse comprises two short pulses. For multiple pulse width modulation, there any finite numbers of pulses (say p) and in this case p and δ are the control parameters. The waveform for such control is shown in Figure.4.3(d).

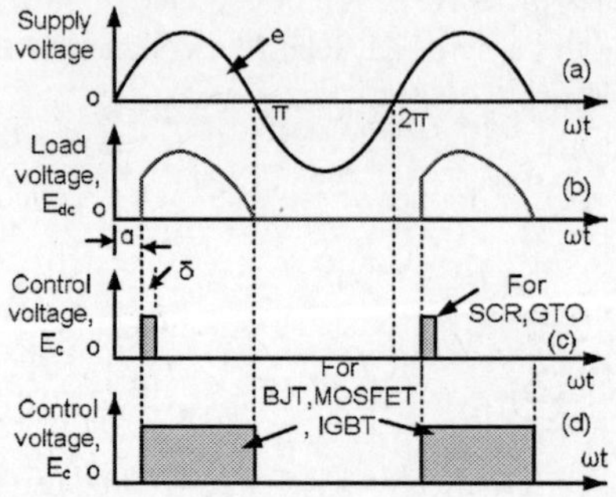

Figure.4.3 (b) Waveform for phase control

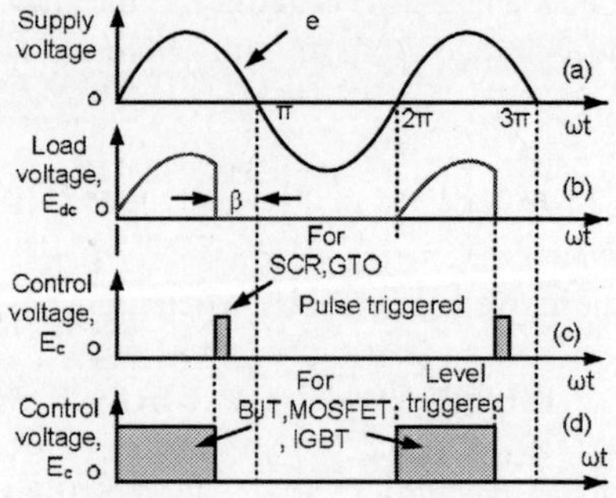

Figure.4.3 (c) Waveform for conduction angle control

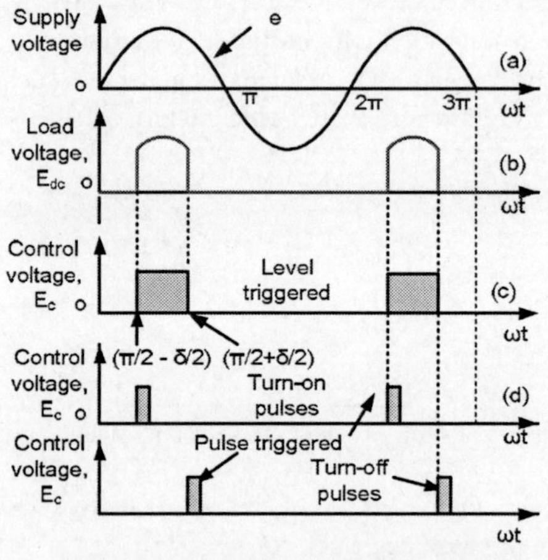

Figure.4.3(d) Waveform PWM control

4.4 PERFORMANCE PARAMETERS IN RECTIFIER CIRCUITS

There are various types of rectifier circuits. The performance of these is evaluated in terms of some parameters. They are given below.

(a) The **average value of the output voltage**. It is denoted by E_{dc} **or** V_{dc} or sometimes V_o.

(b) The **average value of output current** or load current. It is normally denoted by I_{dc} **or** I_o.

(c) The **DC output power**, denoted by P_{dc} is given by

$$P_{dc} = V_{dc} I_{dc} \left(\text{or } V_o I_o \text{ or } E_{dc} I_{dc}\right) \tag{4.2}$$

(d) The **root mean square (RMS) value of output voltage.** It is denoted by E_{rms} **or** V_{rms}

(e) The **rms value of output current** (I_{rms})

(f) The **output AC power.** It is denoted by P_{ac} and is given by

$$P_{ac} = E_{rms}.I_{rms} \text{ or } V_{rms}.I_{rms} \tag{4.3}$$

(g) The **efficiency of rectification**. It is denoted by η and given by

$$\eta = \frac{P_{dc}}{P_{ac}} \tag{4.4}$$

For an ideal rectifier, efficiency η is 100%.

(h) The output voltage is composed of two components. They are (i) **dc component**, (ii) **ac component, or ripple.**

(i) The **effective or rms value of the ac component** of the output voltage. It is normally denoted by ***E****ac* **or** ***V****ac* and expressed as

$$V_{ac} \text{ or } E_{ac} = \sqrt{E_{rms}^{\;2} - E_{dc}^{\;2}} \text{ or } \sqrt{V_{rms}^{\;2} - V_{dc}^{\;2}} \tag{4.5}$$

(j) The **form factor.** It is the measure of the shape of the output voltage. It is denoted by **FF** and given by

$$FF = \frac{V_{rms}(\text{or } E_{rms})}{V_{dc}(\text{or } E_{dc})} \tag{4.6}$$

(k) **Ripple factor.** It is the measure of the content of the ripple in the DC output. It is normally denoted by ***RF***, and given by

$$RF = \frac{V_{ac}(\text{or } E_{ac})}{V_{dc}(\text{or } E_{dc})} = \sqrt{\left(\frac{V_{rms}}{V_{dc}}\right)^2 - 1} = \sqrt{FF^2 - 1} \tag{4.7}$$

For an ideal rectifier, *Vac* is equal to zero. Hence, $RF = 0$.

(l) The **transformer utilization factor**. It is denoted normally by ***TUF***, and defined as follows

$$TUF = \frac{P_{dc}}{V_s I_s (\text{or } E_s I_s)}, \tag{4.8}$$

where *Vs* and *Is* are the RMS value of voltage and current at the secondary of the transformer at the input. For an ideal rectifier, *TUF* is equal to unity.

(m) The displacement factor. It is the cosine of the angle (Φ) between the fundamental component of input voltage and current. It is also known as **displacement power factor (*DPF*).** It is denoted by ***DF*** and given by

$$\textbf{\textit{DPF}} \textbf{ or } \textbf{\textit{DF}} = \text{Cos}\Phi \tag{4.9}$$

(n) The **harmonic factor (*HF*)** of the output current. *HF* measures the distortion of the waveform. It is called total harmonic distortion **(*THD*).** It is expressed as

$$THD \text{ or } HF = \left(\frac{I_s^{\;2} - I_{s1}^{\;2}}{I_{s1}^{\;2}}\right)^{1/2} = \left[\left(\frac{I_s}{I_{s1}}\right)^2 - 1\right]^{1/2} \tag{4.10}$$

where I_{s1} is the fundamental component of input current I_s. Both I_s and I_{s1} are expressed in rms value. For an ideal rectifier, *THD* is equal to zero.

(o) The **power factor** of a rectifier circuit is given by ***PF*** and is expressed as

$$PF = \frac{V_s . I_{s1}}{V_s . I_s}\cos\phi = \frac{I_{s1}}{I_s}\cos\phi \tag{4.11}$$

In the case of purely sinusoidal input current (i_s), $I_s = I_{s1,}$ and hence, the power factor *PF* is equal to the distortion factor. For an ideal rectifier, *PF* is equal to unity.

(p) The **crest factor**. It is normally denoted by ***CF***. It is a measure of peak input current (denoted by $I_{s(peak)}$ in caparison with input current I_s. The CF is expressed as

$$CF = \frac{I_{s(peak)}}{I_s} \tag{4.12}$$

4.5. SINGLE PHASE HALF WAVE CONTROLLED RECTIFIER

There is only one SCR used in the circuit of a single-phase half-wave-controlled rectifier. It is placed between the connected load and the ac source. Controlled rectifier efficiency is extremely sensitive to the nature and settings of the connected load circuit.

4.5.1. Single Phase Half Wave Controlled Rectifier with Resistive (R) Load

The schematic representation of a single-phase half-wave converter with resistive (R) load can be seen in Figure.4.4 (a). The trigger circuit is not portrayed in the diagram. The supply voltage or transformer secondary voltage, $e = E_m \sin\omega t$ supplies the power to the circuit. It is assumed that the forward and reverse blocking ratings of the thyristors are never exceeded by the peak supply voltage. Figure.4.4 (b) depicts the voltage and current waveforms for the circuit shown in Figure.4.4(a). Throughout the Unit, it is assumed that SCR is ideal.

During the positive half-cycle of the supply voltage, the anode of the SCR is positive with respect (w.r.t) to its cathode. The SCR prevents the load current from flowing forward until it is triggered by an appropriate gate pulse. The voltage drop across the SCR is assumed to be zero. Thus, the full supply voltage is applied to the load when the thyristor is fired at an angle of α. Consequently, the load is connected directly to the ac supply. With a source that has no reactance and a load that only has resistance, the waveform of the current after the thyristor is turned on will be the same as the wave of the applied voltage, and the current will depend on the amplitude of the voltage and the value of the load resistance R. The load current will continue to flow until it is commutated at $\omega t = \pi$ by the inversion of the supply voltage, as shown in Figure.4.4 (b). The thyristor conducts during $(\pi - \alpha = \beta)$ is called the conduction angle. By varying the firing angle α, the output voltage can be regulated. During the negative half-cycle, the thyristor prevents current from flowing through the load, so load R is not supplied with power.

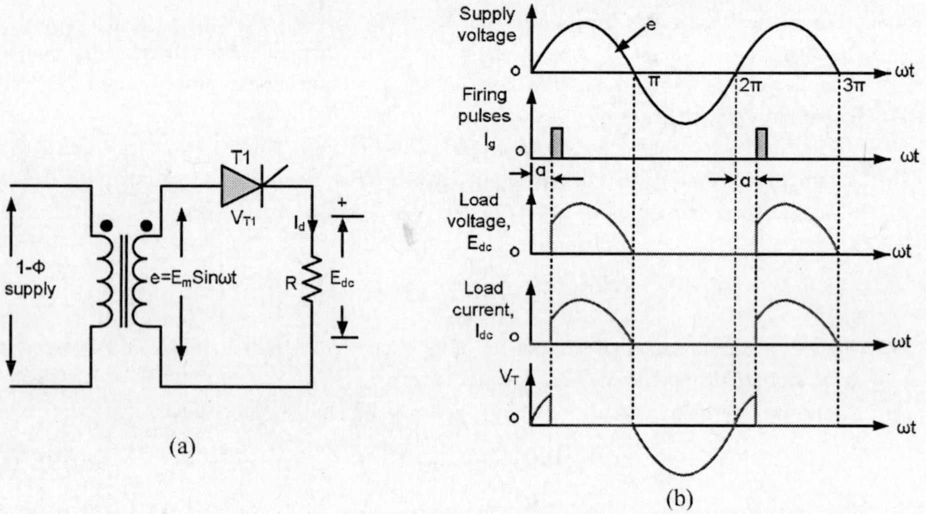

Figure.4.4 Single-phase half-wave rectifier with R load (a) Circuit, (b) waveform

Following are the steps for deriving the average load-voltage value

$$E_{dc} = \frac{1}{2\pi}\int_{\alpha}^{\pi} E_m \sin\omega t \, d(\omega t) \tag{4.13}$$

Where E_m represents the maximum ac input voltage.

$$\begin{aligned} E_{dc} &= \frac{1}{2\pi} E_m [-\cos\omega t]_{\alpha}^{\pi} \\ &= \frac{E_m}{2\pi}(1+\cos\alpha) \end{aligned} \tag{4.14a}$$

The average output voltage becomes maximum when $\alpha = 0^0$. The maximum value of output voltage is given by

$$E_{dm} = \frac{E_m}{2\pi}(1+\cos\alpha) = \frac{E_m}{2\pi}(1+1) = \frac{E_m}{\pi} \quad (4.14b)$$

The average current drawn by the load can be expressed as follows when using the R load:

$$I_d = \frac{E_m}{2\pi R}(1+\cos\alpha) \quad (4.15)$$

The RMS load voltage is calculated as follows for a firing angle of α:

$$E_{rms} = \left[\frac{1}{2\pi}\int_\alpha^\pi (E_m \sin\omega t)^2\, d(\omega t)\right]^{1/2} = \left[\frac{E_m^2}{2\pi}\int_\alpha^\pi \sin^2\omega t\, d(\omega t)\right]^{1/2}$$

$$= \left[\frac{1}{2\pi}\int_\alpha^\pi \left(\frac{1-\cos 2\omega t}{2}\right) d(\omega t)\right]^{1/2} = E_m\left[\frac{1}{4\pi}\left(\omega t - \frac{\sin 2\omega t}{2}\right)_\alpha^\pi\right]^{1/2}$$

$$\therefore E_{rms} = E_m\left[\frac{\pi-\alpha}{4\pi} + \frac{\sin 2\alpha}{8\pi}\right]^{1/2} \quad (4.16)$$

$$\text{Or} \quad E_{rms} = \frac{E_m}{2\sqrt{\pi}}\left[(\pi-\alpha) + \frac{1}{2}\sin 2\alpha\right]^{\frac{1}{2}} \quad (4.17)$$

The RMS value of the output current is given by

$$I_{rms} = \frac{E_{rms}}{R} = \frac{E_m}{2R\sqrt{\pi}}\left[(\pi-\alpha) + \frac{1}{2}\sin 2\alpha\right]^{\frac{1}{2}} \quad (4.18)$$

The power delivered to the load is given by the multiplication of RMS current and RMS voltage, and given by

$$P_{rms} = E_{rms} I_{rms} = \frac{E_{rms}^2}{R} = I_{rms}^2 R \quad (4.19)$$

The input voltampere is the multiplication of the rms value of the source voltage and the total rms value of the line current and is given by

$$\text{Input volt ampere} = E_s I_{rms} = \frac{\sqrt{2}E_s^2}{2R\sqrt{\pi}}\left[(\pi-\alpha) + \frac{1}{2}\sin 2\alpha\right]^{\frac{1}{2}} \quad (4.20)$$

The input power factor is given by

$$\text{Input power factor} = \frac{\text{Power delivered to the load}}{\text{Input VA}} = \frac{E_{rms} I_{rms}}{E_s I_{rms}} = \frac{E_{rms}}{E_s} \quad (4.21)$$

Also, the input power factor can be calculated by putting the value of E_{rms} from (4.17) and is given by

$$\text{Input power factor} = \frac{1}{\sqrt{2\pi}}\left[(\pi-\alpha) + \frac{1}{2}\sin 2\alpha\right]^{\frac{1}{2}} \quad (4.22)$$

4.5.2. Single Phase Half Wave Controlled Rectifier with Inductive (RL) Load

In Figure.4.5 (a), an inductive load is shown connected to a single-phase half-wave-controlled rectifier. Figure.4.5 (b) illustrates the waveforms of both the voltage and the current.

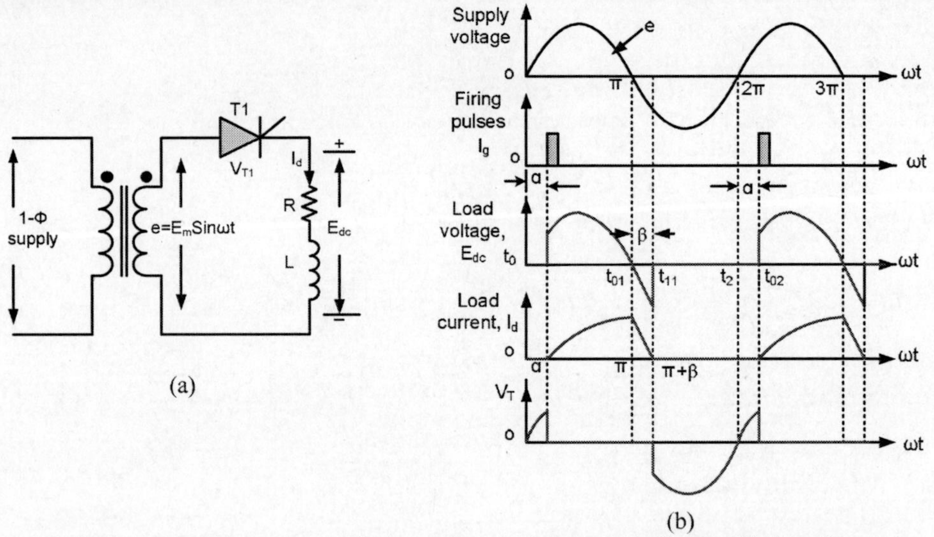

Figure.4.5 Single phase half wave-controlled rectifier with RL load (a) Circuit, (b) Waveforms

The load is considered to be extremely inductive. Under inductive loads, the circuit behaves slightly differently. After the thyristor is triggered at time t_{01}, the load current through the inductive load will rise steadily over some time. Thus, supply voltage appears across the load. The gradual increase in current is caused by the inductive load. The inductor stores energy from time t_{01} to t_1. At time t_1, the supply voltage is inverted, but the thyristor continues to conduct. This is because the current that is passing through the inductance cannot be made to equal zero. During a negative-voltage half-cycle, current flows until the energy that was stored in the inductance is dissipated in the load resistor, at which point some of the energy is returned to its source. Because of the energy that is stored in the inductor, the current will continue to flow up until instant t_{11}. At the time t_{11}, the supply voltage is negative, which causes the thyristor to turn off. This is because the load current is at zero at this point. When the pulse is applied once more at time t_{02}, the process starts all over again. Because of this, the effect of the inductive load lengthens the amount of time that the SCR is conducted. The half-wave circuit is not utilized very often because it is incapable of providing continuous load current and has a large ripple in the output voltage.

The average voltage across the load is

$$E_{dc} = \frac{1}{2\pi}\int_{\alpha}^{\pi+\alpha} E_m \sin\omega t \, d(\omega t) \tag{4.23}$$

Here, it has been assumed that the SCR operates for a duration α during the negative half cycle.

$$\begin{aligned} E_{dc} &= \frac{E_m}{2\pi}[-\cos\omega t]_{\alpha}^{\pi+\alpha} \\ &= \frac{E_m}{\pi}\cos\alpha \end{aligned} \tag{4.24}$$

In terms of conduction angle β, we can write equations (4.5) and (4.12) as follows.

$$E_{dc} = \frac{1}{2\pi}\int_{\alpha}^{\beta} E_m \sin\omega t \, d(\omega t) = \frac{E_m}{2\pi}(\cos\alpha - \cos\beta) \tag{4.25}$$

The average load current is given by

$$I_0 \ or \ I_{dc} = \frac{E_{dc}}{R} = \frac{E_m}{2\pi R}(\cos\alpha - \cos\beta) \tag{4.26}$$

The RMS value of load voltage is given by

$$E_{rms} = \left[\frac{1}{2\pi}\int_{\alpha}^{\beta} E_m^{\,2}\sin^2\omega t.d(\omega t)\right]^{\frac{1}{2}} = \frac{E_m}{2\sqrt{\pi}}\left[(\beta-\alpha) - \frac{1}{2}\{\sin 2\beta - \sin 2\alpha\}\right]^{\frac{1}{2}} \tag{4.27}$$

It can be seen from the above equations that in the case of an inductive load, the average load voltage is decreased with an increase in firing angle. This is because SCR operates in a negative cycle.

4.5.3. Single Phase Half Wave Controlled Rectifier with freewheeling diode

Many circuits, especially uncontrolled or half-controlled ones, employ the use of a diode across the load. This diode can be called either a commutating diode or a bypass diode, or a freewheeling diode depending on the context. When the load voltage goes into reverse, it commutates or transfers the load current flowing through the SCR. This diode has two primary functions. They are

(a) It prevents the load voltage from being inverted except in the case of a small diode voltage.

(b) The main rectifier thyristors can revert to their blocking states after the load current has been diverted away from them.

Following are the advantages of the freewheeling diode in converter circuits

(a) It improves the input power factor.

(b) Improves load current waveforms and thus the performance of the load better

Figure.4.6 (a) depicts a half-wave-controlled rectifier that also includes a freewheeling diode D_F that is connected across the *RL* load. The waveforms are shown in Figure.4.6 (b). The D_F diode prevents the thyristor from conducting beyond π or 180°.

At $\omega t = 0$, the source voltage becomes positive, the SCR is forward-biased, and D_F is reverse-biased. When the SCR is triggered at an angle α, the SCR *T1* conducts, the supply voltage $e = E_m Sin\omega t$ is appeared across the load till $\omega t = \pi$. At $\omega t = \pi$, the source voltage is zero. After $\omega t = \pi$, the source voltage (e) becomes negative and the thyristor is reverse-biased and will not conduct, the freewheeling diode is forward-biased and conducts. Thus, load current transfers the current from SCR to F_D. It is assumed that the load current does not decays to zero till the thyristor is triggered at $(2\pi+\alpha)$. The voltage drop across the F_D is almost zero. Thus, the load voltage during the period for which F_D conduct is zero The voltage across thyristor *T1* during this period is shown in Figure.4.6(b). The turn-off time of the SCR is given by

$$t_c = \frac{\pi}{\omega} \tag{4.28}$$

The source and the SCR currents are the same. The operation of the half-wave rectifier with a freewheeling diode is dived into two modes of operation. They are as follows.

Mode-I: Conduction mode. The range of angles is $\alpha \le \omega t \le \pi$

Here the SCR conducts from α to π, $(2\pi+\alpha)$ to 3π, and so on. The FD is reverse-biased and will not conduct. The time duration is second. Let us assume that at the beginning of this mode load current is I_o. The expression for load current, i_O can be derived as follows. The voltage equation of the circuit is given by

$$Ri_0 + L\frac{di_o}{dt} = E_m \sin\omega t \tag{4.29}$$

Solving for i_O, we will get

$$i_0 = \frac{E_m}{Z}\sin(\omega t - \phi) + Ae^{-\left(\frac{R}{L}\right)t} \tag{4.30}$$

where the angle Φ is the angle by which the source current I_s lags behind the source voltage E_s. The value of constant A can be obtained from the initial condition. At $\omega t = 0,\ i_0 = I_0$. The value time t is expressed by $t = \alpha/\omega$. Putting these values in the above equation, and solving for A, we will get,

$$A = \left[I_0 - \frac{E_m}{Z}\sin(\alpha - \phi)\right]e^{\frac{R\alpha}{\omega L}} \tag{4.31}$$

$$\therefore i_0 = \frac{E_m}{Z}Sin(\omega t - \phi) + \left[I_0 - \frac{E_m}{Z}Sin(\alpha - \phi)\right]e^{-\left(\frac{R}{L}\right)\left(t - \frac{\alpha}{\omega}\right)} \tag{4.32}$$

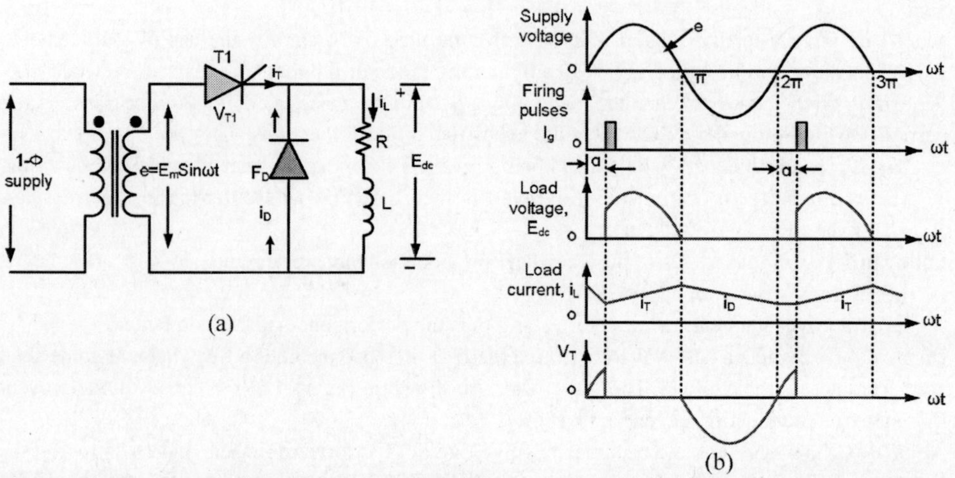

Figure.4.6 Single phase half wave-controlled rectifier with a freewheeling diode (a) circuit, (b) Waveforms

Mode-II: Freewheeling mode. The range of angles is $\pi \le \omega t \le (2\pi + \alpha)$

The range of this mode is π to 2π, 3π to 4π, and so on. The SCR is reversed biased and F_D is forward-biased. Let at the beginning of this mode load current is I_{01} and the load current is continuous. The voltage equation of this loop is given by

$$0 = Ri_0 + L\frac{di_0}{dt} \tag{4.33}$$

The solution for i_0 is given by

$$i_0 = Ae^{-\left(\frac{R}{L}\right)t} \tag{4.34}$$

Putting the initial condition, we can find the value of constant A. At $\omega t = \pi$, $i_0 = I_{01}$. Thus, the expression for A is

$$A = I_{01}e^{\left(R\pi/\omega L\right)} \tag{4.35}$$

$$i_0 = I_{01}e^{-\left(\frac{R}{L}\right)\left(t - \frac{\pi}{\omega}\right)} \tag{4.36}$$

The average load voltage, E_{dc} or V_0 is given by

$$E_{dc} \text{ or } V_0 = \frac{1}{2\pi}\int_{\alpha}^{\pi} E_m \sin \omega t.d(\omega t) = \frac{E_m}{2\pi}(1 + \cos\alpha) \tag{4.37}$$

The average load current is given by

$$I_0 \text{ (or } I_L) = \frac{E_{dc}}{R} = \frac{E_m}{2\pi R}(1 + \text{Cos}\alpha) \tag{4.38}$$

4.6. FULL WAVE-CONTROLLED RECTIFIER

There are different types of controlled rectifiers. Also, many ways are available for the classification of controller rectifiers. Some of them are given below.

(a) Based on the supply phases. Example Single phase, and three phase. The rectifiers that were discussed previously in this unit are single-phase half-wave converters.

(b) Based on the number of current loads current pulses per cycle of the supply voltage. The converters already discussed are single phase half wave-controlled rectifiers are single phase single pulse converters because these converters produce only one pulse of load current during one cycle of the supply voltage. If the converter produces two pulses per cycle, they are called two pulse converters.

The disadvantages of single-phase half-wave or single-phase one-pulse converters are minimized by using single-phase full-wave or single-phase two-pulse converters. Practically there are two basic single-phase full wave or single-phase two pulse-controlled rectifiers. They are

(a) Single phase full wave (or two pulses) midpoint-controlled rectifier

(b) Single phase full wave bridge rectifier or single phase two pulse bridge rectifier.

Both types will be discussed in detail.

4.6.1. Single phase full wave (or two pulse) mid-point-controlled rectifier

A single-phase (1-Φ) full wave-controlled rectifier circuit with a mid-point configuration uses a center-tapped secondary winding 1-Φ transformer along with two thyristors to complete the circuit. This type of converter is sometimes referred to as the two-pulse converter because to trigger the various thyristors, it is necessary to produce either two triggering pulses or two sets of triggering pulses during each supply cycle. Most of the time, these kinds of circuits are used for low-rated outputs.

4.6.1.1. Single phase full wave mid-point-controlled rectifier with resistive (R) load

The fundamental configuration of a single-phase, centre-tap controlled rectifier with a resistive load is illustrated in Figure.4.7(a). Unlike half-wave rectifiers, full-wave rectifiers allow for independent phase control of the positive and negative halves of the AC supply, resulting in a DC voltage of larger amplitude and less ripple.

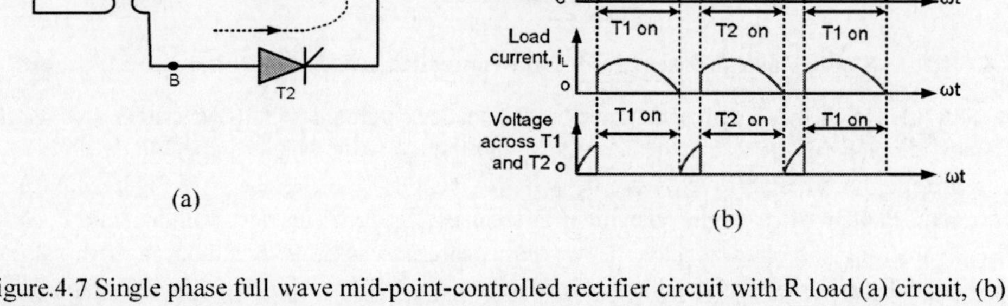

Figure.4.7 Single phase full wave mid-point-controlled rectifier circuit with R load (a) circuit, (b) waveform

Terminal *A* of the transformer has a positive potential with respect to terminal *B* or terminal *A* has a positive potential with respect to *N* when the supply voltage is in its positive half-cycle. As a result, SCR_1 (T_1) is forward biased while SCR_2 (T_2) is reverse biased. Since the SCRs don't get any pulses to turn them on, they are in the off state. When T_1 is triggered at an angle of firing α, current would flow from terminal *A* through T_1 and the resistive load *R*, and then back to the center tap of the transformer. Figure.4.7(a) also shows this current path. Until the line voltage flips polarity and T_1 is turned off, this current will continue to flow up to the angle π. The conduction period of T_1 can be any number between 0 and π, depending on the value of α and the parameters of the load circuit.

The transformer terminal *B* is positive with respect to *N* during the supply voltage negative half-cycle, which causes T_2 to be forward-biased and T_1 to be reverse-biased. When T_2 is triggered at an angle of ($\pi + \alpha$), the current travels from terminal *B* through T_2 and the load (*R*) before going back to the center tap of the transformer. This current is maintained until angle 2π is reached, at which point T_2 is turned off. Since the same firing angle is used to trigger both SCRs, it is reasonable to assume that they each take an equal amount of the load current. As there are two input waves (i.e., positive and negative) applied across the load. Because of this, two current pulses are traveling in the same direction across the load. As a result, the ripple frequency across the load is equal to twice the frequency of the supply. The waveforms of the voltage and current for this configuration can be seen in Figure.4.7 (b). As can be seen in Figure.4.7 (b), the load current will never be continuous when dealing with a purely resistive load.

Following is the derivation of the voltage and current relations:

E_{dc}, the dc voltage across the *R* load, is calculated as follows:

$$E_{dc} = \frac{1}{\pi}\int_{\alpha}^{\pi} E_m \sin\omega t\, d(\omega t) = \frac{E_m}{\pi}[-\cos\omega t]_{\alpha}^{\pi} = \frac{E_m}{\pi}(1+\cos\alpha) \tag{4.39}$$

The formula for the average load current is given by

$$I_{dc} = \frac{E_m}{\pi R}(1+\cos\alpha) \tag{4.40}$$

For a specific firing angle α, the RMS load voltage:

$$E_{rms} = \left[\frac{1}{\pi}\int_{\alpha}^{\pi} E_m^2 \sin^2\omega t\, d(\omega t)\right]^{1/2} = E_m\left[\frac{1}{\pi}\int_{\alpha}^{\pi} \sin^2\omega t\, d(\omega t)\right]^{1/2}$$

$$= E_m\left[\frac{1}{\pi}\int_{\alpha}^{\pi}\left(\frac{1-\cos 2\omega t}{2}\right)d(\omega t)\right]^{1/2} = E_m\left[\frac{1}{2\pi}\left(\omega t - \frac{\sin 2\omega t}{2}\right)_{\alpha}^{\pi}\right]^{1/2}$$

$$= E_m\left[\frac{1}{2\pi}\left(\pi - \alpha + \frac{\sin 2\alpha - \sin 2\pi}{2}\right)\right]^{1/2} = E_m\left[\frac{1}{2\pi}\left(\pi - \alpha + \frac{\sin 2\alpha}{2} - 0\right)\right]^{1/2}$$

$$\text{Thus, } E_{rms} = E_m\left[\frac{\pi-\alpha}{2\pi} + \frac{\sin 2\alpha}{4\pi}\right]^{1/2} \tag{4.41}$$

4.6.1.2. Single phase full wave mid-point-controlled rectifier with inductive (RL) load

Figure.4.8 (a) is a schematic representation of an RL-loaded single-phase full-wave controlled rectifier. Waveforms of both voltage and current are shown in Figure.4.8 (b), which can be found below.

After e_1 becomes positive, T_1 can be switched on at any time, as shown in Figure. 4.8 (b). The current in the inductive load begins to build up as soon as T_1 is turned on. This will maintain T_1 on state until the value of e_1 becomes negative. However, as soon as e_1 begins to exhibit negative values, e_2 begins to exhibit positive values. This causes T_2 to fire immediately, turning it on and consuming the load current while simultaneously applying a reverse voltage to thyristor T_1, which transfers its current to T_2. Figure.4.8 (b) shows the V_T waveform for a thyristor. When V_T is positive, the thyristor can be turned on at any time. The maximum value of the secondary transformer voltage is represented by $2E_m$, which is the peak voltage that is measured across the thyristor in both directions.

The load current can be either continuous or intermittent, depending on the value of the inductance. Continuous load current flow is observed whenever the inductance value is greater than its critical value. It is considered to be discontinuous if the measured inductance value is lower than the critical value. The analysis presented here assumes that the inductance is sufficient to allow each thyristor to conduct for 180°. In addition to this, because both thyristors are triggered with the same delay angle, the load current is divided in half. Because of the large inductance in the circuit and the continuous current conduction, thyristors can keep conducting even when the voltage at their anode is negative in comparison to the voltage at their cathode. This behavior is illustrated in Figure.4.8(b). It has been demonstrated that the load current is constant dc.

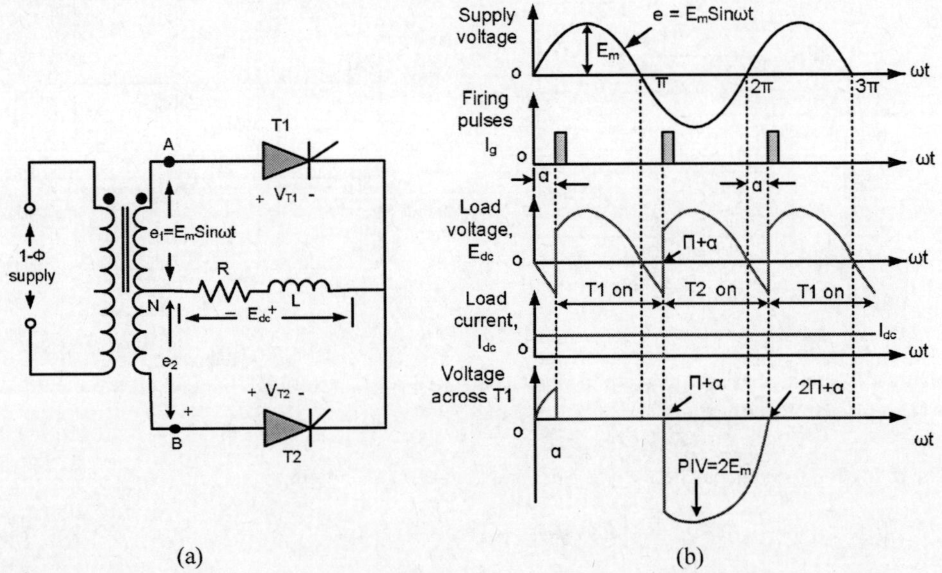

Figure.4.8 Single phase full wave mid-point-controlled rectifier circuit with RL load (a) circuit, (b) Waveform

The output DC voltage can now be calculated as:

$$E_{dc} = \frac{1}{\pi}\int_{\alpha}^{\pi+\alpha} E_m \sin\omega t\, d(\omega t) = \frac{E_m}{\pi}[\cos\alpha - \cos\omega(\pi+\alpha)] = \frac{2E_m}{\pi}\cos\alpha \tag{4.42}$$

This equation has led to some conclusions.

(a) When $\alpha = 0°$, the output voltage will be at its highest point.

(b) At a firing angle $\alpha = 90°$, the voltage output will be zero. It indicates that the output voltage will have equal amounts of positive and negative regions.

(c) The converter operates in inversion mode for firing angles greater than 90°. At $\alpha = 180°$, the voltage will be negative at its maximum.

According to the observations, the output power can be set to any value by adjusting the firing angle, which ranges from 0 to 90 degrees for inductive loads and from 0 to 180 degrees for resistive loads.

4.6.1.3. Single phase full wave mid-point-controlled rectifier with freewheeling diode

Figure.4.9(a) is a representation of the full-wave Centre-tapped phase-controlled circuit, which includes an inductive load with a freewheeling diode. Waveforms of the load voltage and current are also depicted in Figure.4.9(b). The thyristors are triggered at angle α, as seen in Figure.4.9(b). The variable dc voltage at the load is achieved by changing the firing angle. The load voltage cannot be negative as the supply voltage approaches zero at 180 degrees, as shown by Figure.4.9(b), because the freewheeling diode, D_F, begins to conduct. Current freewheeling through the diode maintains a constant load current. Figure.4.9(b) also depicts the conduction period of thyristors and diodes. Due to the energy storage capabilities of the inductive load, current flows through the feedback diode in the direction of the arrow shown in Figure.4.9(a). This current decay rate is influenced by the time constant of the load.

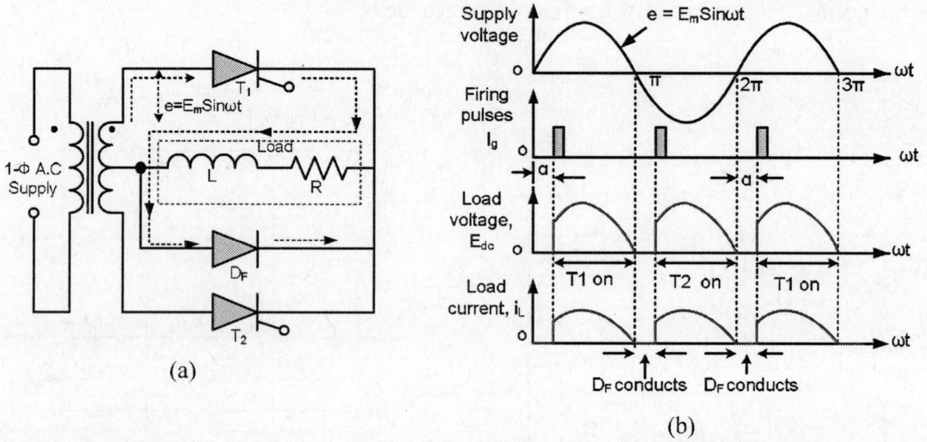

Figure.4.9 Single phase full wave mid-point-controlled rectifier with RL load and freewheeling diode. (a) Circuit, (b) Waveform

The following formula can be used to calculate the average dc output voltage:

$$E_{dc} = \frac{1}{\pi}\int_{\alpha}^{\pi} E_m \sin\omega t \, d(\omega t) = \frac{E_m}{\pi}(1+\cos\alpha) \tag{4.43}$$

The dc load current can be calculated using

$$I_{dc} = \frac{E_m}{\pi R}(1+\cos\alpha) \tag{4.44}$$

The formulas for the average current flow through the D_F are as follows:

$$I_{DF} = I_{dc}\frac{\alpha}{\pi} = \frac{E_m}{\pi R}(1+\cos\alpha)\frac{\alpha}{R} \tag{4.45}$$

4.6.2. Single phase full wave bridge rectifiers

Four thyristors are required for a single-phase bridge converter. This configuration results in operation in two quadrants. This type of converter configuration is known as a two-quadrant converter or a fully controlled converter. Figure.4.1(b) illustrates a two-quadrant converter where voltage polarity can change but the current direction cannot. Two thyristors are frequently replaced by two diodes to modify bridge configurations. This configuration yields one quadrant operation. Such a converter configuration implied a semi converter or one-quadrant converter. A one-quadrant converter has a single output polarity for DC voltage and current, as shown in Figure.4.1(a). Load on the converter can be either resistive (R) or inductive (RL) type.

4.6.2.1. Single phase full wave bridge rectifiers with resistive (R) load

Figure.4.10(a) depicts a circuit diagram of a single-phase fully controlled bridge rectifier. This circuit adheres to the same fundamental rules as the two-pulse midpoint circuit depicted in Figure.4.7(a). In the bridge configuration, diagonally opposite pair of SCRs are made switched on and off simultaneously. SCR T_1 and SCR T_2 are forward biased during the positive half of the supply voltage, and if they are triggered at the same time, current flows the path *L*-SCR T_1-*R* SCR T_2-*N*. During the negative half of the supply voltage, when T_3 and T_4 are forward-biased, current flows through the path *N*- SCR T_3-*R*-SCR T_4 -*L* if they are both turned on at the same time. At the same firing angle α, SCR T_1, *SCR* T_2, and SCR T_3, *SCR* T_4 are triggered during the positive and negative halves of the supply voltage. If the supply voltage drops to zero, the resulting current is also zero. Thus, by natural commutation, SCR T_1, *SCR* T_2, and SCR T_3, *SCR* T_4 are turned off in the positive and negative halves of the cycle, respectively. Figure.4.10(b) depicts the voltage and current waveforms for the circuit configuration shown in Figure.4.10(a). For this bridge configuration, E_{dc}, I_{dc}, and E_{rms} expressions are the same as equations (4.39), (4.40), and (4.41), respectively.

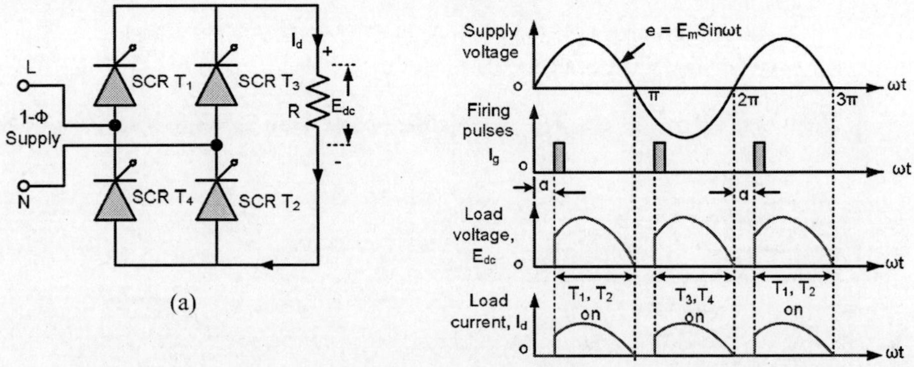

Figure.4.10 Fully controlled bridge circuit with R load (a) Circuit, (b) Waveform

4.6.2.2. Single phase full wave bridge rectifiers with an inductive (RL) load

Figure.4.11 depicts a single-phase fully controlled bridge circuit with an RL load. SCRs T_1 and T_2 must be fired together at the positive half cycle, and thyristors T_3 and T_4 in the negative half cycle of supply voltage, for conduction to occur and current to flow. Both SCRs T_1 and T_2 are fired from the same firing circuit to ensure simultaneous firing. The current ripple in the circuit is dampened by inductance *L*. While a higher value of *L* will cause continuous current flow across the load. A small value of *L* will not be able to support a continuous load current for a large firing angle. Figure.4.12 displays the waveforms with two distinct firing angles.

The DC side voltage output consists of a constant DC component on top of which an AC ripple component with a fundamental frequency twice that of supply is superimposed. The fundamental component of the square waveform of the supply input current has a phase relationship with the input voltage and has an amplitude of I_d. Figure.4.12(a) shows SCR T_1 and SCR T_2 triggered at the firing angle $\alpha = 60°$. The current follows the path *L1*- SCR T_1 - *L* - *R* - SCR T_2 - N. At this instant, supply voltage appears across the load, forcing current to flow through the load. The load current, I_d is considered to be constant. As shown in Figure.4.12(a) alongside the applied voltage, I_d current flows from source to load and has a direction from line to neutral, which is taken as positive. The voltage is now reversed at instant π. The thyristors SCR T_1 and SCR T_2 are conducting and therefore, the negative supply voltage is visible across the output terminals because of the extremely large inductance L, which also ensures that the current is maintained in the same direction and at a constant magnitude. T_3 and T_4 are turned at angle $\pi + \alpha$. Thus, the thyristors SCR T_1 and SCR T_2 are reverse biased by the negative line voltage, as are the SCR T_1, through SCR T_3, and SCR T_2, through SCR T_4. The current takes the path N - SCR T_3 - L - R - SCR T_4 - $L1$. This process is repeated every half cycle, yielding the output voltage depicted in the figure.

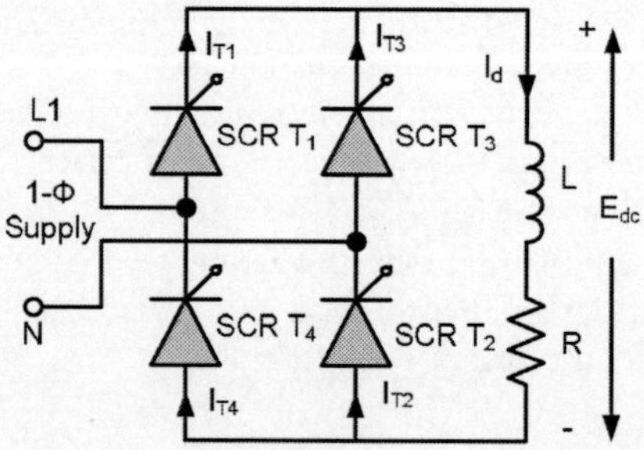

Figure.4.11 Single phase full controlled bridge rectifier with RL load.

The average dc output voltage can be calculated as follows:

$$E_{dc} = \frac{1}{\pi}\int_{\alpha}^{\pi+\alpha} E_m \sin \omega t \, d(\omega t) = \frac{E_m}{\pi}[-\cos\omega t]_{\alpha}^{\pi+\alpha} = \frac{E_m}{\pi}[\cos\alpha - \cos(\pi+\alpha)] E_{dc} = \frac{2E_m}{\pi}\cos\alpha \tag{4.46}$$

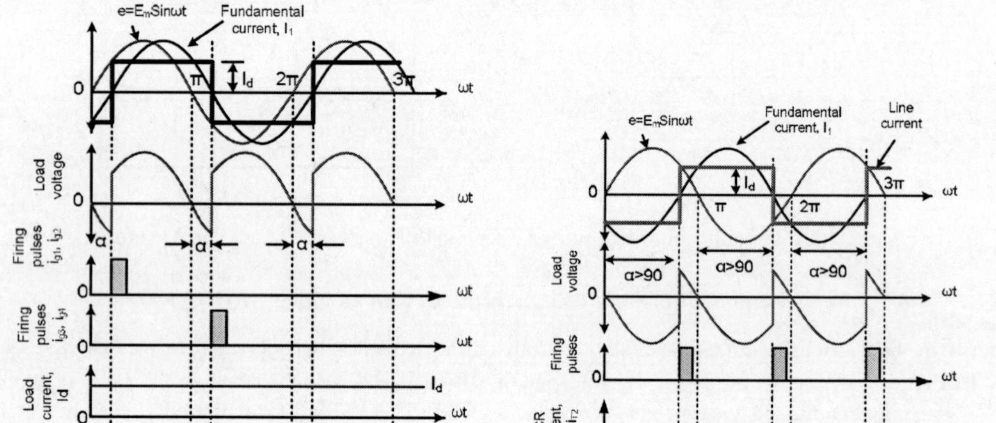

Figure.4.12 Waveform for full wave bridge rectifier with R-L load for (a) α<90^0 (say $\alpha = 60^{\circ}$), (b) α>90^0 (say $\alpha = 135^{\circ}$)

Assuming that there is a continuous flow of current through the DC terminals, the DC voltage output can be continuously varied from a positive maximum to a negative maximum by controlling the firing pulses of the thyristors in a range that extends from 0° to 180°. Even though current flows only one way on the DC side, the converter's power can go either way because the average DC voltage is reversible.

Rectifying mode: While both E_s and I_s in the supply are positive during this period α to π, energy is transferred from the AC source to the load. During the period (π to $\pi + \alpha$), E_s is negative but I_s is positive. This means that the load gives some of its energy back to the AC source. However, the overall direction of the power flow is from the AC source to the load. In addition, as shown by equation (4.46), the converter acts as a rectifier when the $\alpha < 90°$, since the voltage at the DC terminals is positive in this case.

Inverting mode: In Figure.4.12 (b), waveforms for the firing angle greater than 90^0 (say 135°) is presented. There is now an average negative component in the DC voltage, and the AC current fundamental component lags the voltage by 135° (α greater than 90^0). Because the mean DC voltage output is negative for $\alpha > 90°$, the DC power and mean AC power must be negative as well. In other words, the converter is now functioning as a line-commutated inverter, delivering power from the DC side to the AC side.

Since both types of phase-controlled converters have been studied, it is now possible to list the benefits of the single-phase bridge converter over the single-phase mid-point converter:

a) In a midpoint converter, the peak inverse voltage on the SCRs is $2E_m$, while in a fully controlled bridge converter, it is E_m. Therefore, the midpoint configuration handles roughly half the power of the bridge configuration for the same voltage and current ratings of SCRs.
b) Transformer ratings in mid-point converters are typically double the load rating. However, the single-phase bridge converter is an exception to this.

The bridge configuration is therefore preferred over the mid-point configuration. Nonetheless, available source voltage, load-voltage requirements, and component cost largely dictate which of these two types is ultimately chosen. If there is a need to ground the dc terminals, a midpoint converter can be used.

4.7. SINGLE-PHASE HALF CONTROLLED BRIDGE RECTIFIER

Many applications require only a positive voltage in the rectifying mode to operate. In such cases, it is generally preferable to incorporate diodes into the circuit. In this circuit configuration, the mean DC terminal voltage can be continuously controlled from maximum to zero, but the reverse is not possible.

4.7.1. Single phase half-controlled bridge rectifier with resistive (R) load

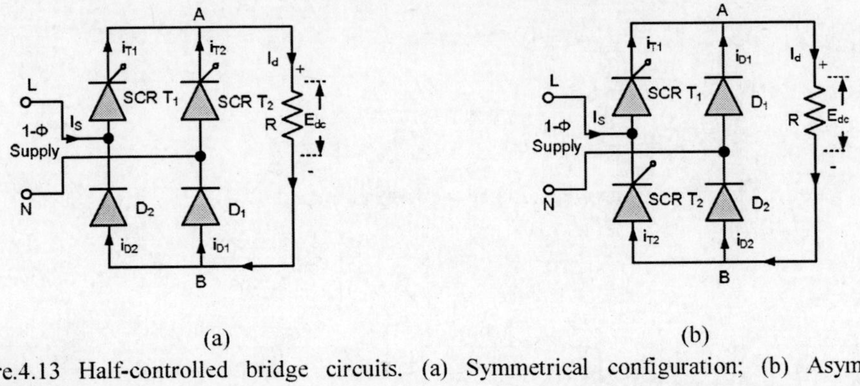

Figure.4.13 Half-controlled bridge circuits. (a) Symmetrical configuration; (b) Asymmetrical configuration.

As can be seen in Figure.4.13, there are two possible configurations namely symmetrical and unsymmetrical configurations for a half-controlled bridge circuit with a resistive load. Semiconverters are another name for half-controlled converters.

A single gate-pulse can be used to trigger either SCR because, in a symmetrical configuration (Figure.4.13(a)), the cathodes of SCR T_1 and SCR T_2 are at the same potential. This allows their gates to be connected. The use of separate-triggering circuits is required for asymmetrical configuration (Figure.4.13(b)). Waveforms for a symmetrical converter circuit are depicted in Figure.4.14.

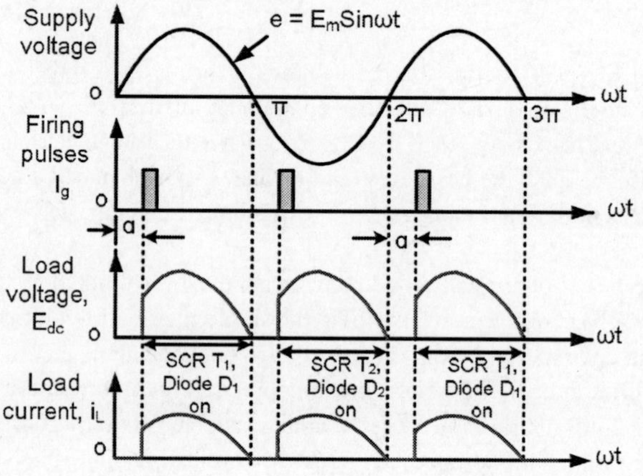

Figure.4.14 Waveforms for symmetrical configuration with R load.

Now let's consider the scenario of a bridge circuit with a half-controlled symmetrical configuration. The SCR T_1 and diode D_1 are forward-biased and operate in the forward blocking mode when the supply voltage is positive. The current follows the path L - SCR T_1 - R - diode D_1 - N when the SCR T_1 is triggered at a firing angle of α. According to Figure.4.14, the load current will continue to flow up until the supply voltage is reversed, which turns it off at time $\omega t = \pi$. Similarly, SCR T_2 and diode D_2 are forward-biased when the supply voltage is negative. The current follows the path N – SCR T_2 - A - R - B - diode D_2 - L, when the SCR T_2 is triggered at a firing angle of $(\pi + \alpha)$ and conducts up to angle 2π. Voltage and current relationships are calculated as follows.

The average dc load voltage is given by

$$E_{dc} = \frac{1}{\pi}\int_{\alpha}^{\pi} E_m \sin \omega t\, d(\omega t) = \frac{1}{\pi} E_m[-\cos \omega t]_{\alpha}^{\pi} = \frac{E_m}{\pi}(1+\cos\alpha) \tag{4.47}$$

The average load current is given by

$$I_d = \frac{E_m}{\pi R}(1+\cos\alpha) \tag{4.48}$$

The RMS load voltage for a firing angle α is derived as follows

$$E_{rms} = \left[\frac{E_m^2}{\pi}\int_{\alpha}^{\pi} \sin^2 \omega t\, d(\omega t)\right]^{1/2} = \left[\frac{1}{\pi}\int_{\alpha}^{\pi}\left(\frac{1-\cos 2\omega t}{2}\right) d(\omega t)\right]^{1/2}$$

$$= E_m\left[\frac{1}{2\pi}\left(\omega t - \frac{\sin 2\omega t}{2}\right)_{\alpha}^{\pi}\right]^{1/2} = E_m\left[\frac{\pi-\alpha}{2\pi} + \frac{\sin 2\alpha}{4\pi}\right]^{1/2} \tag{4.49}$$

4.7.2. Half-controlled bridge rectifier with RL load

There are two possible configurations for a half-controlled bridge circuit with inductive load and both are shown in Figure.4.15. Figure.4.16 displays the voltage and current waveforms for both configurations. Take into consideration the symmetrical configuration of the circuit. Figure.4.15 (a) illustrates that during each positive half-cycle, SCR T_1 is activated at a firing angle α. The current takes the path *L1 - SCR* T_1- point A - *L* - *R* - point B - diode D_1 - *N*. Here, it is assumed that *L* is large enough to support constant load current. The load current I_d is assumed to be constant. Therefore, SCR T_1 and diode D_1 conduct during the positive half of the supply voltage. Since diode D_1 is already conducting, D_2 will be forward-biased when the supply voltage is reversed at $\omega t = \pi$. As a result, D_2 turns on, and the load current flows through diode D_2 and SCR T_1. Since D_1 is reverse-biased by the supply voltage, it is disabled. In this way, during the period π to $(\pi + \alpha)$ in each supply cycle, the load current freely circulates along the route *R* - point *B* - diode D_2 - SCR T_1 - point *A* - *L1*. The forward-biased T_2 is pulsed to trigger at $(\pi + \alpha)$ during the supply voltage negative half-cycle and T_2 is switched on. The supply voltage reverse-biases T_1 and then turns it off as T_2 is turned on. As a result, the load current is conducted through T_2 and D_2, the aforementioned cycle is repeated, and the resulting waveforms, as depicted in Figure.4.15 (a), are analogous to those of a fully-controlled converter with a freewheeling diode. This configuration is referred to as a symmetrical circuit because the conduction periods of the thyristors and the diodes are identical.

Consider the circuit in Figure.4.15 (b), where SCR T_1 and D_1 are forward-biased during the positive half-cycle of the supply voltage. The SCR T_1 is turned ON at firing angle α, as shown in Figure.4.16 (b). The current takes the path *L1 - SCR* T_1 - point *A* - *L* - *R* - *B* - diode D_1 - *N*. As a result, SCR T_1 and diode D_1 operate from α to π. Similarly, T_2 and diode D_2 conduct from $(\pi + \alpha)$ to 2π during each negative half cycle of the supply voltage. The diodes D_1 and D_2 provide the freewheeling action from 0 to α and from π to $(\pi + \alpha)$ in each power cycle. This converter setup makes use of thyristors and diodes whose conduction periods differ. Therefore, this setup is referred to as the asymmetrical configuration.

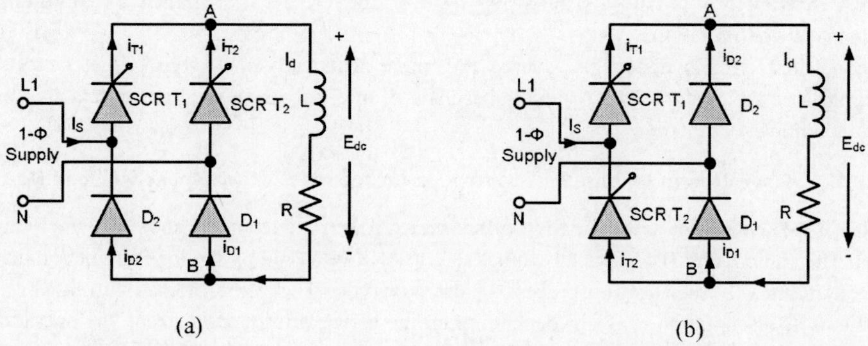

Figure.4.15 Half-controlled bridge rectifier with RL load. (a) Symmetrical configuration; (b) Asymmetrical configuration.

The expression for the dc output voltage is:

$$E_{dc} = \frac{1}{\pi}\int_{\alpha}^{\pi} E_m \sin \omega t \, d(\omega t) = \frac{E_m}{\pi}(1 + \cos\alpha) \qquad (4.50)$$

The following differences can be seen when comparing a 2-pulse half-controlled converter to a full-controlled converter:

(a) Because diodes are used in place of half of the SCRs, half-controlled converters are more affordable than fully-controlled ones.
(b) In the half-controlled converter configuration, the negative voltage at the DC terminals is replaced by freewheeling periods of zero voltage. Getting rid of the negative voltage at the DC terminals is a good thing because it lowers the ripple voltage and, as a result, reduces the need for filtering.
(c) Half-controlled converters have an improved power factor because of the freewheeling behavior of the circuit.

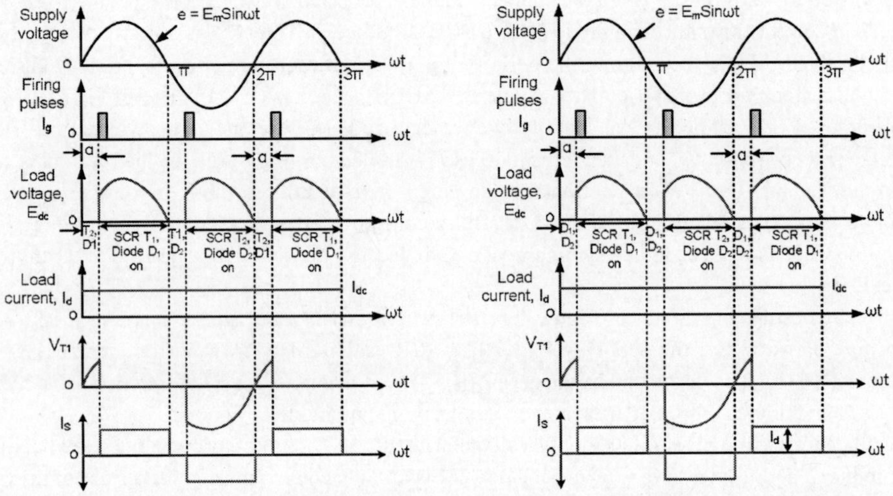

Figure.4.16 Waveform of half-controlled full wave bridge rectifier (a) symmetrical configuration, (b) asymmetrical configuration.

Half-controlled circuits distort the source AC more than fully-controlled bridge circuits. When powering large DC motors from a single-phase AC source, half-controlled converters are commonly used in mainline AC traction.

4.7.2.1. Operation of 1-phase half-controlled bridge rectifier with practical load (R-L load)

Practical loads consist of a combination of resistance R and inductance L, both of which have finite values. The value L and R decides how much energy will be stored by the load and how much will be released during the freewheeling process of the converter. The value of load current thus may be continuous or discontinuous. Thus, depending on the nature of the load current, the operation of the semiconverter for practical load (RL load), is divided into two modes,

(a) Continuous conduction mode.
(b) Discontinuous conduction mode.

Continuous conduction mode: Continuous conduction mode demonstrates a non-zero load current that gradually rises and falls with a finite ripple but does not reach zero because of the larger value of load inductance. Switches in a symmetrical arrangement will continue to have a conduction angle of π. As can be seen in Figure.4.17 (a), the load current drops from I_{max} to I_{min} during the freewheeling intervals. Because it produces better results, this mode is preferred in motor control applications.

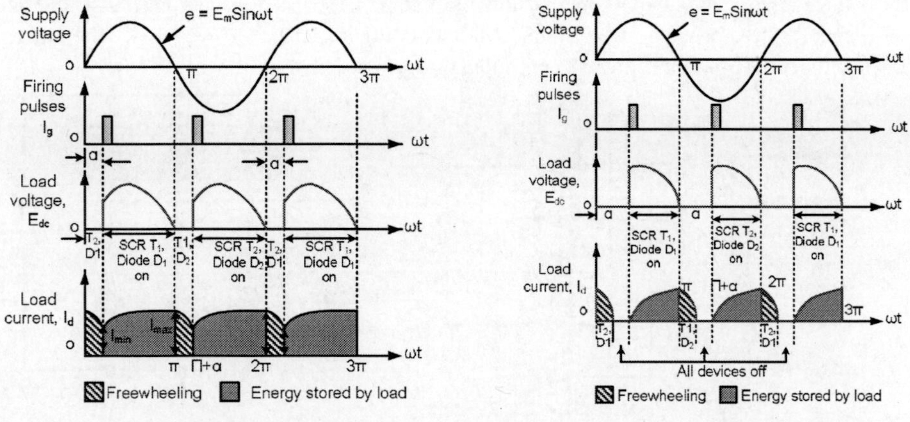

(a) (b)

Figure.4.17 Waveform for the half-controlled symmetrical bridge with practical R-L load (a) Continuous conduction mode, (b) discontinuous conduction mode

Discontinuous condition mode: During the freewheeling interval of the discontinuous mode, the load current drops to zero well before the next SCR is turned on, as shown in Figure.4.17 (b). Increased load current ripple results in a decreased conduction period of devices. This mode is not appropriate for applications such as dc motor speed control.

4.8. Common Anode and Common Cathode for Three-Phase Controlled Rectifier

Before discussing three phase-controlled rectifiers it is necessary to understand the common cathode and common anode configuration of converters using diodes. Figure.4.18 shows such circuits. In Figure.4.18(a), the cathode terminals of the diodes are connected together at a common point and it is called a common cathode arrangement. On the other hand, in Figure.4.18(b) the anode terminals are connected together at a common point and it is called a common anode arrangement. These figures are of three-phase half-wave configurations. In Figure.4.18(a), diodes A_1, B_1, and C_1 are connected in a common load. The other side of the load is connected to the neutral point N of the supply. The load terminal *P* is positive in this figure. On the other hand, the anodes of A_2, B_2, and C_2 are connected to common point *Q*. The terminal of the load is connected to the neutral point of the supply. The terminal *Q* is negative in this case. Thus, the output voltage is positive in the common cathode configuration and negative in the case of the common anode configuration.

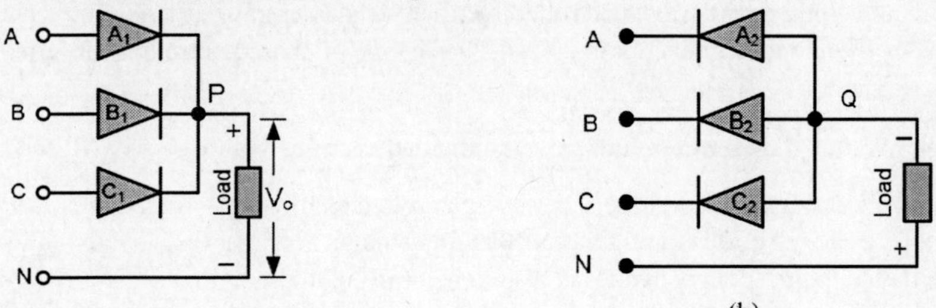

(a) (b)

Figure.4.18 Three phase half wave diode rectifier (a) Common cathode anode configuration, (b) common anode configuration

If the common anode and common cathode circuit are connected in series, we will get six pulse converters. In Figure.4.19(a) the neutral point is shown. But in this series connection, there is a load current if there is no neutral. The same is shown in Figure.4.19(b). After rearrangement, the ultimate six-pulse circuit (three-phase bridge) circuit is shown in Figure.4.19(c). If we replace the diodes with SCRs we will get the 3-phase full wave controlled bridge rectifier.

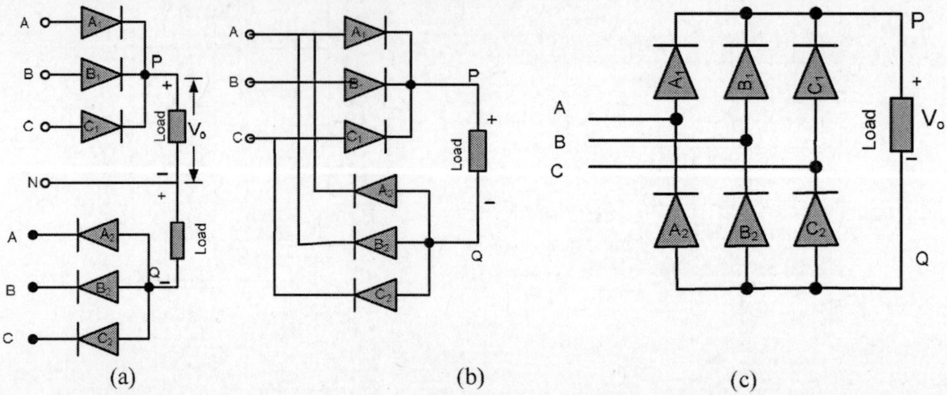

Figure.4.19 (a) &(b) Series connection of 3-phase half wave circuits, (c) Ultimate 3-phase full wave circuit

4.9. THREE-PHASE CONTROLLED CONVERTERS

When powered by a single-phase input, the converter's DC outputs exhibit excessive ripple voltage. The extra heat that is generated by this ripple is usually not desired. As a result, a large inductor is required to smooth out the output voltage and reduce the possibility of discontinuous operation. As the number of pulses increases, the amount of smoothing that is required decreases. A three-phase AC source with an appropriate converter link allows for a rise in the number of pulses. When the converter's number of input pulses is raised, the number of segments that create the output voltage rises as well, and the ripple content reduces. The higher the number of pulses, the better the output. Large power applications typically employ three-phase rectifier circuits. Three-phase controlled converter circuits can be classified as follows:

1. Three-pulse converters configuration.
2. Six-pulse converters configuration.
3. Twelve pulse converters

4.9.1. Three Pulse Converters

The term "three-phase half wave-controlled rectifier" is also used to describe three pulse converters. The three-pulse midpoint converter is the most basic type of phase-controlled converter that runs on a three-phase supply.

4.9.1.1. Three phase half wave-controlled rectifier with resistive (R) load

The circuit diagram of a three-phase half wave-controlled rectifier with *R* load is shown in Figure.4.20 (a). All of the phases of this configuration share a common terminal that is referred to as the neutral point or the midpoint. This converter is also known as the mid-point configuration. The transformer primary is connected in a delta, and the secondary is connected in a star, as shown in the diagram. The load is connected between the three-phase common contact and the neutral point. Three-phase voltage is depicted as a vector in Figure.4.20(b).

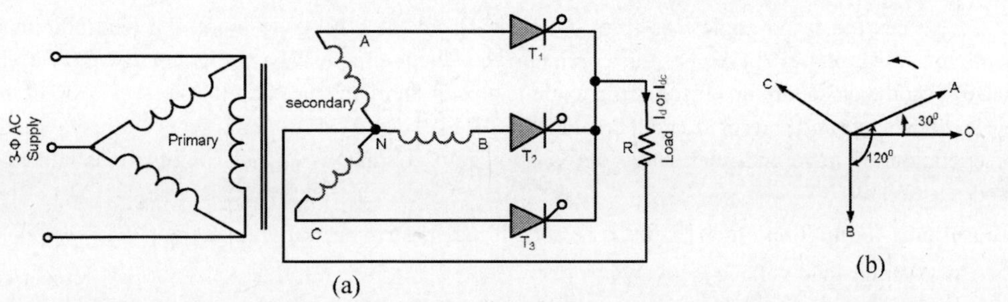

Figure.4.20 Three phase half wave-controlled rectifier with R load (a) circuit, (b) Vector diagram of 3-Φ voltage.

In this configuration, only one SCR that is connected to the phase with the highest instantaneous positive value is active at any given moment. Given that the other conducting phase keeps all SCRs reverse-biased below a phase angle of 30°, none of them can be triggered at lower angles. Therefore, the firing angle is calculated from 30° with respect to the corresponding phase voltage for a particular SCR connected in a specific phase.

Figure.4.21 displays the waveforms for the three-phase half wave-controlled rectifier feeding *R* load. If you look at the vector diagram, you'll see that both phase *A* and phase *C* are positive with respect to neutral. Since the T_1 connected to phase *A* is reverse biased by the conducting T_3, it cannot be triggered at an angle lower than 30°. So, 30° is the minimum angle for firing. The voltage waveforms cross over points, which correspond to the points at which equivalent thyristors would begin to conduct, are used to calculate the firing angle α. Hence the conduction period for each thyristor equal is 120°. T_1, as depicted in Figure.4.21 (a), will conduct from ωt = 30° to 150° because it is more positive than T_2, and T_3 during this time. T_2 will conduct from ωt = 150° to 270° and T_3 from ωt = 270° to 390°. When one SCR is conducting, the other two blocking SCRs are reverse biased, causing the common cathode terminal *p* to rise to the highest positive voltage of that phase. The voltage across the load is shown in Figure.4.21. The *R* load allows for two different modes of conduction.

(a) Continuous conduction (α < 30°).
(b) Discontinuous conduction (α > 30°).

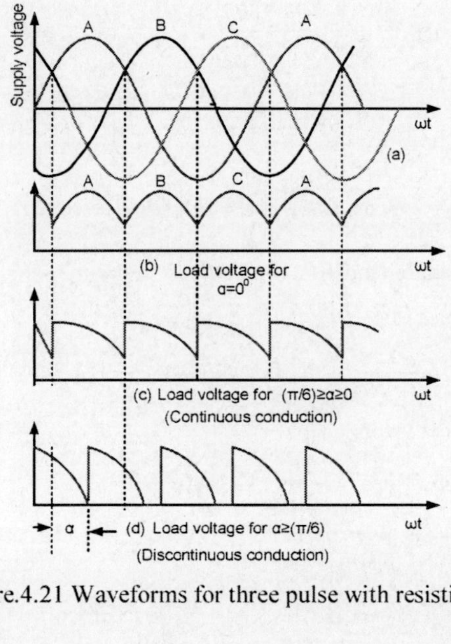

Figure.4.21 Waveforms for three pulse with resistive load.

When the firing angle α is from 0° to 30°, from the cross-over point, the load current is continuous. Figure.4.21 (c) depicts this continuous conduction mode. The output voltage and current pulses become discontinuous if the firing angle is greater than 30°, meaning that for a portion of the cycle, the voltage and current are both zero. Figure.4.21 (d) depicts this intermittent conduction mode of operation. Voltage and current are derived for both continuous and discontinuous conduction modes as follows.

Continuous conduction mode (for $0 \le \alpha \le 30^{o}$) : If the phase to neutral voltage $E_{AN} = E_m\ sin\omega t$, then

(a) Average load voltage is given by

$$E_{dc} = \frac{1}{2\pi/3}\int_{\alpha+30^{o}}^{\alpha+150^{o}} E_m \sin\omega t.d(\omega t) = \frac{3E_m}{2\pi}[-\cos\omega t]_{\alpha+30^{o}}^{\alpha+150^{o}} = \frac{3\sqrt{3}}{2\pi}E_m Cos\alpha \tag{4.51}$$

(b) Average load current is given by

$$I_d = \frac{3\sqrt{3}}{2\pi R}E_m Cos\alpha \tag{4.52}$$

(c) The maximum value of DC output voltage (V_{dm} or E_{dm}), is expressed as

$$V_{dm} = E_{dm} = \frac{3\sqrt{3}}{2\pi}E_m \tag{4.53}$$

(d) The normalized value of DC output voltage (V_n or E_n), is expressed as

$$V_n \text{ or } E_n = \frac{E_{dc}}{E_{dm}} = Cos\alpha \tag{4.54}$$

(e) The RMS load voltage for a given firing angle α is given by:

$$E_{rms} = \left[\frac{3}{2\pi}\int_{\alpha+30^{o}}^{\alpha+150^{o}} E_m^2 \sin^2\omega t\ d(\omega t)\right]^{1/2} = E_m\left[\frac{3}{2\pi}\int_{\alpha+30^{o}}^{\alpha+150^{o}}\left(\frac{1-\cos 2\omega t}{2}\right)d(\omega t)\right]^{1/2}$$

$$= E_m\left[\frac{3}{4\pi}\int_{\alpha+30^{o}}^{\alpha+150^{o}} d(\omega t) - \frac{3}{4\pi}\int_{\alpha+30^{o}}^{\alpha+150^{o}}\cos 2\omega t\ d(\omega t)\right]^{1/2}$$

$$= E_m\left[\frac{3}{4\pi}(\omega t)_{\alpha+30^{o}}^{\alpha+150^{o}} - \frac{3}{4\pi}\left(\frac{\sin 2\omega t}{2}\right)_{\alpha+30^{o}}^{\alpha+150^{o}}\right]^{1/2}$$

$$= E_m\left[\frac{3}{4\pi}[\alpha+150^{o}-30^{o}] - \frac{3}{8\pi}[\sin 2(\alpha+150^{o}) - \sin 2(\alpha+30^{o})]\right]^{1/2}$$

$$= E_m\left[\frac{1}{2} - \frac{3}{8\pi}2(-\cos 2\alpha)\frac{\sqrt{3}}{2}\right]^{1/2} = E_m\left[\frac{1}{2} + \frac{3\sqrt{3}}{8\pi}\cos 2\alpha\right]^{1/2} \tag{4.55}$$

Discontinuous conduction mode (for $30^{o} \le \alpha \le 150^{o}$)

(a) Average load voltage is given by

$$E_{dc} = \frac{1}{2\pi/3}\int_{\alpha+30^{o}}^{180^{o}} E_m \sin\omega t.d(\omega t) = \frac{3E_m}{2\pi}\left[1+\cos(\alpha+30^{o})\right] \tag{4.56}$$

(b) Average load current is given

$$I_d \text{ or } I_O \text{ or } I_{dc} = \frac{3E_m}{2\pi R}\left[1+\cos(\alpha+30^{o})\right] \tag{4.57}$$

(c) The RMS load voltage is given by

$$E_{rms} = \left[\frac{3}{2\pi/3} \int_{\alpha+30^\circ}^{180^\circ} E_m^2 \sin^2 \omega t \, d(\omega t) \right]^{1/2} = \frac{\sqrt{3}E_m}{2\sqrt{2}} \left[\frac{5\pi - 3\alpha}{3\pi} + \frac{\sin(2\alpha + \pi/3)}{\pi} \right]^{1/2} \quad (4.58)$$

4.9.1.2. Three phase half wave-controlled rectifier with inductive load (RL)

The circuit diagram of a three-phase half-wave-controlled rectifier with RL load is shown in Figure.4.22. The transformer primary is connected in a delta, and the secondary is connected in a star, as shown in the diagram. Figure.4.23 displays the waveforms for the three-phase half wave-controlled rectifier feeding RL load assuming continuous load current. Figure.4.23 (a) shows the waveforms for the firing angle of 0°. Each thyristor conducts for a period of 120° of supply voltage, and the output has three pulses because the ripple voltage fundamental frequency is three times the supply frequency. Each thyristor current waveform is unidirectional, rectangular in shape, and has a 120° duration of conduction.

Figure.4.23 (b) shows the waveforms for the firing angle of 45°. Because of this change, the fundamental component of the ac input current now lags the voltage by 45°, and the average voltage at the dc output has been lowered.

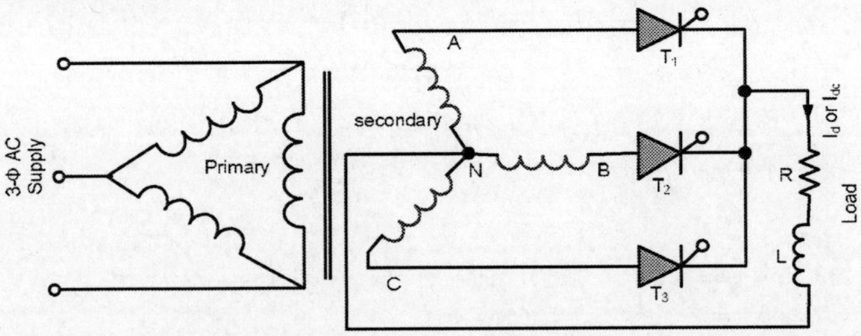

Figure.4.22 Three phase half wave-controlled rectifier with RL load.

Figure.4.23(c) shows the waveforms for the firing angle of 90°. The fundamental component of the current waveform of the ac input lags the voltage by 90°, so the average voltage at the dc output is zero. Even when the instantaneous voltage is in the negative range, the continuous flow of current is maintained because of the large smoothing reactor. Therefore, to achieve rectifier operation, the firing angle must be changed from 0° to 90°.

Figure.4.24 (a) shows the waveforms for the firing angle of 135°. The dc terminals have a negative mean voltage when measured with a dc voltmeter. Because of this, the rectifier connection acts as a load within the dc circuit and transfers energy from the dc circuit to the ac system.
Figure.4.24 (b) shows the waveforms for the firing angle of 180°. The maximum value of the mean voltage at the dc terminals is negative, which is almost equal to and the opposite of the maximum value of the positive value that can be obtained when the angle is set to 0°. As a result, the converter is now in "inversion mode," with a nearly 180° phase shift between the fundamental component of the ac input current waveform and the voltage.

The graph of the average output voltage versus the firing angle is displayed in Figure.4.24. The following can be used to derive the expression for the average dc output voltage:

$$E_{dc} = \frac{1}{2\pi/3} \int_{\alpha+30^\circ}^{\alpha+150^\circ} E_m \sin \omega t \, d(\omega t) = \frac{3\sqrt{3}}{2\pi} E_m \cos\alpha \quad (4.59)$$

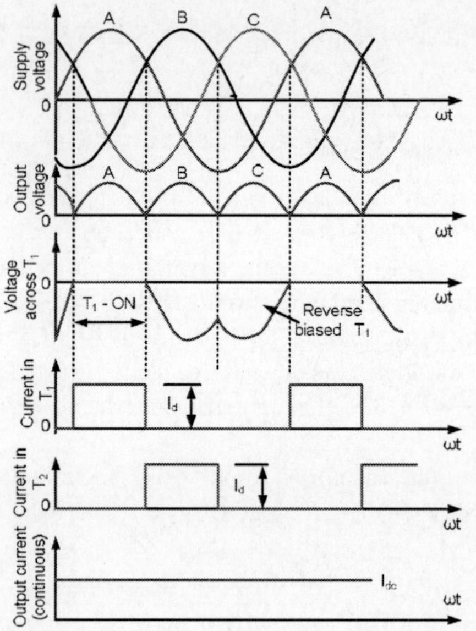

Figure.4.23 (a) Voltage and current waveforms for $\alpha = 0°$.

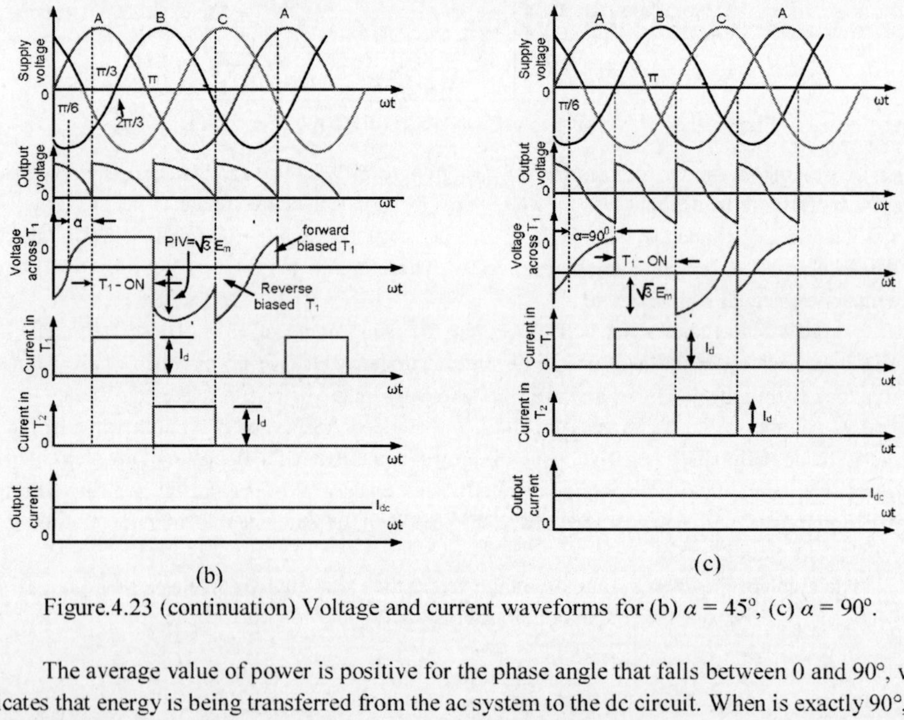

Figure.4.23 (continuation) Voltage and current waveforms for (b) $\alpha = 45°$, (c) $\alpha = 90°$.

The average value of power is positive for the phase angle that falls between 0 and 90°, which indicates that energy is being transferred from the ac system to the dc circuit. When is exactly 90°, both the positive and negative values of power cancel out, and the effective power is zero. In the case where α is greater than 90°, the average power is negative, and energy is transferred from the dc circuit to the ac system.

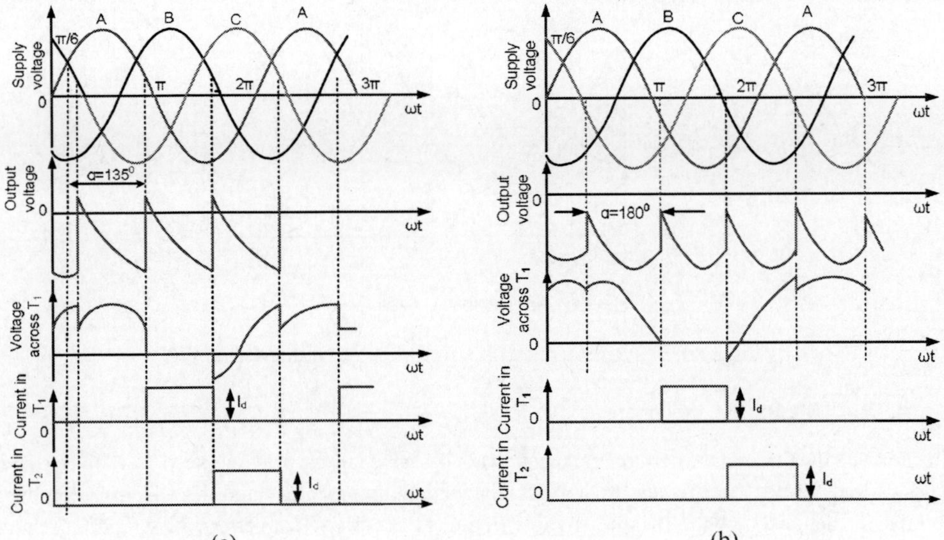

(a) (b)

Figure.4.24 Voltage and current waveforms for (a) $\alpha = 135°$, (b) $\alpha = 180°$

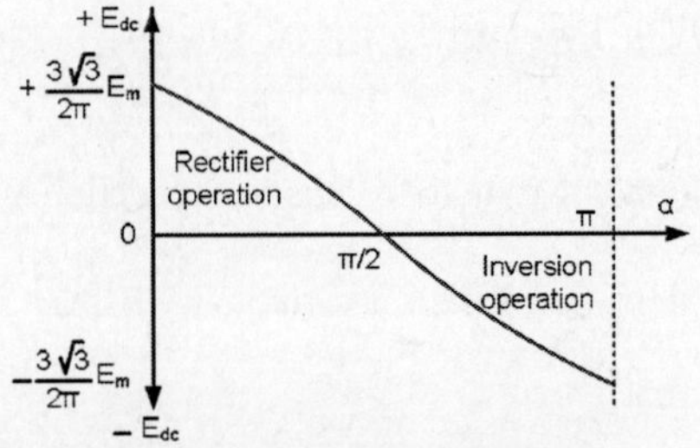

Figure.4.25 Graph of E_{dc} vs. α.

4.9.1.3. Three phase half wave-controlled rectifier with inductive load (RL) and freewheeling diode

The schematic for a three-pulse mid-point converter that includes a freewheeling diode is shown in Figure.4.26. Figure.4.27 depicts the resulting waveforms. The dc terminal voltage of the converter is always positive for firing angles that are less than 30°, and the freewheeling diode does not come into operation during these circumstances. When the firing angle exceeds this threshold, the diode begins to allow the load current to freewheel through it for brief intervals, interrupting the input line current and preventing the dc terminal load voltage from falling below zero. This circuit has a maximum firing angle range of 150°.

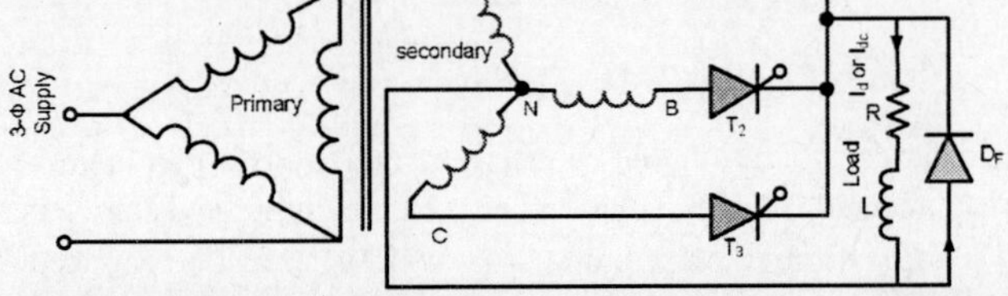

Figure.4.26 Three phase half wave controlled rectifier with RL load and freewheeling diode.

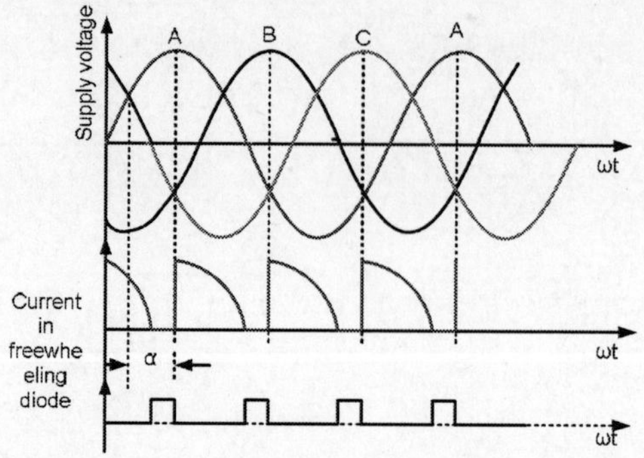

Figure.4.27 Waveforms with a freewheeling diode for $\alpha = 60^{o}$.

4.9.2. Six Pulse Converters

The fact that the three-pulse converters, which were described in the previous section, require particular types of transformer arrangement to stop DC magnetization has prevented them from becoming very popular. To achieve both high output and low ripple, a three-phase converter with a high pulse number has been designed. Connections with six pulses are the most common type utilized in industrial applications; however, connections with twelve pulses are favored when it comes to transmission lines.

When compared to three-pulse converters, six-pulse converters have several advantages, including the following:

(a) Easy commutation.

(b) The decrease in lower-order harmonics results in a reduction in the amount of distortion on the AC side.

(c) The amount of inductance needed in a series on the load side reduces the ripple greatly diminished.

4.9.2.1 Basics of three-phase fully controlled bridge converter

Figure.4.27 (a) depicts a 6-pulse bridge converter, which can be formed by connecting the DC terminals of two 3-pulse converters in series. This type of converter is referred to as a bridge converter. This converter sees extensive use in industrial applications that necessitate a two-quadrant operation.

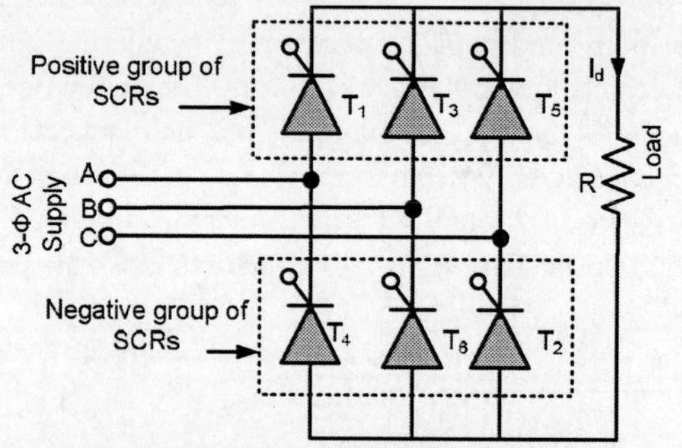

Figure.4.27 (a) Three-phase full converter.

In this particular configuration, the connection of the transformer is not required. However, the transformer is to be connected if better isolation between the output and source is needed, or if a higher output is required. If a transformer is used, one of its windings must be connected in the delta because this configuration provides a path for third-harmonic current to circulate. As a result, the third-harmonic current does not appear in phase, which is an improvement. In this bridge, six numbers of SCRs are used. They are denoted as T_1, T_2, T_3, T_4, T_5, and T_6 A positive group is formed here by using T_1, T_3, and T_5, whereas a negative group is formed here by using T_4, T_6, and T_2. When the supply voltages are positive, the SCRs belonging to the positive group are activated, and when the supply voltages are negative, the SCRs belonging to the negative group are activated. When the SCRs are connected to the AC supply, the firing gate pulses are applied to them in the appropriate order to ensure proper operation. However, if only a single SCR is triggered, there will be no flow of current because the other SCR that is in the path of the current will be in the off-state. Therefore, to get the circuit to work, we have to simultaneously fire two SCRs, one is of the positive group and the other is of the negative group. This allows current to start flowing through the circuit. To provide an accurate description of how the circuit works, one must keep in mind the following points:

(a) It is important that each SCR be triggered at the appropriate firing angle.
(b) The maximum angle of conduction for each SCR is 120°.
(c) It is necessary to trigger SCRs in the order T_1, T_2, T_3, T_4, T_5, and T_6.
(d) The phase shift that occurs between the triggering of the two SCRs that are adjacent to one another is 60°.
(e) At any instant, two SCRs can conduct and there are six pairs.
(f) At any given time, two of the SCRs have the potential to conduct, and there are a total of six pairs (i.e., (T_6, T_1), (T_1, T_2), (T_2, T_3), (T_3, T_4), (T_4, T_5) and (T_5, T_6)).
(g) Each pair has a 60° angle of operation.
(h) Incoming SCRs are used to commutate outgoing SCRs; for example, T_1 is used to commutate T_5; T_2 is used to commutate T_6, and so on.
(i) It is the line voltage that is applied across the load when both SCRs are conducting, one from the positive group and one from the negative group.
(j) The first half-bridge (T_1/T_4) is connected to Phase A, the second half-bridge (T_3/T_6) is connected to Phase B, and the third half-bridge (T_5/T_2) is connected to Phase C.
(k) When a half-bridge top (positive group) SCR conducts, the corresponding phase current is positive; when the bottom (negative group) SCR conducts, the corresponding phase current is negative.
(l) It is evident from Figure.4.27 (b) that when the phase voltage makes a 30° angle with neutral, the upper SCR in that half-bridge is forward biased. Similarly, the lower SCR is considered to have a forward bias when its phase forms an angle of 210° with neutral. As a result, when ωt = 30°, α = 0°.

The conducting pair, as well as the incoming and outgoing SCRs, are shown in Table.4.1.

Table.4.1 Firing sequence of SCRs

S.No.	ωt	**Incoming SCR**	**Conducting pair**	**Outgoing SCR**	**Line voltage across the load**
1	$30^\circ + \alpha$	T_1	(T_6, T_1)	T_5	E_{AB}
2	$90^\circ + \alpha$	T_2	(T_1, T_2)	T_6	$EACH$
3	$150^\circ + \alpha$	T_3	(T_2, T_3)	T_1	E_{BC}
4	$210^\circ + \alpha$	T_4	(T_3, T_4)	T_2	E_{BA}
5	$270^\circ + \alpha$	T_5	(T_4, T_5)	T_3	E_{CA}
6	$330^\circ + \alpha$	T_6	(T_5, T_6)	T_4	E_{CB}

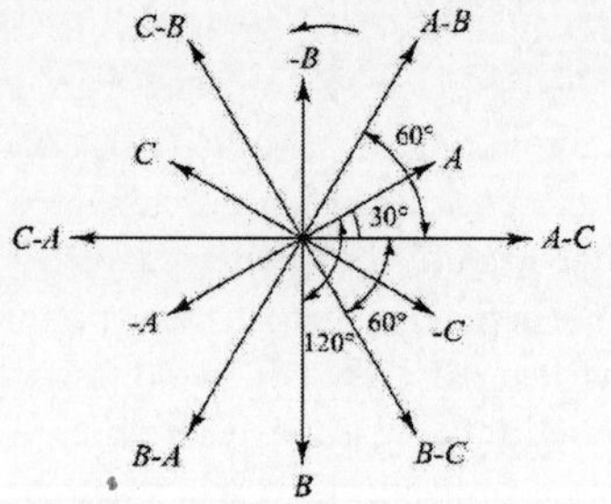

Figure.4.27 (b) Vector diagram.

The following are the defining equations for phase and line voltages:

$E_{AN} = E_m \sin(\omega t)$ $E_{AB} = \sqrt{3}\, E_m \sin(\omega t + 30^\circ)$ $E_{BA} = \sqrt{3}\, E_m \sin(\omega t - 150^\circ)$

$E_{BN} = E_m \sin(\omega t - 120^\circ)$ $E_{AC} = \sqrt{3}\, E_m \sin(\omega t - 30^\circ)$ $E_{CA} = \sqrt{3}\, E_m \sin(\omega t + 150^\circ)$

$E_{CN} = E_m \sin(\omega t + 120^\circ)$ $E_{BC} = \sqrt{3}\, E_m \sin(\omega t - 90^\circ)$ $E_{CB} = \sqrt{3}\, E_m \sin(\omega t + 90^\circ)$

4.9.2.2 Three-phase fully controlled bridge rectifier with resistive (R) load

Figure.4.28 depicts a three-phase fully controlled bridge rectifier supplying an R load. Each SCR must be pulsed twice in its conduction cycle, or at 60° intervals, for a six-pulse operation. There are six line-to-line phasors, with a maximum conduction angle of 60°, as shown in Figure.4.27 (b). Figure.4.29 (a) displays the output voltage waveform for various values of α = 0°, 30°, 60°, 90°, and 120°.

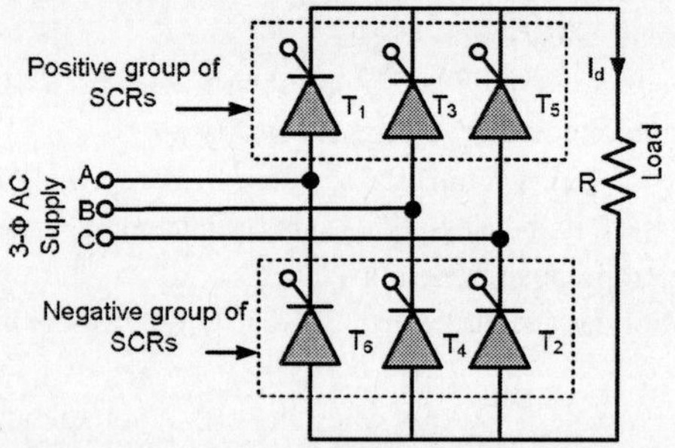

Figure.4.28 Three-phase fully controlled bridge rectifier with R load.

From Figure.4.29 (a), the following points can be made:

(a) For all values of α, the voltage output follows a six-pulse wave with a 300 Hz ripple frequency.

(b) **Continuous conduction mode:** $(0 \le \alpha \le \pi/3)$: When the phasor (A-B) is allowed to conduct at an angle between 0 and $\pi/3$, it will continue to conduct by 60° after the phasor (A-C) has been fired. T_6 conduction is shifted to T_2 conduction. T_6 is turned off as a result of the reverse voltage that is applied across it from phases C and B. After conducting for 60°, the phasor (A-C) is switched out for the phasor (B-C) when the voltage of phase B becomes higher than that of the phases C and A. As a result, the load current remains continuous for firing angles ranging from 0 to $\pi/3$.

(c) **Discontinuous conduction mode:** $(\pi/3 \le \alpha \le 2\pi/3)$: When $\pi/3 \le \alpha \le 2\pi/3$, the phasor ($A$-$B$) conducts up to an angle of 180°, after which both the T_1 and T_6 are turned off and after 60°, when T_2 and T_1 are triggered, phase (A-C) conducts also up to angle 180°, therefore the load current remains zero (discontinuous) from angle 180° to the next firing pulse, therefore the fully-controlled bridge circuit produces a ripple frequency of six-times the supply.

(d) The output voltage is zero when the firing angle is 120°, so the maximum firing angle is also 120°.

Following is the derivation of the voltage and current expressions for both modes:

(i) Continuous conduction mode: $(\alpha \le 60^\circ)$

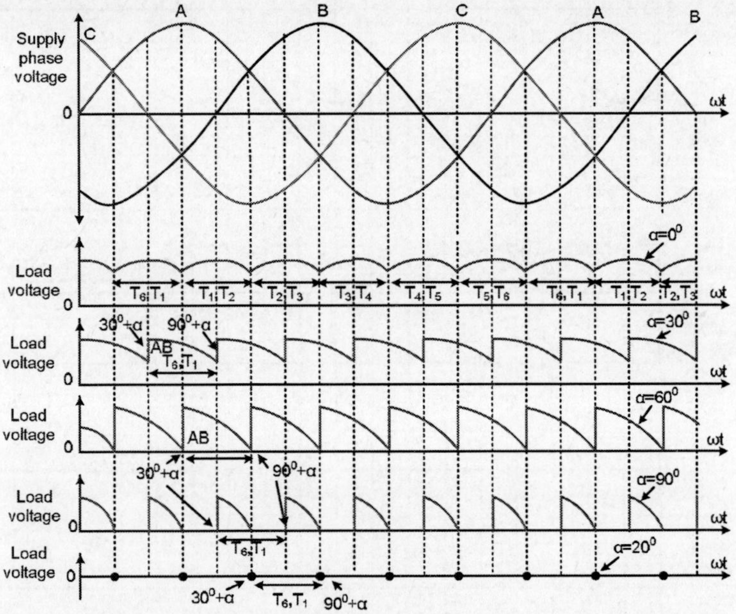

Figure.4.29 (a)Waveform for Voltage of 3-Φ full wave bridge rectifier with R load at various α

Average load voltages are given by

$$E_{dc} = \frac{1}{2\pi}\int_{0}^{2\pi} E_{dc}\omega t\, d(\omega t) \tag{4.60}$$

$$E_{dc} = 6 \times \frac{1}{2\pi}\int_{\frac{\pi}{6}+\alpha}^{\frac{\pi}{2}+\alpha} E_{AB}\,\omega t\, d(\omega t)$$

$$E_{dc} = \frac{3}{\pi}\int_{\frac{\pi}{6}+\alpha}^{\frac{\pi}{2}+\alpha} \sqrt{3}E_m \sin\left(\omega t + \frac{\pi}{6}\right) d(\omega t) = \frac{3\sqrt{3}E_m}{\pi}\int_{\frac{\pi}{3}+\alpha}^{\frac{2\pi}{2}+\alpha} \sin \omega t\, d(\omega t)$$

$$= \frac{3\sqrt{3}E_m}{\pi}(\cos \omega t)_{2\pi/3+\alpha}^{\pi/3+\alpha} = \frac{3\sqrt{3}E_m}{\pi}\left[\cos\left(\frac{\pi}{3}+\alpha\right) - \cos\left(\frac{2\pi}{3}+\alpha\right)\right]$$

$$= \frac{3\sqrt{3}E_m}{\pi}\left[\cos\left(\frac{\pi}{3}+\alpha\right) + \cos\left(\frac{\pi}{3}-\alpha\right)\right]$$

$$= \frac{3\sqrt{3}E_m}{\pi}\left(2.\cos\left(\frac{\pi}{3}\right).\cos\alpha\right) = \frac{3\sqrt{3}E_m}{\pi}\cos\alpha \tag{4.61}$$

The average load current is given by

$$I_d = \frac{3\sqrt{3}E_m}{\pi R}\cos\alpha \tag{4.62}$$

(ii) Discontinuous conduction mode: $(\alpha > 60^{o})$

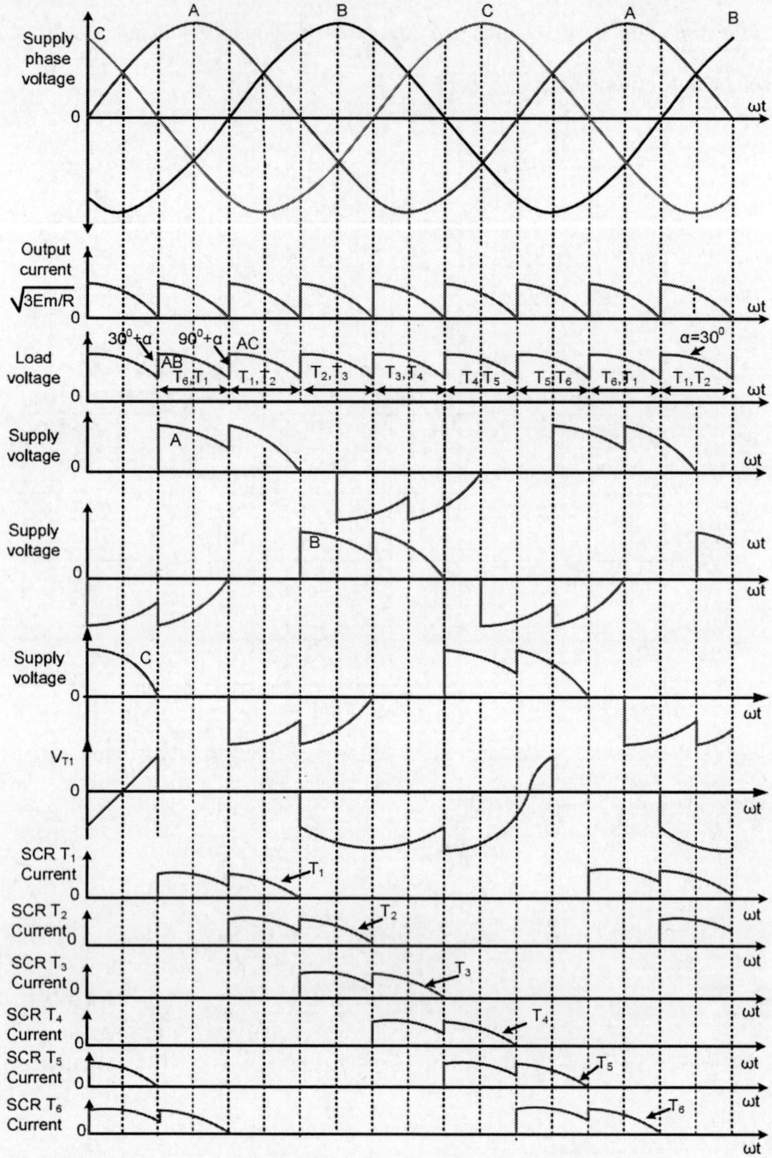

Figure.4.29 (b) Waveforms at firing angle 30°. in 3-Φ full wave bridge rectifier with R load

Form Figure.4.29 (a)

$$E_{dc} = 6 \times \frac{1}{2\pi} \int_{\frac{\pi}{6}+\alpha}^{\frac{5\pi}{2}} \sqrt{3}E_m \sin\left(\omega t + \frac{\pi}{6}\right) = \frac{3\sqrt{3}E_m}{\pi} \int_{\frac{\pi}{3}+\alpha}^{\pi} \sin \omega t \, d(\omega t)$$

$$\frac{3\sqrt{3}E_m}{\pi} (\cos \omega t)_{\pi}^{\pi/3+\alpha} = \frac{3\sqrt{3}E_m}{\pi}\left[1 + \cos\left(\alpha + \frac{\pi}{3}\right)\right] \tag{4.63}$$

For α_{max}, $E_{dc} = 0°$

$$\therefore \frac{3\sqrt{3}E_m}{\pi}\left[1+\cos\left(\alpha+\frac{\pi}{3}\right)\right]=0$$

Hence, $\alpha_{max} = 120°$

$$I_d = \frac{3\sqrt{3}E_m}{\pi R}\left[1+\cos\left(\alpha+\frac{\pi}{3}\right)\right] \tag{4.64}$$

In the Figure.4.29(a) only load voltage waveforms at various values of α are shown. The other waveform say current in the load, through the SCRs, input current in various phases, and the voltage across SCRs (only for T_1) are shown in Figure.4.29(b) for $\alpha = 30^0$. The following can be deduced from these waveforms:

(a) Waveforms of the output voltage and current are six pulses with a ripple frequency of 300 Hz.

(b) The waveforms of the supply current are 120° in each half cycle when is less than 60°, but when is greater than 60° (which is not shown in the figure), the waveforms are less than 120° in each half cycle.

(c) PIV rating of SCR is $\sqrt{3}E_m$.

4.9.2.3 Three phase fully controlled bridge rectifier with highly inductive load

Figure.4.30 (a) shows the three-phase fully controlled bridge rectifier with a highly inductive load. The SCRs are triggered at $\pi/3$. The frequency of the output ripple is $6f_s$ where f_s is the switching frequency. The requirement of the filter at the output is less than that of half wave rectifier. At angle $\pi/6+\alpha$, the SCR T_6 is already conducting and SCR T_1 is fired. In the interval $(\pi/6+\alpha) \le \omega t \le (\pi/2+\alpha)$, T_6 and T_1 conduct and line voltage is E_{AB} appear across the load. At $\omega t = \pi/2+\alpha$, the SCR T_6 becomes reverse biased and turned off and SCR T_2 is fired. In the interval $(\pi/2+\alpha) \le \omega t \le (5\pi/6+\alpha)$, T_2 and T_1 conduct and line voltage are E_{AC} appears across the load. With the numbering of SCRs numbered as shown, the triggering sequence is 12, 23, 34, 45 56, and 61. The waveforms for voltage and current are depicted in Figure.4.30(b).

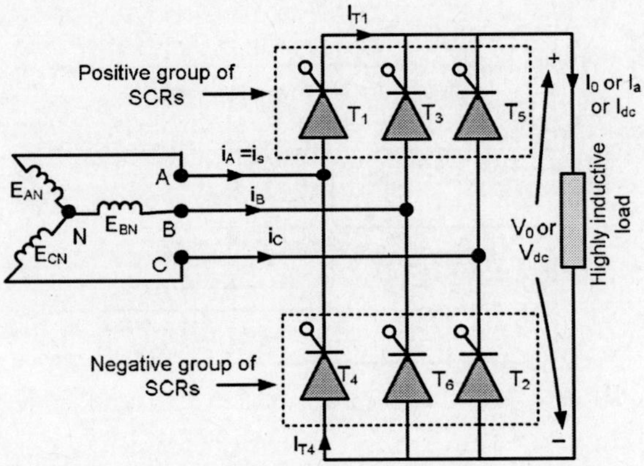

Figure.4.30(a) Circuit diagram for three-phase full converter

The average output voltage is obtained as

$$E_{dc} = 6\times\frac{1}{2\pi}\int_{\frac{\pi}{6}+\alpha}^{\frac{\pi}{2}+\alpha} E_{AB}\,\omega t\; d(\omega t) = \frac{3}{\pi}\int_{\frac{\pi}{6}+\alpha}^{\frac{\pi}{2}+\alpha} \sqrt{3}\,E_m \sin(\omega t + 30)\; d(\omega t) = \frac{3}{\pi}\int_{\frac{\pi}{3}+\alpha}^{\frac{2\pi}{3}+\alpha} \sqrt{3}\,E_m \sin(\omega t)\; d(\omega t)$$

$$= \frac{3\sqrt{3}E_m}{\pi}[\cos(\omega t)]_{120+\alpha}^{60+\alpha} = \frac{3\sqrt{3}}{\pi}E_m[\cos(60+\alpha)-\cos(120+\alpha)].$$

$$E_{dc} = \frac{3\sqrt{3}E_m}{\pi}\cos\alpha \text{ for } 0 \le \alpha \le 180^\circ \tag{4.65}$$

The maximum value of output voltage is obtained when $\alpha = 0^0$. It is given by E_{dcm} or V_{dcm} and given by

$$E_{dcm} \text{ or } V_{dcm} = \frac{3\sqrt{3}E_m}{\pi} \tag{4.66}$$

The normalized average output voltage is given by

$$E_n \text{ or } V_n = \frac{E_{dc}}{E_{dcm}} = Cos\alpha \tag{4.67}$$

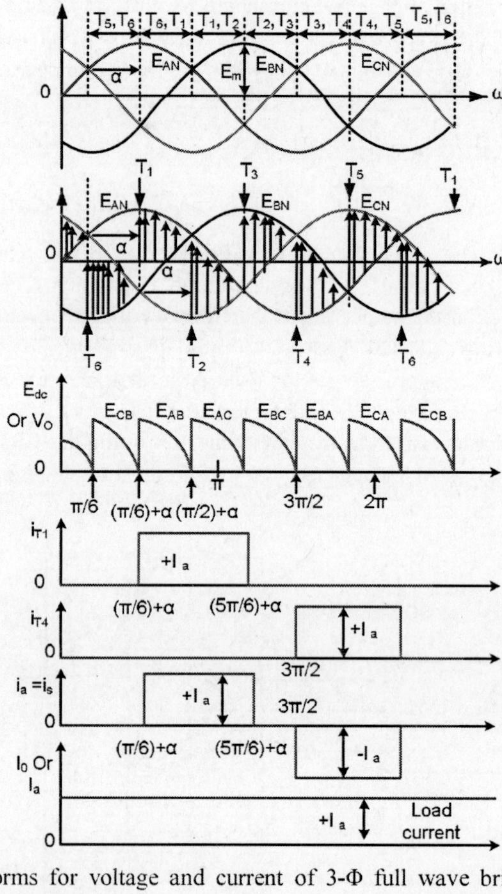

Figure.4.30(b) Waveforms for voltage and current of 3-Φ full wave bridge converter with highly inductive load

The converter operation is said to be in **rectification mode** when the firing angle goes from 0° to 90°, at which point the voltage drops from its maximum to its minimum. When the firing angle is between 90° and 180°, the converter is operating in the **inversion mode**, and the voltage will range from 0 to a maximum negative value.

The expression for the RMS value of output voltage is obtained as follows

$$E_{rms} = \left[\frac{1}{2\pi}\int_0^{2\pi} E_{dc}^2(\omega t)\, d(\omega t)\right]^{1/2} = \left[6\times\frac{1}{2\pi}\int_{30+\alpha}^{90+\alpha} E_{AB}^2(\omega t)\, d(\omega t)\right]^{1/2}$$

$$= \left[\frac{9E_m^2}{2\pi}\left\{60^\circ + \frac{1}{2}(\sin(120+2\alpha) - \sin(240+2\alpha))\right\}\right]^{1/2}$$

$$= \left[\frac{9E_m^2}{2\pi}\left\{60^\circ + \frac{1}{2}\left(\sqrt{3}\cos 2\alpha\right)\right\}\right]^{1/2}$$

$$= \frac{3E_m}{2}\left[\frac{2}{3} + \frac{\sqrt{3}}{\pi}\cos 2\alpha\right]^{1/2}, \text{ for } 0 \le \alpha \le 180^\circ$$

$$= E_{rms} = \sqrt{3}E_m\left(\frac{1}{2} + \frac{3\sqrt{3}}{4\pi}\cos 2\alpha\right)^{1/2} \tag{4.68}$$

4.9.2.4 Three phase fully controlled bridge rectifier with R-L load for various values of α

Figure.4.31 (a) illustrates the waveforms of the output voltage along with the firing angles. It is presumed that the load inductance has a very large value to flow a constant load current.
Figure.4.31 (a) Output voltage waveform with RL load.
The following observations can be made based on Figure.4.31.

(a) Waveforms with R-load for $\alpha = 0^\circ$, 30°, and 60° are comparable.
(b) Due to the inductive nature of the load, the voltage turns negative for $\alpha > 60^\circ$.
(c) When α is equal to 90°, the area under the positive cycle and the area under the negative cycle are the same size, which means that the average voltage is 0.
(d) When α is less than or equal to 90°, the output voltage is (+), but when α is greater than or equal to 90°, the average output voltage is (-).
(e) The maximum firing angle $\alpha = 180^\circ$.

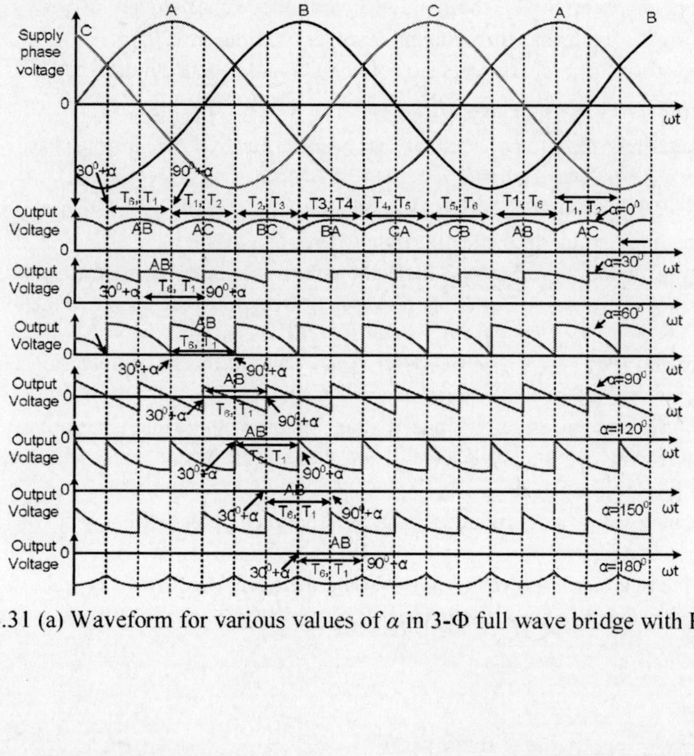

Figure.4.31 (a) Waveform for various values of α in 3-Φ full wave bridge with R-L load

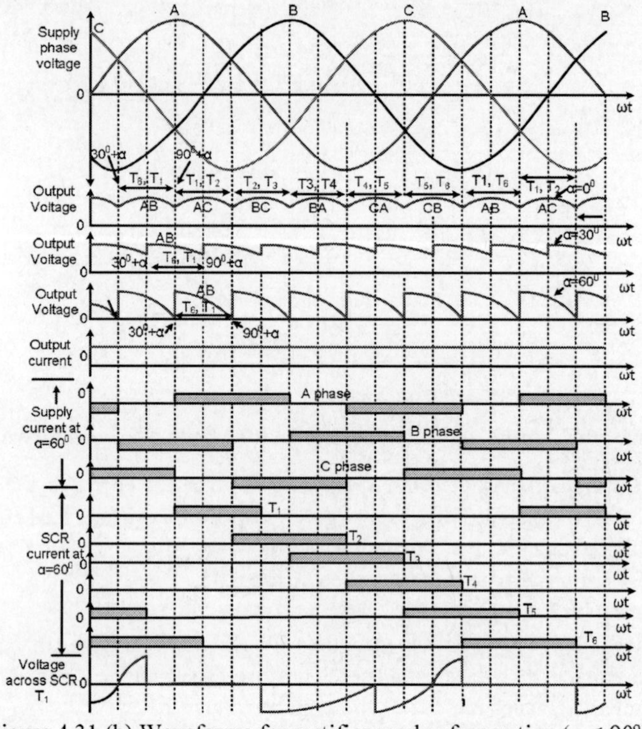

Figure.4.31 (b) Waveforms for rectifier mode of operation ($\alpha < 90°$).

Rectifier Operation (α < 90°) Rectifier mode waveforms are depicted in Figure.4.31 (b). The following observations can be made based on these waveforms:

a) For firing angles up to 90°, the supply waveforms are quasi-square waves.
b) SCR current waveforms are rectangular waves of duration 120°.
c) The waveforms of SCR currents are rectangular and have a duration of 120°.
d) Thyristor has a PIV value of $\sqrt{3}E_m$.
e) The phase of the three supply currents has been shifted by 120°, and the phase of the six thyristor currents has been shifted by 60°.
f) The waveform of the load current is continuous and free of ripples; consequently, the average, RMS, and peak values are all equivalent to the value of I_{dc}.
g) Because E_{dc} and I_{dc} are both positive, the circuit works in the first quadrant. This is because positive $P_{dc}(av)$ values indicate that power is flowing from the source to the load.

Inverter Operation (90° < α < 180°) Inverter mode waveforms are depicted in Figure.4.31 (c). The following observations can be made based on these waveforms:

a) The inverting operation results in an average output voltage that is negative.
b) The waveforms for supply currents have the appearance of quasi-square waves and have a mutual phase shift of 120°.
c) The current waveforms produced by an SCR are rectangular with a width of 120° and a mutual phase shift of 60°.
d) Since average output power $P_{dc}(av)$ is in the negative, the flow of power is typically from the load to the source (since E_{dc} is in the negative and I_{dc} is in the positive).

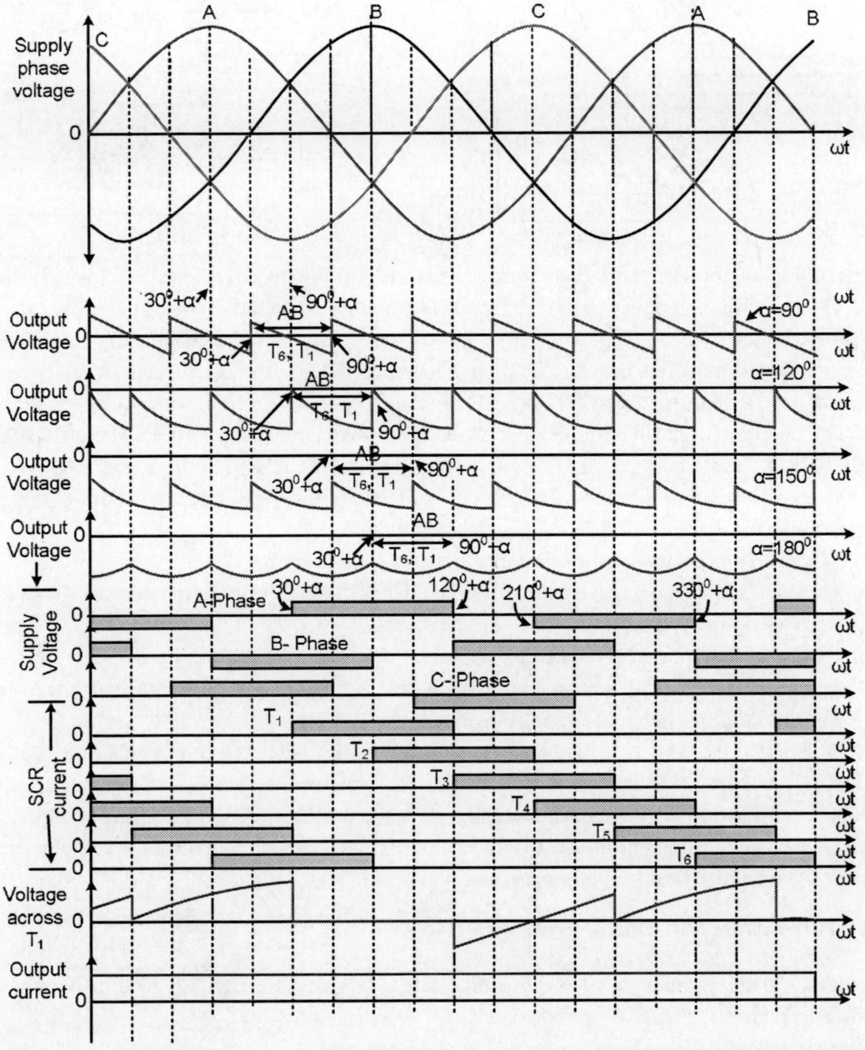

Figure.4.31 (c) Waveforms for inversion operation ($\alpha > 90^{o}$) of 3-Φ full wave bridge with R-L load.

4.10. THREE PHASE HALF CONTROLLED BRIDGE CONVERTER

It is possible to put bridge-connected rectifiers into a freewheeling mode of operation by exchanging half of their SCRs for diodes in the circuit. As a result, the circuit of a three-phase half-controlled bridge converter includes three SCRs in three of the arms and three diodes in the remaining three arms. Figure.4.32 (a) presents the schematic representation of the three-phase symmetrical-half controlled bridge rectifier that can be constructed. In this instance, the asymmetrical configuration is not utilized because doing so would result in an imbalance in the line currents on the ac side.

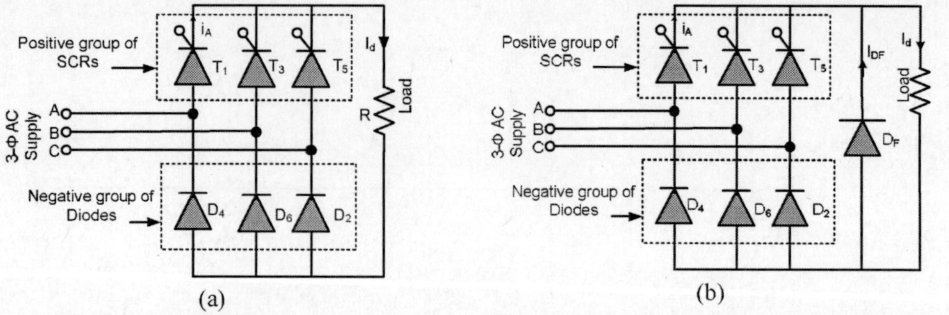

Figure.4.32 Half-controlled three-phase bridge converter (a) R load, (b) R load, and freewheeling diode

In industrial applications that require only one quadrant of operation, three-phase semiconverters are typically the device of choice. However, compared to three-phase half-wave converters, this one performs better, even though its power factor drops as the delay angle grows larger. This converter has built-in freewheeling action, which contributes to an improvement in its power factor. The action of freewheeling occurs between the SCR and diode in the same arm, specifically between SCR T_1 and diode D_4; SCR T_3 and diode D_6, and SCR T_5 and diode D_2. The voltage drop across the load ($V_{TH} + V_D$) becomes approximately $2V$ due to this freewheeling action.

As a result, this results in the following disadvantages:

(a) The average output voltage will be reduced as a result of this voltage drop.

(b) Conduction losses increase while the converter is in the free-wheeling period, which leads to a decrease in the converter's overall efficiency.

(c) Since SCRs continue to conduct even when the line voltage is in its negative cycle, we can deduce that the period of conduction of every SCR is 180°. As a result, both the average and the rms current ratings of the SCRs will increase.

Putting an external freewheeling diode across the load, as demonstrated in Figure.4.32 (b), is one way to mitigate all of these potential drawbacks. This results in a lower on-state voltage drop across the load as well as lower conduction losses. Additionally, it results in a more continuous flow of current through the load. The free-wheeling diode also ensures that each SCR will commutate at the end of its respective half cycle. As a result, the rating of the devices will decrease.

4.10.1 Three Phase half control bride with R load

The voltage and current waveforms for the three-phase semiconverter with R load (of Figure.4.32 (a)) are shown in Figure.4.33 (a).

The following details are essential for grasping the procedure.

(a) Once a diode is forward-biased, it will begin conducting.

(b) The order in which the line voltages are conducted is as follows: E_{AB}, E_{AC}, E_{BC}, E_{BA}, E_{CA} and E_{CB}.

(c) A line voltage with the highest value in comparison to others will conduct, i.e. when it forms a 60° angle with neutral.

(d) Devices that are conducting at each line voltage are listed in Table.4.2.

(e) There are six devices, and the phase voltage of each device is considered forward-biased when it is at an angle of 30° with the neutral. Table.4.3 gives $\alpha = 0$ for each device.

(f) Each line phasor and, consequently, each pair conducts when α is equal to 60°. The instant that each phasor begins conducting is listed in Table.4.4, along with the incoming device, the outgoing device, the pair conduction, and the conducting period of each pair.

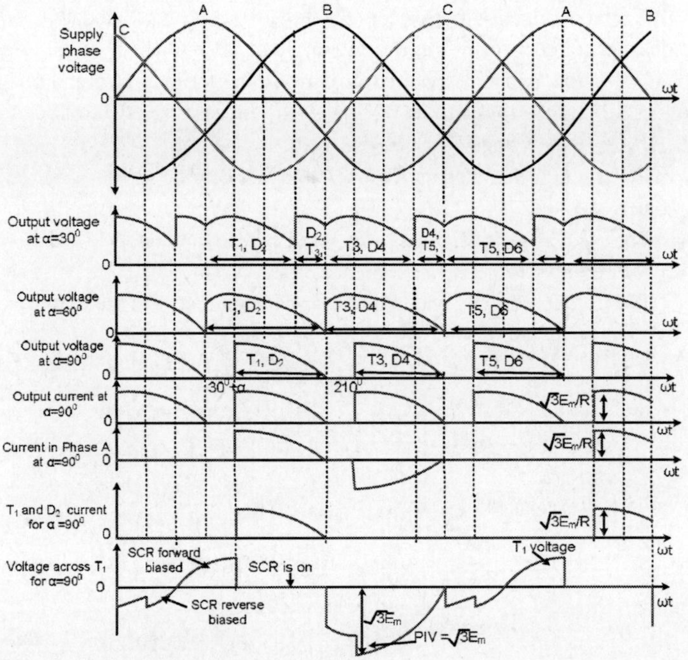

Figure.4.33 (a) Voltage and current waveforms for three-phase semiconverter with R load ($\alpha = 30^o$, 60°, and 90°).

Table.4.2

S.No	Conducting line voltages	Conducting devices
1	E_{AB}	(D_6, T_1)
2	E_{AC}	(T_1, D_2)
3	E_{BC}	(D_2, T_3)
4	E_{BA}	(T_3, D_4)
5	E_{CA}	(D_4, T_5)
6	E_{CB}	(T_5, D_6)

Table.4.3

S.No.	Devices	Phase voltage at 30° with neutral	Angle from $\omega t = 0$ concerning phase A
1	T_1	E_A	$\omega t = 30^o$
2	D_2	E_C	$\omega t = 90^o$
3	T_3	E_B	$\omega t = 150^o$
4	D_4	$-E_A$	$\omega t = 210^o$
5	T_5	$-E_C$	$\omega t = 270^o$
6	D_6	$-E_B$	$\omega t = 300^o$

(g) The following can be deduced from the information presented in Table.4.4

a) The phasors E_{AC}, $E_{BA.}$ and E_{CB} start conducting at $\omega t = 90^0$. 220°, and 330^0 respectively. The phasors E_{AB}, $E_{CB.}$ and E_{CA} start conducting at angle $\omega t = \alpha + 30^0$, $\alpha+150°$, and $\alpha+270^0$ respectively. Each SCR and diodes conduct for 120^0.

b) The phasors E_{AB}, E_{CB}, and E_{CA} conducts for 60-α. The phasors E_{AC}, E_{BA}, and E_{CB} for 60+α. The total conduction period for (E_{AB} + E_{AC}), (E_{BC} + E_{BA}), and (E_{CA} + E_{CB}) is 120°.

(h) All phasors conduct for 60° at a firing angle of 0°, but at any angle less than or equal to 60°, phasors E_{AB}, E_{CB}, and E_{CA} conduct for periods shorter than 60°, and phasors E_{AC}, E_{BA}, and E_{CB} conduct for periods longer than 60°. It is still the case that the total conduction period of (E_{AB} + E_{AC}), (E_{BC} + E_{BA}), and (E_{CA} + E_{CB}) is 120°.

a) For $\alpha = 60^0$, the phasors E_{AB}, E_{CB}, and E_{CA} conduct 0^0. and E_{AC}, E_{BA} and E_{CB} conduct for period 120^0.

b) For $\alpha > 60^0$, the phasors E_{AB}, E_{CB}, and E_{CA} do not conduct, and E_{AC}, E_{BA}, and E_{CB} conduct for period < 120^0.

c) If α is equal to 180°, then no phasor can conduct, and the output voltage will be 0.

Expression for output voltage:

Case-1 $\alpha \le 60^\circ$.

$$E_{dc} = 3\times\frac{1}{2\pi}\left[\int_{30+\alpha}^{90} E_{AB}(\omega t)\, d(\omega t) + \int_{90}^{150+\alpha} E_{AC}(\omega t)\, d(\omega t)\right]$$

Putting the value of $E_{AB}(\omega t)$ and $E_{AC}(\omega t)$

$$E_{dc} = \frac{3}{2\pi}\left[\int_{30+\alpha}^{90} \sqrt{3}\, E_m \sin(\omega t + 30)\, d(\omega t) + \int_{90}^{150+\alpha} \sqrt{3}\, E_m \sin(\omega t - 30)\, d(\omega t)\right]$$

$$= \frac{3\sqrt{3}E_m}{2\pi}[1 + \cos(60+\alpha) - \cos(120+\alpha)] = \frac{3\sqrt{3}E_m}{2\pi}[1+\cos\alpha] \quad (4.69)$$

Case-2 $\alpha \ge 60^\circ$

$$E_{dc} = 3\times\frac{1}{2\pi}\left[\int_{30+\alpha}^{210} E_{AC}(\omega t)\, d(\omega t)\right]$$

Substituting the value of $E_{AC}(\omega t)$, we get,

$$E_{dc} = \frac{3}{2\pi}\int_{30+\alpha}^{210} \sqrt{3}\, E_m \sin(\omega t - 30)\, d(\omega t) = \frac{3\sqrt{3}E_m}{2\pi}[\cos(\omega t - 30)]_{210}^{30+\alpha}$$

$$= \frac{3\sqrt{3}E_m}{2\pi}[\cos(\alpha) - \cos(180))] = \frac{3\sqrt{3}E_m}{2\pi}(1+\cos\alpha) \quad (4.70)$$

Expression for rms output voltage:

Case-1 $\alpha \le 60^\circ$,

The RMS output voltage is given by

$$E_{rms} = \left\{\frac{3}{2\pi}\left[\int_{30+\alpha}^{90} E_{AB}^2(\omega t)\, d(\omega t) + \int_{90}^{150+\alpha} E_{AC}^2(\omega t)\, d(\omega t)\right]\right\}^{1/2}$$

From which we will get,

$$E_{rms} = \left\{\frac{9E_m^2}{4\pi}\left[\frac{2\pi}{3} + \frac{\sqrt{3}}{2}(1+\cos 2\alpha)\right]\right\}^{1/2} = \frac{3}{2}E_m\left[\frac{2}{3} + \frac{\sqrt{3}}{2\pi}(1+\cos 2\alpha)\right]^{1/2} \quad (4.71)$$

Case-1 $\alpha > 60^\circ$,

The RMS output voltage is given by

$$E_{rms} = \left[\frac{3}{2\pi}\int_{30+\alpha}^{210} E_{AC}^2(\omega t)\, d(\omega t)\right]^{1/2} = \left[\frac{3}{2\pi}\int_{30+\alpha}^{210} \left(\sqrt{3}E_m \sin(\omega t - 30)\right)^2 d(\omega t)\right]^{1/2}$$

From which we will get,

$$E_{rms} = \frac{3E_m}{2}\left[\frac{\pi - \alpha + 1/2\sin 2\alpha}{\pi}\right]^{1/2} \quad (4.72)$$

Table.4.4 Incoming, and outgoing conducting devices

S.No	Firing instant	Income device	Outgoing device	Pair conducting and phasor	Conducting period of each pair	Conducting period of outgoing device
1	$\omega t = 30° + \alpha$	T_1	T_5	D_6 and T_1 (E_{AB})	$90° - (30° + \alpha) = 60° - \alpha$	$360° + 30° + \alpha - (270° + \alpha) = 120°$
2	$90°$	D_2	D_6	T_1 and D_2 (E_{AC})	$(150° + \alpha - 90°) = 60° + \alpha$	$360° + 90° - 330° = 120°$
3	$150° + \alpha$	T_3	T_1	D_2 and T_3 (E_{BC})	$210° - (150° + \alpha) = 60° - \alpha$	$150° + \alpha - (30° + \alpha) = 120°$
4	$210°$	D_4	D_2	T_3 and D_4 (E_{BA})	$270° + \alpha - 210° = 60° + \alpha$	$210° - 90° = 120°$
5	$270° + \alpha$	T_5	T_3	D_4 and T_5 (E_{CA})	$330° - (270° + \alpha) = 60° - \alpha$	$270° + \alpha - (150° + \alpha) = 120°$
6	$330°$	D_6	D_4	T_5 and D_6 (E_{CB})	$360° + (30° + \alpha) - 330° = 60° + \alpha$	$330° - 210° = 120°$

4.10.2 Operation 3-Φ half control bridge rectifier (Semiconverter) with inductive load

Figure.4.33 (b) illustrates the waveforms for a semiconverter operating with a large inductive load. It is possible to deduce the following information from the data presented in Figure.4.33 (b):

(a) The waveform of the voltage remains the same when a resistive (R) load is present. As a result, the average and rms values of the output voltage waveform are the same as those given by equations (4.69), (4.70), (4.71), and (4.72).

(b) **Continuous conduction mode:** When α is less than 60°, the waveform of the output current is continuous, and when is equal to 30°, there are no ripples visible. As a result, the current waveform has a form factor (FF) of unity, and the ripple factor has a value of zero.

(c) **Discontinuous conduction mode:** When the firing angle is greater than 60°, a discontinuous mode will occur, as will be shown for an angle of 90°. It is possible to deduce from the waveforms that the freewheeling action causes the output voltage to drop to zero for a portion of the time.

(d) SCRs and diodes both have a PIV rating of $\sqrt{3}E_m$.

(e) The maximum value of α is 180°.

(f) The highest possible value for α is 180°.

4.11. PHASE-CONTROLLED RECTIFIER WITH INPUT IMPEDANCE

Practically there are resistance and inductance present in the source. These parameters affect the change in current and for this change, some time is required. As a result the commutation of the outgoing SCRs delay. To commutate some fine time required. The current will also rise at the same rate in the case of incoming SCRs. The commutation process takes a significant period during which incoming and outgoing SCRs conduct simultaneously. This period is called the overlap period. The angle for which incoming and outgoing SCRs conducts is called the overlap angle or commutation angle. The commutation angle is denoted in general by the symbol μ. The waveform for voltage and current is also different from the circuit in which the input inductance is neglected. In the output, the effect of inductance is reflected in terms of reduction average voltage and distortion of total harmonic term. On the input side, the overlap angle slightly reduces the displacement factor and modifies the distortion terms. In the case of AC supply, the inductive reactance is greater than the resistance. The inductance is responsible for delaying the current change. If the source resistance is considered there is an excess voltage drop in it and average output power reduces. In the analysis of the power electronics converters, the input resistance is normally neglected because its value is very as compared to input inductive reactance. Hence, the effect on commutation angle is due to negligible resistance. In the case of fully controlled converters, the commutation overlap is more than that of semiconverters. The commutation In this section, the analysis of two circuits namely single-phase and three-phase fully controlled bridge rectifiers.

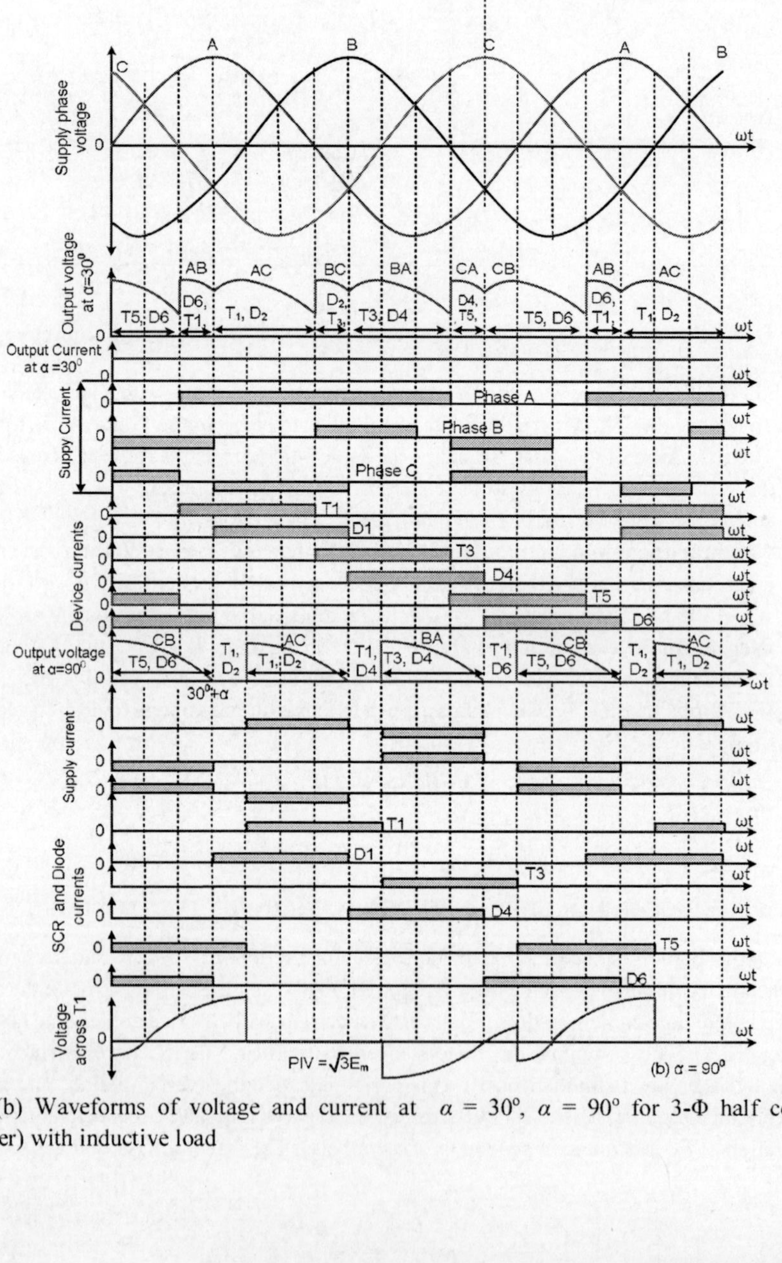

Figure.4.33 (b) Waveforms of voltage and current at $\alpha = 30^\circ$, $\alpha = 90^\circ$ for 3-Φ half controlled (semiconverter) with inductive load

4.11.1 Single phase fully controlled bridge rectifier considering source inductance

The single-phase full-wave bridge circuit considering source inductance is shown in Figure.4.34(a). L_s is the source inductance in this circuit. The major source of this converter is transformer leakage inductance. The input source voltage is e_s. The terminals of e_s are *P* and *N*. The load comprises resistance *R* and inductance *L*. The load current is considered constant. The circuit of Figure.4.34(a) can be represented by its equivalent circuit as shown in Figure.4.34(b). In this equivalent circuit, two paths are shown. One path represents the flow of current when terminal *P* is positive (during the positive half cycle of e_s) and the other is the current path when terminal *N* is positive (during the negative half cycle of e_s). During the positive half cycle, the flow of current (i_{t1}) is in the path *P* - L_s - *T1* - Load (*R* and *L*) - *T2* - *N* of Figure.4.32(a). Let this path is called Path1. Similarly, during the negative half cycle the path of current is *N* - L_s - *T3* - Load (*R* and *L*) - *T4* - *P*, and say this path is named Path2. In the equivalent circuit Path1 comprise e_1 - L_s - *T1* - *T2* and load whereas Path2 comprises e_2 - L_s -*T3* - *T4* - and load. The corresponding waveform for output voltage and current are shown in Figure.4.34 (c).

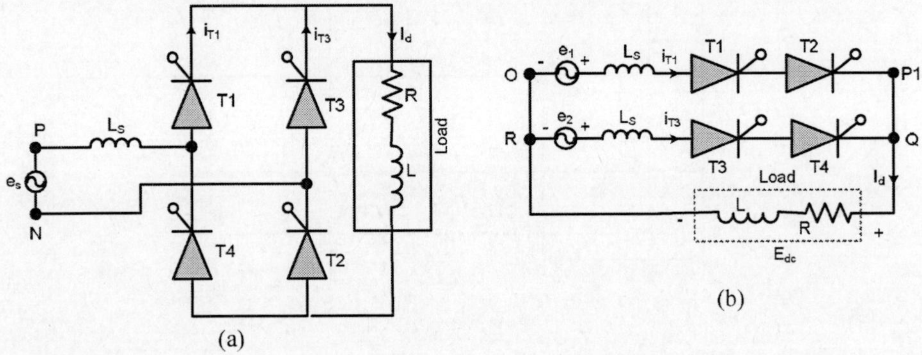

Figure.4.34 The circuit and equivalent circuit of 1-Φ fully controlled bridge rectifier (a) Circuit, (b) Equivalent circuit.

The SCRs *T1* and *T2* fired at angle α. The commutation of SCRs *T3*, and *T4* starts at α and the current gradually decreases from the load current Id and finally to zero at an angle ($\alpha+\mu$). This is because of input inductance L_S. Because of the same reason, the current through SCRs *T1* and *T2* built up gradually from zero at angle α and the final load current at an angle ($\alpha+\mu$). At the commutation period (μ), the conduction of all the SCRs (*T1, T2, T3,* and *T4*) overlap. Application of KVL in the loops of Figure.4.32(b), (neglecting device voltage drops) The obtained equation is

$$e_1 - L_S\frac{di_{T1}}{dt} = e_2 - L_S\frac{di_{T3}}{dt}$$

$$\text{Or } e_1 - e_2 = L_S\left(\frac{di_{T1}}{dt} - \frac{di_{T3}}{dt}\right) \tag{4.73}$$

The equation for e_1 and e_2 are

$$e_1 = E_m Sin\omega t,\ \ e_2 = -E_m Sin\omega t$$

Substituting for e_1 and e_2 in (4.73), the obtained equation is

$$2E_m Sin\omega t = L_S\left(\frac{di_{T1}}{dt} - \frac{di_{T3}}{dt}\right) \tag{4.74}$$

Since we have assumed that I_d is constant, hence we can write

$$i_{T1} + i_{T3} = I_d \tag{4.75}$$

Differentiating (4.75) with respect to time t, the resultant equation is

$$\frac{di_{T1}}{dt} = -\frac{di_{T3}}{dt} \tag{4.76}$$

From equations (4.74) and (4.76), we will get

$$\frac{di_{T1}}{dt} = \frac{E_m}{L_s} Sin\omega t \tag{4.77}$$

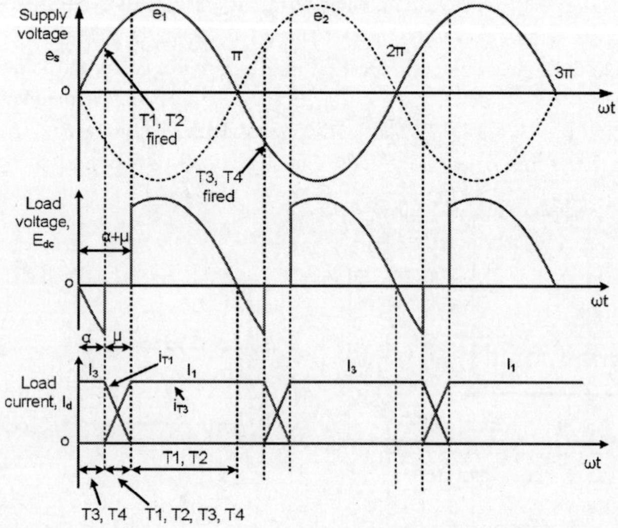

Figure.4.34 (Continue) (c) The waveform for output voltage and current of 1-Φ fully control bridge considering source inductance

During the overlap angle (μ), the current through SCRs $T1$ and $T2$ built up from zero to I_d. At $\omega t = \alpha$, $i_{T1} = 0$, and at $\omega t = \alpha+\mu$, $i_{T1} = I_d$. Thus from (4.77), we can find the expression for I_d as follows.

$$\int_0^{I_d} di_{T1} = \frac{E_m}{L_S} \int_{\alpha/\omega}^{(\alpha+\omega)/\omega} \sin \omega t d(\omega t) \tag{4.78}$$

From this equation, we can find the expression for I_d and the expression is

$$I_d = \frac{E_m}{\omega L_s}\left[\cos\alpha - \cos(\alpha+\mu)\right] \tag{4.79}$$

Thus,

$$\left[\cos\alpha - \cos(\alpha+\mu)\right] = I_d \frac{\omega L_s}{E_m} \tag{4.80}$$

Or

$$\cos\alpha - I_d \frac{\omega L_s}{E_m} = \cos(\alpha+\mu) \tag{4.81}$$

Or

$$\cos\alpha = \cos(\alpha+\mu) + I_d \frac{\omega L_s}{E_m} \tag{4.82}$$

The expression for E_{dc} is obtained as follows

$$E_{dc} = \frac{E_m}{\pi} \int_{\alpha+\mu}^{\pi+\alpha} Sin\omega t d(\omega t) = \frac{E_m}{\pi}[\cos(\alpha+\mu) - \cos(\alpha+\pi)] = \frac{E_m}{\pi}[\cos\alpha + \cos(\alpha+\mu)] \tag{4.83}$$

From equations (4.81) and (4.83), we will get

$$E_{dc} = \frac{E_m}{\pi}\left[\cos\alpha + \cos(\alpha+\mu)\right] = \frac{E_m}{\pi}\left[2\cos\alpha - I_d\frac{\omega L_s}{E_m}\right] = \frac{2E_m}{\pi}\cos\alpha - I_d\frac{\omega L_s}{\pi} \quad (4.84)$$

We can also express E_{dc} from (4.82) and (4.83) as follows.

$$E_{dc} = \frac{E_m}{\pi}\left[\cos\alpha + \cos(\alpha+\mu)\right] = \frac{2E_m}{\pi}\cos(\alpha+\mu) + I_d\frac{\omega L_s}{\pi} \quad (4.85)$$

From equation (4.84), we can conclude that if we consider the source inductance, the output D.C voltage is reduced by $I_d\frac{\omega L_s}{\pi}$. The equation (4.84), can be represented by an equivalent circuit as shown in Figure.4.34(d) in which the diode symbol represents unidirectional, source voltage is represented by $\frac{2E_m}{\pi}\cos\alpha$, $\frac{\omega L_s}{\pi}$ is the series resistance.

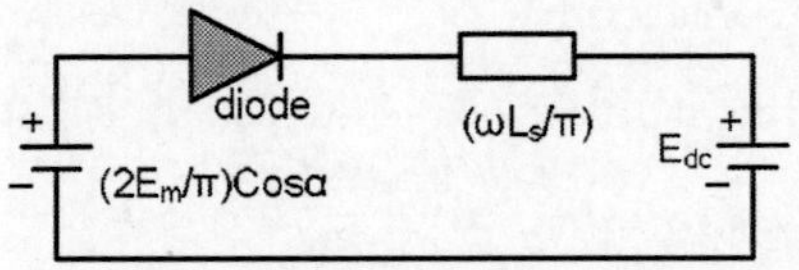

Figure.4.34 (Continue) (d) DC Equivalent circuit for the single-phase full converter considering source inductance.

4.11.2 Three phase fully controlled bridge rectifier considering source inductance

A 3- phase fully controlled rectifier with source inductance L_S is shown in Figure.4.35. In this analysis, we assumed that the load is constant. A 3-Φ supply is shown in Figure.4.36(a). The output voltage of the 3-Φ fully controlled bridge rectifier is shown in Figure.4.36(b).

When both the firing angle (α) and conduction angle (μ) are zero, the conduction of SCRs shown in Figure.4.36(c). Here, SCRs T_5, and T_6 conduct up to 30^0. The SCRs T_1 and T_6 conduct for the period of 60^0 (from $\omega t = 30^0$ to 90^0), The SCRs T_1 and T_2 conduct from $\omega t = 90^0$ to 150^0) and in this way other SCRs also conduct as shown in the figure. It is observed that only two SCRs conduct at a time.

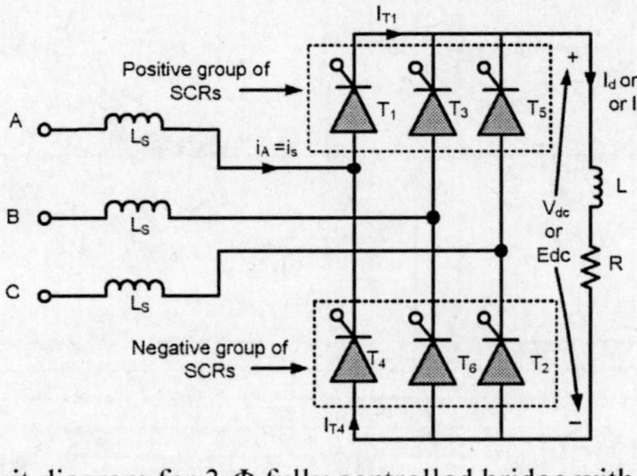

Figure.4.35 Circuit diagram for 3-Φ fully controlled bridge with source inductance

To explain the effect of overlap, we assume that the firing angle $\alpha = 0^0$ and the overlap angle is less than 60^0. The phase voltages of source voltages are denoted by V_a, V_b, and V_c. The effect of overlap is shown in Figure.4.36(d). It is seen that T_5 and T_6 conduct from $\omega t = 0^0$ to 30^0. At $\omega t = 30^0$, the SCR T_5 is outgoing SCR and SCR T_1 is incoming. When T_1 is triggered the current through T_5 is start decaying and through T_1 the current rises. At $\omega t = 30^0+\mu$, the current through T_5 is zero, and through T_1, it is the output current, I_d. From $\omega t = 30^0$ to $30^0+\mu$, the SCR T_5, SCR T_6, and SCR T_1 conduct. After $30^0+\mu$ SCR

T_1 and T_6 conducts. At $\omega t = 90^0$, the SCR T_2 is fired. The current through SCR T_6 decays and SCR T_2 rises. In the period $\omega t = 90^0$ to $90^0+\mu$, the SCR T_6, SCR T_1, and SCR T_2 conducts. At $\omega t = 90^0+\mu$, the current through T_6 is zero, and through T_2, it is the output current, I_d. After $90^0+\mu$ SCR T_1 and T_2 conduct. The operation sequence of the SCR repeats with other SCRs also. Thus, when an SCR from the positive SCR group (upper SCRs T_1, T_3, and T_5) is undergoing commutation, two from this group and one from the negative group conduct. After the commutation of an SCR from the +Ve group is completed, only two SCRs conduct, one from the +ve group and another from the -Ve group conducts. For this when an SCR from the -Ve group undergoes commutation, three SCRs one from the +Ve group and the other two from the -Ve group conducts for the period equal to μ. After that, two SCRs, one from the +Ve group and another from the -Ve group conducted. This operation is shown in Figure.4.36(d). It is also observed that there are six shaded areas per cycle of input voltage (in Figure.4.36(d)), which indicates six commutations in one cycle. During the commutation of T_5 and T_1, the output voltage is the average of V_c and V_a i.e. the output voltage from $\omega t = 30^0$ to $30^0+\mu$, is $\frac{V_c+V_a}{2}$. It is indicated by pq in Figure.4.36(b). Similarly, during the commutation of SCR T_6 and SCR T_4 (of -Ve group), the output voltage is the average voltage is V_b and V_c i.e $\frac{V_b+V_c}{2}$. It is indicated by p_1q_1 in Figure.4.36(b). In this way, During the commutation of T_1 and T_3, the output voltage is the average of V_a and V_b i.e $\frac{V_a+V_b}{2}$. It is indicated by p_2q_2 in Figure.4.36(b). Like a single-phase full-wave converter, the output voltage is reduced due to the introduction of source inductance. This reduction is due to the triangular area (pqr, $p_1q_1r_1$, p_2,q_2,r_2) shown in Figure.4.36(b).

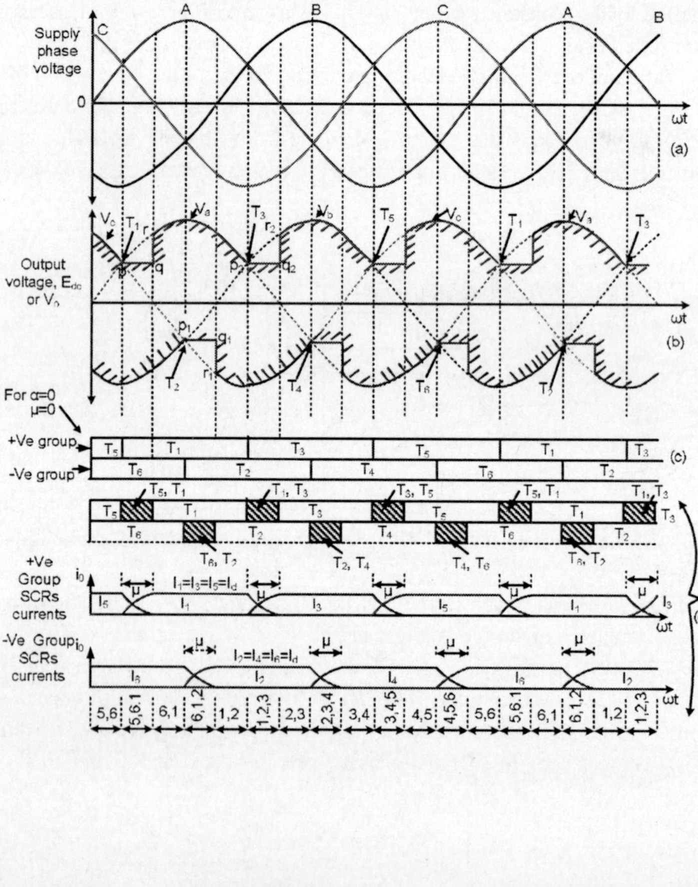

Figure.4.36 Waveforms for voltage and current (a) Supply voltage, (b) Output voltage, (c) sequence of SCRs when $\alpha = 0^0$, $\mu=0^0$, (d) sequence SCR conduction and output currents for a positive and negative group of SCRs

The output voltage during the overlap period is obtained by dividing the triangular area by periodicity (π/3) as follows.

$$= \frac{3}{\pi}\int_0^{\mu} v_L d(\omega t) = \frac{3}{\pi}\int_0^{\mu} L_s \frac{di}{dt} d(\omega t) = \frac{3L_S}{\pi}\int_0^{\mu/\omega} \omega \frac{di}{dt} dt = \frac{3\omega L_S}{\pi}\int_0^{I_d} di = \frac{3\omega L_S}{\pi} I_d \qquad (4.86)$$

The output voltage without overlap is given by

$$E_{dc} = \frac{3E_m}{\pi}\cos\alpha \qquad (4.87)$$

The output voltage with overlap is given by

$$E_{dc} = \frac{3E_m}{\pi}\cos\alpha - \frac{3\omega L_s}{\pi} I_d \qquad (4.88)$$

In general, the voltage drop in the overlap portions for the m pulse converter is given by

$$= \frac{m\omega L_S . I_d}{2\pi} \qquad (4.89)$$

In terms of conduction angle μ, the output voltage considering overlap due to input inductance for 3 phase fully controlled converter (6 pulse) is given by

$$E_{dc} = \frac{3E_m}{\pi}\cos(\alpha + \mu) + \frac{3\omega L_s}{\pi} I_d \qquad (4.90)$$

Unit Summary

This Unit explores various phase-controlled rectifiers. Various circuits and their operation have been presented with related waveforms for various voltages and currents. Following is the summary of this unit.

1. There are two types of converters namely uncontrolled and controlled (phase-controlled). Power diodes are used for uncontrolled converters and thyristors are used for controlled converters. In between there is a converter called a semi converter. Semi converters used both diodes and thyristors. The converters in which thyristors are present are called phase-controlled converters (PCCs).
2. Depending on the type of input voltage phase controlled rectifies are divided into two. They are (i) single phase and (ii) three phase. Each type is again classified into (i) semi converter, (ii) full converter, and (iii) dual converter. Semi converters are one quadrant and it has one polarity of the output voltage and current. Full converters are two-quadrant converters. The output voltage may be either positive or negative polarity. Output current is of positive polarity. Dual converters are four-quadrant converters. Both output voltage and current are either positive or negative.
3. The converters can also be classified based on the number of pulses per cycle of the supply voltage. The single-phase half-wave-controlled rectifiers are single-phase single-pulse converters because these converters produce only one pulse of load current during one cycle of the supply voltage. If the converter produces two pulses per cycle, they are called two pulse converters.
4. The firing angle or delay angle (α) of a thyristor is an angular difference between the time it is triggered and the time it would conduct as a diode. As a result, the firing angle is the angle that is measured from the instant that produces the highest average output voltage to the instant that it is triggered. The extinction angle (γ) is the angular distance from the start of the positive half cycle of the input supply to the instant the load current drops to zero. Conduction angle (β) is

defined as the angle that measures the SCR period of conduction (how long it conducts in one cycle of time).

5. Various parameters measure the performance of a converter. They are the average value of the output voltage, the average value of output current or load current, DC output power, the root mean square (RMS) value of output voltage, the RMS value of output current, output AC power, the efficiency of rectification, effective or rms value of ac component, form factor (*FF*), Ripple factor (*RF*), transformer utilization factor (*TUF*), displacement power factor (*DPF* or *DF*). harmonic factor (*HF*), or total harmonic distortion (*THD*), power factor, and crest factor.
6. Single phase half wave-controlled rectifier contains only one SCR. Many converter circuits employ a diode called a commutating or bypass or freewheeling diode. When the load voltage goes into reverse, it commutates or transfers the load current flowing through the SCR. This diode has two primary functions. They are (i) it prevents the load voltage from being inverted except in the case of a small diode voltage, (ii) the main rectifier thyristors can revert to their blocking states after the load current has been diverted away from them. The advantages of the freewheeling diode in converter circuits are (i) it improves the input power factor, (ii) improves load current waveforms and thus the performance of the load better.
7. The full wave converter uses a minimum of two semiconductor devices. In the case of a single-phase full-wave converter, there are two constructions. Single phase full wave midpoint-controlled converter and single-phase full wave bridge converter. The former uses two SCRs and the latter uses four SCRs for the fully controlled converter. In the case of semi controlled bridge converter, two diodes and two thyristors are used. Single-phase full converters are also two pulse converters. The 3-Φ phase-controlled converters use a minimum of three semiconductor devices. The 3-Φ half wave-controlled converter uses three SCRs. The three-phase full wave-controlled bridge converter uses six SCRs. On the other hand, three diodes and three SCRs are used in 3-Φ semi controlled or semi converter.
8. The common cathode and common anode connection are used to construct the 3-Φ bridge converter.
9. The source impedance affects the output voltages and currents. In general, two device conducts at a time in the case of a three-phase converter. If the source impedance is not considered, there is no overlapping of the device conduction and two devices (SCRs in case of fully controlled) conducts. But when the same is considered, there is an overlapping period on which three devices conduct one at a time.

Exercises

Example.4.1

A DC load resistance of R = 10 Ω is powered from a 1-Φ transformer operating at 230V, 50 Hz by a half wave-controlled rectifier circuit. Determine the (a) rectification efficiency (b) form factor (FF) (c) voltage ripple factor (VRF) (d) transformer utilization factor (TUF) and (e) pick inverse voltage (PIV) of the SCR for a firing angle of 60°.

Solution: Given $E_s = 230$ V, f = 50 Hz, $R = 10\ \Omega$, $\alpha = 60°$

The average output voltage is obtained as

$$E_{dc} = \frac{E_m}{2\pi}(1+\cos\alpha) = \frac{\sqrt{2}\times 230}{2\pi}(1+\cos 60°) = 77.64\,\text{V}$$

The average output current is obtained as follows.

$$I_{dc} = \frac{77.64\,\text{V}}{10} = 7.76\,\text{A}$$

The RMS voltage is calculated as follows.

$$E_{rms} = E_m\left[\frac{\pi-\alpha}{4\pi}+\frac{\sin 2\alpha}{8\pi}\right]^{1/2} = \frac{\sqrt{2}\times 230}{2\sqrt{\pi}}\left[\left(\pi-\frac{\pi}{3}\right)+\frac{1}{2}\sin 120^{\circ}\right]^{1/2} = 145.87\,\text{V}$$

The RMS current is calculated as follows

$$I_{rms} = \frac{145.87}{10} = 14.58\,\text{A}$$

The DC power is obtained as follows.

$$P_{dc} = E_{dc}I_{dc} = 77.64\times 7.76 = 602.8\,\text{W}$$

The AC power is obtained as follows.

$$P_{ac} = E_{rms}I_{rms} = 145.87\times 14.58 = 2127.85\,\text{W}$$

From the above, the required values are obtained as follows.

(a) Rectification efficiency, $\eta = \frac{P_{dc}}{P_{ac}} = \frac{602.8}{2127.85} = 0.2833 \text{ or } 28.33\%$

(b) Form factor, $\text{FF} = \frac{V_{rms}}{V_{dc}} = \frac{145.87}{77.64} = 1.87$

(c) Voltage ripple factor, $\text{VRF} = \sqrt{\text{FF}^2 - 1} = \sqrt{1.87^2 - 1} = 1.59$

(d) Transformer utilization factor, $\text{TUF} = \frac{E_{dc}I_{dc}}{E_sI_s} = \frac{E_{dc}I_{dc}}{E_sI_{rms}} = \frac{602.8}{230\times 14.58} = 0.1797$

(e) Pick inverse voltage,

Example.4.2

A 1-Φ, 230 Volt, 50 Hz source fed power to a 1 kW, 230 Volt heater through an SCR. Determine the power draw by the heating element for $\alpha = 45^{\circ}$ and 90°.

Solution: Heater resistance is $\text{R} = \frac{(230)^2}{1000} = 52.9\Omega$

The RMS voltage for firing angle 45° is obtained as

$$E_{rms} = E_m\left[\frac{\pi-\alpha}{4\pi}+\frac{\sin 2\alpha}{8\pi}\right]^{1/2} = \frac{\sqrt{2}\times 230}{2\sqrt{\pi}}\left[\left(\pi-\frac{\pi}{4}\right)+\frac{1}{2}\sin 90^{\circ}\right]^{1/2} = 155.07\,\text{V}$$

The power consumed by the heater element at a firing angle of 45° is calculated as follows.

$$\frac{E_{rms}^2}{R} = \frac{(155.07)^2}{52.9} = 454.57\,\text{watts}$$

The RMS voltage for firing angle 90° is obtained as

$$E_{rms} = E_m\left[\frac{\pi-\alpha}{4\pi}+\frac{\sin 2\alpha}{8\pi}\right]^{1/2} = \frac{\sqrt{2}\times 230}{2\sqrt{\pi}}\left[\left(\pi-\frac{\pi}{2}\right)+0\right]^{1/2} = 115\,\text{V}$$

The power consumed by the heater element at a firing angle of 90° is calculated as follows.

$$\frac{E_{rms}^2}{R} = \frac{(115)^2}{52.9} = 250\,\text{watts.}$$

Example.4.3

A load resistance of $R = 10\ \Omega$ is powered from a 1-Φ, 120 V, 50Hz source through a single-phase half wave-controlled converter. Find the (a) α (b) average and rms values of currents output (c) average and rms value of SCR currents whereas average voltage output is 25% of the maximum average output voltage.

Solution: (a) The average output voltage is given by $E_{dc} = \frac{E_m}{2\pi}(1+\cos\alpha)$

Firing angle of 0° results in maximum average output voltage. Thus

$$E_{dc\max} = \frac{E_m}{\pi}$$

It is given that

$$E_{dc} = 25\%\left(\frac{E_m}{\pi}\right) = 0.25\left(\frac{E_m}{\pi}\right)$$

$$0.25\left(\frac{E_m}{\pi}\right) = \frac{E_m}{2\pi}(1+\cos\alpha)$$

Thus

$$\alpha = \frac{\pi}{3} \text{ and } E_{dc} = 0.238E_m$$

(b) The average output current is given by $I_{dc} = \frac{E_{dc}}{R}$

$$= \frac{0.238E_m}{R} = \frac{0.238\times 120\times\sqrt{2}}{10} = 4.04 \text{ A}$$

The RMS voltage is given by

$$E_{rms} = E_m\left[\frac{\pi-\alpha}{4\pi} + \frac{\sin 2\alpha}{8\pi}\right]^{1/2}$$

$$= 0.448E_m = 0.448\times 120\times\sqrt{2} = 76.09 \text{ V}$$

The RMS current is given by $I_{rms} = \frac{E_{rms}}{R} = 7.61 \text{ A}$

(c) The average and rms SCR current will be identical to the average and rms load current.

Example.4.4

SCRs of 1-Φ mid-point converter and bridge converters with rating, peak forward voltage, and average on-state current are 1000V and 40A respectively. Use a safety factor of 2.5 to figure out how much power these two converters can handle.

Solution: Single-phase, midpoint converter with peak voltage across the SCR, $2E_m$

Thus, the midpoint converter has a maximum voltage of $\frac{1000}{2\times 2.5} = 200\text{ V}$

The mid-point converter's maximum allowable average power is given by

$$= \left(\frac{2E_m}{\pi}\cos\alpha\right)I_{Tav} = \frac{2\times 200}{\pi}\times 40\times\frac{1}{1000} = 5.09 \text{ kW}$$

Single-phase bridge converter with peak voltage across the SCR, E_m

Thus, the bridge converter has a maximum voltage of $\frac{1000}{2.5} = 400\text{ V}$

The bridge converter's maximum allowable average power: $= \frac{2\times 400}{1000\times\pi}\times 40 = 10.18 \text{ kW}$

Example.4.5

A 230V 1-Φ AC source supplies 12A at 150V to a DC inductive load. Provide specifics on meeting this specification in either the (a) midpoint or (b) bridge configuration, with $\alpha = 30°$. The voltage drop across each SCR is 1.5 volts, and compare this to the two configurations.

Solution: (a) For midpoint configuration:

$$E_{dc} = \frac{2E_m}{\pi}\cos\alpha$$

With a voltage drop across each SCR of 1.5 volts, we can write,

$$E_{dc} = \frac{2E_m}{\pi}\cos\alpha - 1.5$$

$$150 = \frac{2E_m}{\pi}\cos 30 - 1.5$$

$$E_m = \frac{\pi}{2\cos 30}(151.5) = 274.8 \text{ V}$$

The RMS voltage of each section of the T/F secondary is

$$E_{rms} = \frac{274.8}{\sqrt{2}} = 194.31 \text{ V}$$

And carrier rms current is

$$I_{rms} = \frac{12}{\sqrt{2}} = 8.49 \text{ A}$$

kVA rating of T/F $= 2 \times 194.31 \times 8.49 = 3.3 \text{ kVA}$

T/F primary current $= 12 \times \left(\frac{194.31}{230}\right) = 10.14 \text{ A}$

PIV of SCR $= 2E_m = 2 \times 274.8 = 549.6 \text{ V}$

(b) For bridge configuration $E_{dc} = \frac{2E_m}{\pi}\cos\alpha$

With a voltage drop across each SCR of 1.5 volts, we can write, $E_{dc} = \frac{2E_m}{\pi}\cos\alpha - 3$

$= 150 = \frac{2E_m}{\pi}\cos 30 - 3$

From which $E_m = \frac{\pi}{2\cos 30}(153) = 277.52 \text{ V}$

The T/F secondary rms voltage is: $E_{rms} = \frac{277.52}{\sqrt{2}} = 196.24 \text{ V}$

As T/F secondary of square wave thus rms current: $I_{rms} = 12$ A.

T/F kVA rating $= 196.24 \times 12 = 2.35 \text{ kVA}$

T/F primary current $= 12 \times \left(\frac{196.24}{230}\right) = 10.24 \text{ A}$

PIV of SCR $= E_m = 277.52 \text{ V}$

(c) Mid-point configuration: SCR loss $= \frac{1.5}{150} \times 100 = 1\%$.

Bridge configuration: SCR loss $= \frac{3}{150} \times 100 = 2\%$.

In terms of transformer size and SCR voltage rating, the bridge configuration outperforms the mid-point configuration.

Example.4.6

The firing angle, α for a half-wave-controlled rectifier is 60^0. The load is purely resistive (R). Find (a) rectification efficiency, (b) Form factor (FF), (c) Ripple factor (RF), (d) TUF, (e) PIV for SCR

Solution:

(a) The rectification efficiency is given by, where P_{dc} is the load DC power output, and P_{ac} is the rms load power. The P_{dc} is given by and The P_{ac} is given by $P_{ac} = \frac{E_{rms}^2}{R}$. Hence, to find the η, we have to find the E_{dc} and E_{rms}.

The E_{dc} is obtained as follows

$$E_{dc} = \frac{E_m}{2\pi}(1+\cos\alpha) = \frac{E_m}{2\pi}(1+\cos 60^0) = 0.239E_m$$

The E_{rms} is obtained as follows

$$E_{rms} = E_m\left[\frac{\pi-\alpha}{4\pi}+\frac{\sin 2\alpha}{8\pi}\right]^{1/2} = E_m\left[\frac{\pi-\frac{\pi}{3}}{4\pi}+\frac{\sin(2x60^0)}{8\pi}\right]^{1/2} = 0.485E_m$$

Thus, the η is obtained as

$$\eta = \frac{P_{dc}}{P_{ac}} = \frac{(0.239E_m)^2}{(0.485E_m)^2} = 24.28\%$$

(b) The form factor (FF) is $\quad FF = \frac{E_{rms}}{E_{dc}} = \frac{0.485E_m}{0.239E_m} = 203.3\%$

(c) The RF is $\quad RF = (FF^2-1)^{1/2} = (2.03^2-1)^{1/2} = 1.77 \text{ or } 177\%$

(d) The TUF is given by $TUF = \frac{P_{dc}}{E_s I_s}$, where $E_s = \frac{E_m}{\sqrt{2}} = 0.707E_m$, I_s = rms current = $\frac{E_{rms}}{R} = \frac{0.485E_m}{R}$

Thus, TUF is obtained as follows.

$$TUF = \frac{P_{dc}}{E_s I_s} = \left(\frac{\frac{E_{dc}^2}{R}}{E_s I}\right) = \left(\frac{\frac{(0.239E_m)^2}{R}}{0.707E_m \text{x} \frac{0.485E_m}{R}}\right) = 0.166 \text{ or } 16.6\%$$

(e) The PIV is equal to E_m

Example.4.7

A load resistance of R = 10 Ω series with a large inductance powered from 230V, 50Hz 1-Φ supply through a single-phase full converter. Find the performance parameters of the converter for a 45° firing angle.

Solution: Average voltage output is given by, $E_{dc} = \frac{2E_m}{\pi}\cos\alpha = \frac{2\sqrt{2}\times 230}{\pi}\cos 45^\circ = 146.42 \text{ V}$

The average output current is given by $\quad I_{dc} = \frac{E_{dc}}{R} = \frac{146.42}{10} = 14.64 \text{ A}$

The RMS voltage is $\quad E_{rms} = \frac{E_m}{\sqrt{2}} = E_s = 230 \text{ V}$

As of load current is ripple free, $\quad I_{rms} = I_s = I_{dc} = 14.64 \text{ A}$

Output dc power, $\quad P_{dc} = E_{dc} I_{dc} = 146.42 \times 14.64 = 2143.97 \text{ W}$

Output ac power, $P_{ac} = E_{rms} I_{rms} = 230 \times 14.64 = 3367.73$ W

Rectification efficiency, $= \dfrac{P_{dc}}{P_{ac}} = \dfrac{2143.97}{3367.73} \times 100 = 63.66\%$

Form factor (FF), $\text{FF} = \dfrac{E_{rms}}{E_{dc}} = \dfrac{230}{146.42} = 1.57$

Voltage ripple factor (VRF), $\text{VRF} = \sqrt{\text{FF}^2 - 1} = \sqrt{1.57^2 - 1} = 1.21$

As the load current is ripple-free, the current ripple factor (CRF) = 0.

Fundamental rms input current (I_{s1}), $I_{s1} = \dfrac{2\sqrt{2}}{\pi} \times 14.64 = 13.18$ A

$$\theta_1 = -\alpha = -45^\circ$$

Displacement factor (DF), $\text{DF} = \cos\theta_1 = \cos(+45) = 0.707$

Current distortion factor (CDF), $\text{CDF} = \dfrac{I_{s1}}{I_s} = \dfrac{13.18}{14.64} = 0.90$

Input power factor (PF), $\text{PF} = \text{CDF} \times \text{DF} = 0.90 \times 0.707 = 0.636$ lag

Total harmonic distortion (THD) obtained as $\text{THD} = \left[\dfrac{1}{\text{CDF}^2} - 1\right]^{1/2} = \left[\dfrac{1}{0.90^2} - 1\right]^{1/2} = 0.483$

Active power input (P_i) is obtained as $P_i = V_o \times I_o = 146.42 \times 14.64 = 2143.97$ W

Reactive power input (Q_i) is obtained as

$$P_i = \frac{2E_m}{\pi} I_{dc} \sin\alpha = \frac{2\sqrt{2} \times 230}{\pi} \times 14.64 \times \sin 45^\circ = 2143.96 \text{ VAr}$$

Example.4.8

A load resistance of R = 10 Ω series with a large inductance powered from 230V, 50Hz, 1-Φ supply through a 1-Φ semiconverter. Find the performance parameters of the converter for a 45° firing angle.

Solution: Average output voltage, $E_{dc} = \dfrac{E_m}{\pi}(1 + \cos\alpha) = \dfrac{\sqrt{2} \times 230}{\pi}(1 + \cos 45^\circ) = 176.72$ V

The average output current is $I_{dc} = \dfrac{E_{dc}}{R} = \dfrac{176.72}{10} = 17.67$ A

The output rms voltage is obtained from $E_{rms} = E_s \left[\dfrac{1}{\pi}\left\{(\pi - \alpha) + \dfrac{\sin 2\alpha}{2}\right\}\right]$

$$= 230\left[\frac{1}{\pi}\left\{(\pi - \frac{\pi}{4}) + \frac{\sin 90}{2}\right\}\right] = 219.3 \text{ V}$$

As of load current is ripple free hence, $I_{rms} = I_{dc} = 17.67$ A

The output dc power is obtained as

$$P_{dc} = E_{dc} I_{dc} = 176.72 \times 17.67 = 3122.99 \text{ W}$$

The output ac power is obtained as follows

$$P_{ac} = E_{rms} I_{rms} = 219.3 \times 17.67 = 3875.47 \text{ W}$$

Rectification efficiency, $= \dfrac{P_{dc}}{P_{ac}} = \dfrac{3122.99}{3875.47} \times 100 = 80.58\%$

Form factor (FF), $\text{FF} = \dfrac{E_{rms}}{E_{dc}} = \dfrac{219.3}{176.72} = 1.241$

Voltage ripple factor (VRF), $\text{VRF} = \sqrt{\text{FF}^2 - 1} = \sqrt{1.24^2 - 1} = 0.735$

Current ripple factor (CRF) = 0.

Fundamental rms input current (I_{s1}), $I_{s1} = \frac{2\sqrt{2}}{\pi} \times 17.67 \times \cos\frac{45}{2} = 14.69\ A$

Total rms input current (I_s), $I_s = 17.67\sqrt{\frac{3\pi}{4\pi}} = 15.30\ A$

$$\theta_1 = -\alpha = -45^\circ$$

Displacement factor (DF), $\text{DF} = \cos\theta_1 = \cos\left(-\frac{\alpha}{2}\right) = 0.923$

Current distortion factor (CDF), $\text{CDF} = \frac{I_{s1}}{I_s} = \frac{14.69}{15.30} = 0.96$

Input power factor (PF), $\text{PF} = \text{CDF} \times \text{DF} = 0.96 \times 0.923 = 0.887$ lag

Total harmonic distortion (THD), $\text{THD} = \left[\frac{1}{\text{CDF}^2} - 1\right]^{1/2} = \left[\frac{1}{0.96^2} - 1\right]^{1/2} = 0.290$

Active power input (P_i), $P_i = V_o \times I_o = 176.72 \times 17.67 = 3122.99$ W

Reactive power input (Q_i), $Q_i = \frac{E_m}{\pi} I_{dc} \sin\alpha = \frac{\sqrt{2} \times 230}{\pi} \times 17.67 \times \sin 45^\circ = 1293.79$ VAr

Example.4.9

A 120 V, 50 Hz AC supply is utilized to power a 1-Φ semiconverter. Find (a) α (b) average and rms value of current output (c) average and rms value of SCR currents whereas average voltage output is 25% of the maximum average output voltage.

Solution: (a) The average output voltage is given by

$$E_{dc} = \frac{E_m}{\pi}(1 + \cos\alpha)$$

Firing angle of 0° results in maximum average output voltage. Thus

$$E_{dc\max} = \frac{E_m}{\pi}$$

Given $E_{dc} = 25\%\left(\frac{2E_m}{\pi}\right) = \frac{0.25 \times 2 \times \sqrt{2} \times 120}{\pi} = 27$ V

$$27 = \frac{120 \times \sqrt{2}}{\pi}(1 + \cos\alpha), \text{ Thus } \alpha = 120^\circ$$

(b)The RMS output voltage, $E_{rms} = E_m\left[\frac{\pi - \alpha}{2\pi} + \frac{\sin 2\alpha}{4\pi}\right]^{1/2} = 53\text{V}$

The RMS current output is $I_{rms} = \frac{E_{rms}}{R} = 5.3$ A

Average current output, $I_{dc} = \frac{E_{dc}}{R} = \frac{27}{10} = 2.7$ A

(c) Average SCR current is obtained from

$$I_{TH(av)} = \frac{1}{2\pi}\int_{\alpha}^{\pi} I_m \sin\omega t\ d(\omega t) = \frac{I_m}{2\pi}(1 + \cos\alpha)$$

$$I_m = \frac{E_m}{R} = 16.97 \text{ A Thus } I_{m(av)} = \frac{16.97}{2\pi}\left(1 + \cos\frac{2\pi}{3}\right) = 1.35 \text{ A}$$

The RMS SCR current

$$I_{TH(rms)} = \left\{ \frac{1}{2\pi} \int_{\alpha}^{\pi} I_m^2 \sin^2 \omega t \, d(\omega t) \right\}^{1/2}$$

$$= \left\{ \frac{I_m^2}{2\pi} \int_{\alpha}^{\pi} \frac{1 - \cos 2\omega t}{2} d(\omega t) \right\}^{1/2}$$

$$= \frac{16.97}{2} \left[\frac{1}{\pi} \left(\frac{\pi}{3} - 0.433 \right) \right]^{1/2} = 3.75 \text{ A}$$

Example.4.10

Load resistance of 10Ω is powered from a three-phase star-connected 220V, 50 Hz supply through the three-phase half-wave converter. Find: (i) α (ii) rms and the average value of currents output (iii) average and rms value of SCR currents (iv) rectification efficiency (v) UTF and (vi) input power factor if the average voltage output = 25% of maximum average voltage.

Solution: Source line voltage, , frequency, $f = 50$ Hz, load resistance, $R = 10\ \Omega$

Source phase voltage, $E_s = 220 / \sqrt{2} = 127$ V

$$E_m = \sqrt{2} E_s = 179.63 \text{ V}$$

(i) The DC output voltage is given by $E_{dc} = \dfrac{3\sqrt{3}E_m}{2\pi} \cos\alpha$

Firing angle of 0° results in maximum average output voltage. Thus,

$$E_{dc\max} = \frac{3\sqrt{3}E_m}{\pi} = \frac{3\sqrt{3} \times 179.63}{2\pi} = 148.55 \text{ V}$$

Given $E_{dc} = 25\% E_{dc\max} = 0.25 \times 148.55 = 37.14$ V

$$E_{dc\,normalize} = \frac{E_{dc}}{E_{dc\max}} = \cos\alpha$$

As if for continuous conduction $\alpha \le \pi/6$. i.e., if

$$E_{dc\,normalize} \ge \cos(\pi/6) = 86.6\%$$

As here $E_{dc\,n} = 25\%$, hence load current is discountinous

Hence with R load $\alpha \ge \pi/6$,

$$E_{dc} = \frac{3E_m}{2\pi}[1 + \cos(\alpha + 30^\circ)]$$

$$37.14 = \frac{3 \times 179.63}{2\pi}[1 + \cos(\alpha + 30^\circ)]$$

$$\alpha = 94.5^\circ$$

(ii) Average output current is obtained as $I_{dc} = \dfrac{E_{dc}}{R} = \dfrac{37.14}{10} = 3.714$ A

The RMS output voltage is $E_{rms} = \dfrac{\sqrt{3}E_m}{2\sqrt{2}} \left[\dfrac{5\pi - 3\alpha}{3\pi} + \dfrac{\sin(2\alpha + \pi/3)}{\pi} \right]^{1/2} = 62.12$V.

The RMS output current is $I_{rms} = \dfrac{E_{rms}}{R} = \dfrac{62.12}{10} = 6.21$ A.

(iii) Average SCR current, $I_{T_{avg}} = \dfrac{I_{dc}}{3} = \dfrac{3.714}{3} = 1.24$ A

The RMS SCR current, $I_{T_{rms}} = \dfrac{I_{rms}}{\sqrt{3}} = \dfrac{6.21}{\sqrt{3}} = 3.59\ \text{A}$

(iv) The rectification efficiency is obtained as $\eta = \dfrac{P_{dc}}{P_{ac}} = \dfrac{37.14 \times 3.714}{62.1 \times 6.21} = 35.75\%$

(v) Input VA rating, $3E_s I_s = 3 \times 127 \times 3.59 = 1367\ \text{W}$

$$\text{TUF} = \frac{E_{dc} I_{dc}}{E_s I_s} = \frac{37.14 \times 3.714}{1367} = 10.09\%$$

(vi) Power output, $P_o = I_{rms}^2 R = 385\ \text{W}$

Power factor, $= \dfrac{P_o}{E_s I_s} = \dfrac{385}{1367} = 0.28\ \text{(lag)}$

Example.4.11

(a) A resistive load of 10 Ω takes 5 kW from a 3-Φ full converter for α = 30°. Calculate the magnitude of the per-phase input supply voltage. (b) Repeat part (a) if a large reactor connected in series with the load eliminates ripple in the load current.

Solution:

(a) For restive load, the current output wave shape is identical to the output voltage and for $\alpha \leq 60°$ output voltage and current are continuous.

Here, $\dfrac{E_{rms}^2}{R} = 5000\ \text{W}$ and $E_m = \sqrt{2}\, V_s$

$$2V_s^2 \frac{3}{2\pi} \frac{E_{rms}^2}{R}\left[\frac{\pi}{3} + \frac{\sqrt{3}}{2}\cos 60\right] = 5000 \times 10$$

$$V_s = 188.08\ \text{V}$$

Phase voltage, $V_{ph} = \dfrac{V_s}{\sqrt{3}} = 108.591\ \text{V}$

(b) For ripple free load current; average load current (I_{dc}) = rms load current (I_{rms}).

$$I_{rms} = \frac{E_{dc}}{R} = \left(\frac{3E_m}{\pi}\cos\alpha\right)\frac{1}{R}$$

$$I_{rms}^2 \times R = \left[\frac{3E_m}{\pi}\cos\alpha\right]^2 \frac{1}{R} = 5000\ \text{W}$$

$$V_s = \sqrt{50000} \times \frac{\pi}{\sqrt{2} \times 3\cos 30°} = 191.22\ \text{V}$$

$$V_{ph} = 110.40\ \text{V}$$

Example.4.12

(a) A resistive load of 10 Ω takes 5 kW from a 3-Φ semiconverter for α = 30°. Calculate the magnitude of the per-phase input supply voltage. (b) Repeat part (a) if a large reactor connected in series with the load eliminates ripple in the load current.

Solution:

(a) For $\alpha < 30°$ the voltage output is continuous. And for restive load current is continuous. Output rms voltage also continuous for $\alpha < 30°$.

For $\alpha = 30°$, $$\frac{2V_s^2}{4}\frac{3}{\pi}\left[\frac{2\pi}{3}+\frac{\sqrt{3}}{2}(1+\cos 60)\right] = 5000\times 10$$

From which $V_s = 175.67$ V

Phase voltage, $V_{ph} = \frac{V_s}{\sqrt{3}} = 101.43$ V

(b) For ripple free load current; average load current (I_{dc}) = rms load current (I_{rms}).

$$I_{rms} = \frac{E_{dc}}{R} = \frac{3E_m}{2\pi}(1+\cos\alpha)\frac{1}{R}$$

$$I_{rms}^2 \times R = \left[\frac{3E_m}{2\pi}(1+\cos 30°)\right]^2 \frac{1}{10} = 5000 \text{ W}$$

$$V_s = \sqrt{50000}\times\frac{\sqrt{2}\pi}{3\times 1.866} = 177.44 \text{ V}$$

$$V_{ph} = 191.2 \text{ V}$$

Example.4.13

The supply voltage of a 3-Φ converter is 415V, 50Hz 3-Φ. A load resistor of 100Ω is connected in series with a large smoothing inductor. Find the SCR ratings.

Solution:

The rating of the SCRs is $I_{T(average)}$, $I_{T(RMS)}$, and $I_{T(peak)}$. To find these ratings, the given parameters are Supply voltage = 414V, frequency of supply voltage = 50Hz, load resistance, R =100Ω

The $I_{T(average)}$ is given by

$$I_{T(average)} = \frac{1}{2\pi}\int_0^{2\pi} I_T(\omega t)d(\omega t) = \frac{1}{2\pi}\int_{30+\alpha}^{150+\alpha} I_d d(\omega t) = \frac{I_d}{3}$$

The $I_{T(RMS)}$ is given by

$$I_{T(\text{rms})} = \left[\frac{1}{2\pi}\int_0^{2\pi} i_T^2(\omega t)d(\omega t)\right]^{1/2} = \left[\frac{1}{2\pi}\int_{30+\alpha}^{150+\alpha} I_d^2 d(\omega t)\right]^{1/2} = \left[\frac{I_d^2}{2\pi}.\frac{2\pi}{3}\right]^{1/2} = \frac{I_d}{\sqrt{3}}$$

In these equations, I_d is the DC output current.

The $I_{T(peak)}$ is equal to I_d. The value of the I_d is obtained from $I_d = \frac{E_{dc}}{R}$

Considering, α = 0°, the E_{dc} is obtained from

$$E_{dc} = \frac{3\sqrt{3}E_m}{\pi}\cos\alpha = \frac{3\sqrt{3}\text{x}\sqrt{2/3}\text{x}415}{\pi}\cos 0^0 = 560.5V$$

Thus, $$I_d = \frac{560.5}{100} = 5.605A$$

Now, we can find the ratings as follows.

$$I_{T(average)} = \frac{5.605}{3} = 1.87A;\ I_{T(\text{rms})} = \frac{5.605}{\sqrt{3}} = 3.24A; I_{T(\text{peak})} = 5.605A$$

Example.4.14

The source impedance of a 1-Φ full wave midpoint converter is 0.33 mH. The converter is supplied from 120V, 50Hz supply. A freewheeling diode is connected across the load. A continuous load current of 4A is obtained. Determine the overlap angle for (a) transfer current from the SCR to the commutating diode, and (b) transfer current from the commutating diode to SCR.

Solution:

(a) Let commutation from SCR to diode starts at t = 0, the instant at which load voltage begins to reverse. The SCR voltage drop was also neglected. The figure is shown in Figure.4.37(a).

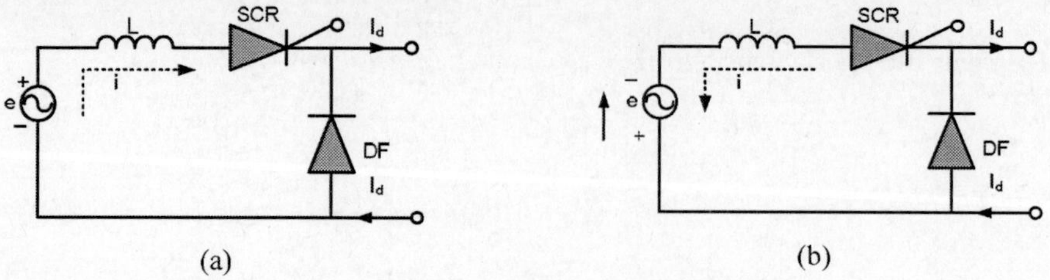

Figure.4.37 Circuits related to example 4.14

Thus, we can write

$$e = -E_m \sin \omega t = -L\frac{di}{dt}$$

$$\text{Hence, } di = E_m Sin\omega t \text{ x } \frac{di}{L}$$

From,

$$i = \frac{E_m}{L}\int_0^t \sin \omega t d(\omega t) = \frac{E_m}{L}(1-\cos \omega t)$$

The commutation is completed when $i = I_d$ and $\omega t = \mu_1$

Thus,

$$I_d = \frac{E_m}{\omega L}(1-\cos \mu)$$

Putting the values, we will get

$$\frac{120}{2\text{x}\pi\text{x}50\text{x}0.33\text{x}10^{-3}}(1-\cos \mu_1) = 4$$

Hence, $\mu_1 = 6.786^0$

(b) The commutation from the freewheeling diode to SCR begins at t = 0, the time at which SCR is triggered. Figure.4.37(b) shows the circuit.

The voltage equation can be written as

$$e = E_m Sin(\omega t + \alpha) = L\frac{di}{dt}$$

From this, we will get an expression for I as follows.

$$i = \frac{E_m\left[\cos\alpha - \cos(\omega t - \alpha)\right]}{\omega L}$$

When $i = I_d$ and $\omega t = \mu_2$, the commutation will be completed. Putting the values, we will get

$$I_d = \frac{E_m\left[\cos\alpha - \cos(\mu_2 - \alpha)\right]}{\omega L}$$

Putting the values of I_d and α, we will get $\mu_2 = 0.54^0$

Example.4.15

A 3-Φ fully controlled bridge rectifier is supplied from a 3-Φ, 400V, 50Hz supply. The triggering angle is 45^0. A continuous load current of 10A is maintained at a load voltage equal to 360V. Determine (a) Source inductance, (b) load resistance, and (c) overlap angle.

Solution:

Given, E_{rms} = 400V (Line), frequency, f = 50Hz, α = 45^0, load current, I_d = 10A, Load voltage, E_{dc} = 360V.

(a) From the equation,

$$E_{dc} = \frac{3\sqrt{3}E_m}{\pi}\cos\alpha - I_d\frac{3\omega L_s}{\pi}$$

The expression for E_m is

$$E_m = 400\text{x}\frac{\sqrt{2}}{\sqrt{3}}$$

Putting the given values in the equation for E_{dc}

$$360 = \frac{3\sqrt{3}}{\pi}\text{x}400\text{x}\frac{\sqrt{2}}{\sqrt{3}}\cos 45^0 - \frac{3\text{x}2\pi\text{x}50\text{x}L_s\text{x}10}{\pi}$$

We will get

$$L_s = 7.3mH$$

Thus, source inductance per phase is 7.3 mH.

(b) The load resistance is calculated as

$$R = \frac{E_{dc}}{I_d} = \frac{360}{10} = 36\Omega$$

(c) Overlap angle is calculated from

$$E_{dc} = \frac{3\sqrt{3}E_m}{\pi}\cos(\alpha + \mu) + I_d\frac{3\omega L_s}{\pi}$$

Substituting the given values, and solving for μ, we will get

$$\mu = 6^0$$

Example.4.16

A 3-Φ fully controlled bridge rectifier is supplied from a 3-Φ, 230V per phase, 50Hz supply having source inductance of 4mH. A continuous load current of 20A is maintained. The load consists of a voltage source of voltage 400V having internal resistance 1Ω. Determine (a) firing angle, and (b) overlap angle.

Solution:

(a) The output voltage of the converter, E_{dc} = 400+1x20 = 420V

The output voltage equation of the converter is

$$E_{dc} = \frac{3\sqrt{3}E_m}{\pi}\cos\alpha - I_d\frac{3\omega L_s}{\pi}$$

Substituting the given values, we will get

$$420 = \frac{3\sqrt{3}\text{x}\sqrt{2}\text{x}230}{\pi}\cos\alpha - 20\text{x}\frac{3(2\pi x50)\text{x}4\text{x}10^{-3}}{\pi}$$

Solving for α, we will get, **α =34.38⁰**

(b) To find the overlap angle (μ), we will use the output voltage equation having μ

$$E_{dc} = \frac{3\sqrt{3}E_m}{\pi}\cos(\alpha + \mu) + I_d\frac{3\omega L_s}{\pi}$$

Substituting the given data and α obtained from (a), we will get

$$420 = \frac{3\sqrt{3}\text{x}\sqrt{2}\text{x}230}{\pi}\cos(34.38^0 + \mu) + 20\text{x}\frac{3(2\pi x50)\text{x}4\text{x}10^{-3}}{\pi}$$

Solving for μ, we will get $\mu = 8.22^0$

Multiple Choice Questions

1. The firing angle is utilized for which of the following purposes?
   (a) To activate an SCR and ignite a device at any desired time
   (b) To regulate the on-off timing of an SCR
   (c) To control the turning-off timing of a general transistor
   (d) To manage the turning-on timing of a diode without the need for a controller
2. When a highly inductive load is supplied by a single-phase diode bridge rectifier, assuming ripple-free load current, the input current at the AC side of the rectifier will be?
   (a) Square wave
   (b) Purely sinusoidal
   (c) Pure DC
   (d) Triangular wave
3. In a three-phase (50Hz) full converter, what is the frequency of the ripple in the output?
   (a) 50 Hz
   (b) 100 Hz
   (c) 150 Hz
   (d) 300 Hz
4. The function of a freewheeling diode in a phase-controlled rectifier is to:
   (a) Contribute to additional harmonics
   (b) Contribute to additional reactive power
   (c) Facilitate inverter operation
   (d) Enhance the line power factor
5. A single-phase full-wave rectifier built with thyristors has a peak value of the sinusoidal input voltage at V_m and a delay angle of $\pi/3$ radians. In this case, the average value of the output voltage is?
   (a) $0.48\ V_m$
   (b) $0.32\ V_m$
   (c) $0.71\ V_m$
   (d) $0.54\ V_m$
6. In a single-phase half-wave controlled rectifier with a resistive load supplied from a 230 V (RMS), 50 Hz source, and a firing angle of 60°, the average output voltage would be:
   (a) 77.5 V
   (b) 52 V
   (c) 155 V
   (d) 104 V
7. In a line-commutated phase-controlled inverter operating at its inverter limit, a commutation failure can occur if:
   (a) The frequency decreases
   (b) The voltage increases
   (c) The frequency increases
   (d) Both the voltage and frequency change in such a way that the ratio of voltage to frequency (v/f) remains constant
8. Among the following options, which one results in a larger peak inverse voltage for the thyristor, while providing the same voltage output?
   (a) Single-phase full-wave centre tapped circuit

(b) Single-phase full-wave bridge circuit
(c) Three-phase full-wave bridge circuit
(d) Three-phase full-wave center tapped circuit

9. In a single-phase fully controlled rectifier feeding a constant DC into the load, if the firing angle α is 30°, what is the displacement factor of the rectifier?
(a) 1
(b) 0.5
(c) 1.73
(d) 0.865

10. In a thyristor-controlled rectifier, the firing angle of the thyristor is to be controlled within which range?
(a) 0° to 180°
(b) 0° to 90°
(c) 90° to 270°
(d) 90° to 180°

11. The frequency of the ripple in the output voltage of a three-phase controlled bridge rectifier depends on which of the following factors?
(a) Load inductance
(b) Firing angle
(c) Supply frequency
(d) Load resistance

12. In a three-phase controlled bridge rectifier, what is the maximum conduction angle for each thyristor?
(a) 60°
(b) 90°
(c) 120°
(d) 150°

13. A three-phase full converter is providing power to a purely resistive load at a 220 V DC voltage with a firing angle of 0°. Determine the output voltage when the firing angle is 90°.
(a) 30 V
(b) 0 V
(c) 90 V
(d) 120 V

14. What is the name of a converter that can operate in both 3-pulse and 6-pulse modes?
(a) Three-phase full-wave converter
(b) Three-phase half-wave converter
(c) Three-phase semi converter
(d) Single-phase semi converter

15. A three-phase full-controlled converter can function as a:
(a) Converter for $\alpha = 0$ to 120^0
(b) Converter for $\alpha = 0$ to 90^0
(c) Converter for $\alpha = 0$ to 180^0
(d) Converter for $\alpha = 0$ to 60^0

16. A fully controlled converter utilizes:
(a) Diodes only
(b) Thyristors only
(c) Both diodes and thyristors
(d) None of the mentioned

17. A single-phase full converter with an R load is a __________ quadrant converter, while a single-phase full converter with an RL load without freewheeling diode (FD) is a __________ quadrant converter.
    (a) One, one
    (b) Two, one
    (c) One, two
    (d) Two, two
18. A single-phase full-controlled bridge converter (B-2) utilizes:
    (a) 4 SCRs and 2 diodes
    (b) 4 SCRs
    (c) 6 SCRs
    (d) 4 SCRs and 2 diodes
19. In a B-2 type full-controlled bridge converter:
    (a) One SCR conducts at a time
    (b) Two SCRs conduct at a time
    (c) Three SCRs conduct at a time
    (d) Four SCRs conduct at a time
20. In a three-phase half-wave rectifier, the DC output voltage is 230 V. The peak inverse voltage across each diode is:
    (a) 481.7 V
    (b) 460 V
    (c) 345 V
    (d) 230 V
21. In a three-phase semi-converter, when the firing angle is less than or equal to 60°, the freewheeling diode conducts for:
    (a) 30°
    (b) 60°
    (c) 90°
    (d) 0°
22. In a single-phase phase-controlled rectifier with a freewheeling diode across the load:
    (a) The instantaneous output voltage V_0 is always positive
    (b) V_0 may be positive or zero
    (c) V_0 may be positive, zero, or negative
    (d) V_0 is always zero or negative
23. The average output voltage is maximized when the SCR is triggered at ωt equals:
    (a) π
    (b) 0
    (c) $\pi/2$
    (d) $\pi/4$
24. In a single-phase thyristor circuit with an R load and a firing angle α, the conduction angle can be expressed as:
    (a) $\pi+\alpha$
    (b) $2\pi+\alpha$
    (c) $\pi-\alpha$
    (d) α
25. In a single-phase half-wave controlled rectifier with a firing angle α and an angular frequency of ω, the circuit turn-off time in seconds can be expressed as:
    (a) α/π

(b) π/α
(c) ω/π
(d) π/ω

26. In a single-phase half-wave thyristor circuit with an R load, the input power factor can be determined by:
(a) RMS source voltage/total RMS line current
(b) RMS input power/power delivered to the load
(c) Cos α
(d) Power delivered to the load/input VA
27. In the case of a single-phase half-wave circuit with an RL load, where the firing angle is α and the extinction angle is β, the conduction angle γ can be expressed as:
(a) $\gamma = \beta+\alpha$
(b) $\gamma = \beta-\alpha$
(c) $\gamma = \beta/\alpha$
(d) $\gamma = \alpha/\beta$
28. Select the accurate statement:
(a) M-2 type connection requires SCRs with higher PIV as compared to those in a B-2 type
(b) M-2 type connection requires SCRs with lower PIV as compared to those in a B-2 type
(c) The average output voltage in the M-2 type is more than that obtained from a B-2 type configuration of the same rating
(d) The average output voltage in the M-2 type is less than that obtained from a B-2 type configuration of the same rating
29. In controlled rectifiers, the nature of the load current (continuous or discontinuous) depends on the:
(a) Type of load and firing angle
(b) Only on the type of load
(c) Only on the firing angle
(d) It is independent of all parameters
30. A single-phase symmetrical semi-converter utilizes:
(a) One SCR and one diode in each leg
(b) Two SCRs and two diodes in each leg
(c) Two SCRs in each leg
(d) Two diodes in each leg
31. In a three-phase, three-pulse, M-3 type controlled converter, the number of SCRs used is:
(a) 1
(b) 2
(c) 3
(d) 4
32. In a three-phase, three-pulse, M-3 type controlled converter, where the firing angle for one of the SCRs is set to 15°, this SCR would initiate conduction at:
(a) 0°
(b) 15°
(c) 30°
(d) 45°
33. The impact of source inductance on the performance of a three-phase controlled converter is to:
(a) Increase the average load voltage
(b) Decrease the average load voltage
(c) Make the load current continuous
(d) Eliminate ripples from the load current

34. In a single-phase full converter, when the output voltage has peak and average values of 325 V and 133 V respectively, the firing angle can be determined as follows:
    (a) 40°
    (b) 140°
    (c) 50°
    (d) 130°
35. In a single-phase semiconverter, when the output voltage has peak and average values of 325 V and 133 V respectively, the firing angle can be determined as follows:
    (a) 40°
    (b) 140°
    (c) 73.40°
    (d) 80°
36. A fully controlled line commutated converter operates as an inverter when the firing angle (α) falls within the range of:
    (a) 0° - 90°
    (b) 90° - 180°
    (c) 90° - 180°, but only when there is a suitable DC source in the load
    (d) 90° - 180°, but only when it supplies a back EMF load
37. In a single-phase half-wave circuit with RL load and a freewheeling diode across the load, if the extinction angle β is greater than π, the conduction periods for the SCR and freewheeling diode, respectively, can be determined as:
    (a) π-α, β
    (b) β-α, π-α
    (c) π-α, β-π
    (d) π-α, π-β
38. In a single-phase one-pulse circuit with RL load and a freewheeling diode, if the extinction angle β is less than π, the conduction periods for the SCR and freewheeling diode, respectively, can be determined as:
    (a) β-α, 0°
    (b) π-α, & π-β
    (c) α, β-α
    (d) β-α, α
39. One of the significant benefits of employing dual converters is that
    (a) it offers cost-effectiveness
    (b) it provides an improved power factor
    (c) it eliminates the need for mechanical switches to switch between operation modes
    (d) it operates at high frequencies
40. In a single-phase full-bridge converter with a freewheeling diode and an inductive load, the load resistance is 15.53 Ω and the inductance is large enough to provide constant and ripple-free DC. The converter is supplied by an ideal 230 V, 50 Hz single-phase source. When the firing delay angle is set to 60°, what is the average value of the diode current?
    (a) 8.1655 A
    (b) 10 A
    (c) 3.33 A
    (d) 5.774 A

Answers to multiple-Choice Questions

(a) (b)	11. (c)	21. (d)	31. (c)
(b) (a)	12. (c)	22. (b)	32. (d)
(c) (d)	13. (b)	23. (b)	33. (b)
(d) (d)	14. (c)	24. (c)	34. (c)
(e) (b)	15. (c)	25. (d)	35. (c)
(f) (a)	16. (b)	26. (d)	36. (d)
(g) (c)	17. (c)	27. (b)	37. (c)
(h) (d)	18. (b)	28. (a)	38. (a)
(i) (d)	19. (b)	29. (a)	39. (c)
(j) (a)	20. (a)	30. (a)	40. (c)

Short and Long Answer Type Questions

1. What do you mean by a controlled rectifier?
2. What is phase controlled rectifier?
3. Define converter.
4. What is the delay angle?
5. Differentiate between uncontrolled and controlled rectifiers.
6. What do you mean by delay or firing angle, extinction angle, and conduction angle? Explain with the necessary diagram.
7. What do you mean by semi converter? Draw two such circuits.
8. What do you mean by half controlled, fully controlled, and semi controlled converters?
9. What are the three methods of control for phase-controlled rectifiers? Explain each with the necessary waveform.
10. What are the various performance parameters of the phase controller rectifier? Write the formulae for each.
11. Draw the circuit diagrams for single phase half wave-controlled rectifier with resistive (R), resistive and inductive (R-L) load. Explain the operation of these circuits with the necessary waveform for various voltages and currents.
12. Draw the circuit diagram for single phase half wave-controlled rectifier with resistive and inductive (R-L) load considering a freewheeling diode. Explain the operation of this circuit with the necessary waveform for various voltages and currents.
13. What do you mean by full converter? Draw two such circuits.
14. Give the principle of phase control.
15. Classify the single-phase full-wave-controlled rectifier. Draw the circuit diagram for each.
16. Explain the operation of single-phase midpoint-controlled rectifiers (with-R load, with R-L load, and R-L load with freewheeling diode) with the necessary waveform for various voltages and currents.
17. Draw the circuit diagram for single-phase full-wave fully controlled bridge rectifiers with various loads.
18. Explain the operation of single-phase full wave fully controlled bridge rectifiers (with R load, with R-L load, and R-L load) with the necessary waveform for various voltages and currents.
19. Draw the circuit diagram for a single-phase half-controlled bridge rectifier with various loads.
20. Explain the operation of a single-phase half-controlled bridge rectifier with various loads with the help of necessary waveform for various voltages and currents.
21. What are the effects of removing a freewheeling diode in semi controlled single phase circuit?

22. Explain the operation of the common anode and common cathode for three phase-controlled rectifiers.
23. Classify 3-Φ phase-controlled rectifiers.
24. Draw the circuit diagrams of a three-phase half-wave-controlled rectifier with resistive (R), and resistive-inductive (R-L) loads.
25. Explain the operation of three phase half wave-controlled rectifier with resistive (R), resistive inductive (R-L) loads. Provide necessary waveforms for various currents and voltages of each converter.
26. Draw the circuit diagram for three phase fully controlled bridge converter with various types of loads.
27. Explain the operation of three phase fully controlled bridge converter with various types of loads. Provide necessary waveforms for various voltages and currents in each of them.
28. Explain the operation of single-phase fully controlled bridge converters considering the source impedance. Provide necessary waveforms for various voltages and currents.
29. Explain the operation of three phase fully controlled bridge converters considering the source impedance. Provide necessary waveforms for various voltages and currents.
30. Explain the effect of input impedance in output voltage and currents in single-phase and three phase converters.

Numerical problems

1. A source of 120V supplies a half-wave-controlled rectifier. Find the voltage across load and power output for the following firing angles if the load resistance is 10 Ω: (a) 0° (b) 45° (c) 90° (d) 135° (e) 180°.
2. A DC load resistance of R = 20 Ω is powered from a 1-Φ transformer operating at 220V, 50 Hz by a half wave-controlled rectifier circuit. Determine the (a) rectification efficiency (b) form factor (FF) (c) voltage ripple factor (VRF) (d) transformer utilization factor (TUF) and (e) pick inverse voltage (PIV) of the SCR for a firing angle of 30°.
3. The firing angle, α for a half wave-controlled rectifier is 30^0. The load is purely resistive (R). Find (a) rectification efficiency, (b) Form factor (FF), (c) Ripple factor (RF), (d) TUF, and (e) PIV for SCR.
4. The firing angle, α for a half-wave-controlled rectifier is 45^0. The load is purely resistive (R). Find (a) rectification efficiency, (b) Form factor (FF), (c) Ripple factor (RF), (d) TUF, and (e) PIV for SCR.
5. A 1-Φ, 220 Volt, 50 Hz source fed power to a 2 kW, 220 Volt heater through an SCR. Determine the power draw by the heating element for $\alpha = 30°$ and 60°.
6. A 220V 1-Φ AC source supply 12A at 160V to a DC inductive load. Provide specifics on meeting this specification in either the (a) midpoint or (b) bridge configuration, with $\alpha = 45°$. The voltage drop across each SCR is 1.6 volts, and compare this to the two configurations.
7. A load having a resistance of R = 10 Ω is supplied by a half-wave-controlled rectifier from a 150 V, 60 Hz source. Find: (a) peak load current (b) average load voltage and load current (c) rms load current (d) load power input (e) conduction angle (f) ripple frequency (h) power factor if the firing angle is 30°.
8. A load resistance of $R = 20\ \Omega$ is powered from a 1-Φ, 110 V, 50Hz source through a single-phase half wave-controlled converter. Find the (a) α (b) average and rms values of currents output (c) average and rms value of SCR currents whereas average voltage output is 30% of the maximum average output voltage.

9. SCRs of 1-Φ mid-point converter and bridge converters with rating, peak forward voltage, and average on-state current are 1100V and 30A respectively. Use a safety factor of 2.6 to figure out how much power these two converters can handle.
10. The load resistance of a single-phase half-wave converter is R = 12 Ω and it is powered by a 230 V, 50 Hz source. Determine the (a) rectification efficiency (b) FF (c)VRF (d) TUF (e) PIV of the SCR for a firing angle of 30°.
11. A load resistance of R = 20 Ω series with a large inductance powered from 220V, 50Hz 1-Φ supply through a single-phase full converter. Find the performance parameters of the converter for a 30° firing angle.
12. A load resistance of R = 20 Ω series with a large inductance powered from 220V, 50Hz, 1-Φ supply through a 1-Φ semiconverter. Find the performance parameters of the converter for a 60° firing angle.
13. A 120 V, 50 Hz AC supply is utilized to power a 1-Φ semiconverter. Find (a) α (b) average and rms value of current output (c) average and rms value of SCR currents whereas average voltage output is 20% of the maximum average output voltage.
14. An inductive load is supplied by a single-phase fully-controlled bridge converter. Find the (i) average output voltage, (ii) supply rms current, (iii) fundamental supply rms current, (iv) DF, (v) supply power factor, (vi) THD, and (vii) VRF for a supply voltage of 230 V and firing angle of 30° under the assumption that the output current is constant.
15. A 120 V, 50 Hz ac supply is used to power a single-phase semiconverter. The load current is continuous and free of ripple at a firing angle of π /2. Find (a) DF (b) THD (c) input power factor.
16. The three-phase, 230 V, 50Hz supply is used to power the 3-Φ half wave converter, with a load resistance R = 10 Ω. The average output voltage must be 50 percent of the maximum output voltage. Find (a) α (b) average and rms load current (c) rectification efficiency.
17. A load resistance of 20Ω is powered from a three-phase star-connected 230V, 50 Hz supply through the three-phase half-wave converter. Find: (i) α (ii) rms and the average value of currents output (iii) average and rms value of SCR currents (iv) rectification efficiency (v) UTF and (vi) input power factor if the average voltage output = 25% of maximum average voltage.
18. A resistive load of 10 Ω takes 6 kW from a 3-Φ full converter for α = 60°. Calculate the magnitude of the per-phase input supply voltage. (b) Repeat part (a) if a large reactor connected in series with the load eliminates ripple in the load current.
19. A three-phase, 400 V, 50 Hz source powers a three-phase, half-wave-controlled converter, which is connected to a load taking a constant current of 36 A. Where SCR has a voltage drop of 1.4 V. Determine: (a) average load voltage for α = 30° and 60°. (b) average and RMS current rating and PIV of SCRs (c) power dispassion in each SCR.
20. A three-phase, three-phase, 400 V, 50 Hz supply a three-phase full converter delivers a ripple-free load current of 10A with an α = 45°. (a) Find the DF, CDF, THD, and PF, (b) Also determine the active and reactive powers input.
21. The supply voltage of a 3-Φ converter is 400V, 50Hz 3-Φ. A load resistor of 120Ω is connected in series with a large smoothing inductor. Find the SCR ratings.
22. The source impedance of a 1-Φ full wave midpoint converter is 0.44 mH. The converter is supplied from 130V, 50Hz supply. A freewheeling diode is connected across the load. A continuous load current of 5A is obtained. Determine the overlap angle for (a) transfer current from the SCR to the commutating diode, and (b) transfer current from the commutating diode to SCR.

23. A 3-Φ fully controlled bridge rectifier is supplied from a 3-Φ, 430V, 50Hz supply. The triggering angle is 30^0. A continuous load current of 15A is maintained at a load voltage equal to 370V. Determine (a) Source inductance, (b) load resistance, and (c) overlap angle.
24. A 3-Φ fully controlled bridge rectifier is supplied from a 3-Φ, 4400V (line), 50Hz supply having source inductance of 5mH. A continuous load current of 25A is maintained. The load consists of a voltage source of voltage 380V having an internal resistance of 1Ω. Determine (a) firing angle, and (b) overlap angle.

Know More

(a) In the case of a 1-Φ full converter considering inductive loads permits only a two-quadrant operation. If two full converters are connected back-to-back both output voltage and currents can be revered. The system provides four-quadrant operation and the converter is called a dual converter. Dual converters are normally used in high-power variable speed drives. The circuit diagram of a single-phase dual converter is shown in Figure.4.38(a). Three-phase dual converters are available and find application in variable speed drives. A circuit for a 3-Φ dual converter is shown in Figure.4.38(b).

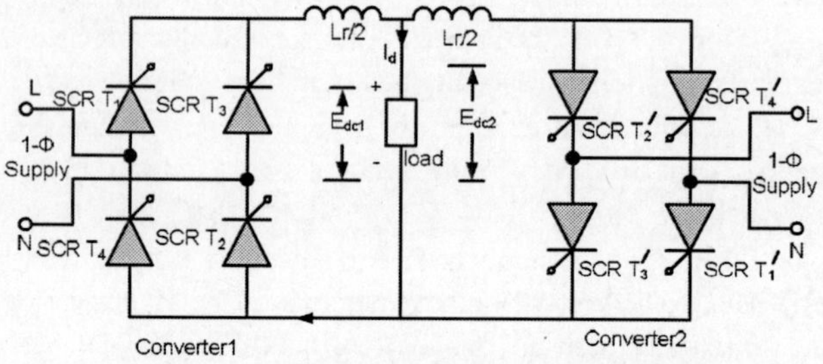

Figure.4.38(a) 1- phase dual converter

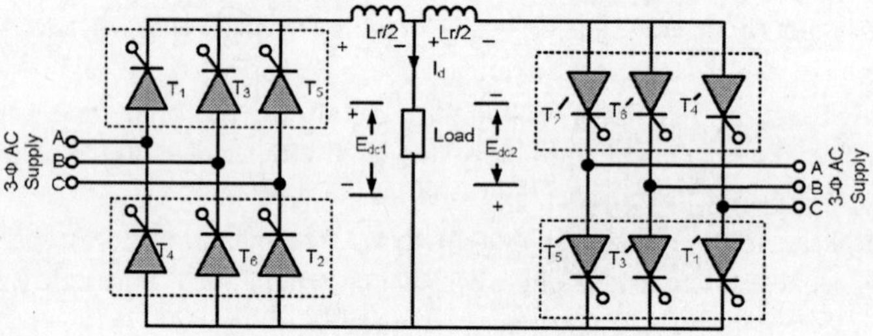

Figure.4.38(b) 3-phase dual converter

(b) In this Unit the stress is given to SCRs. The SCR is the basic device of the thyristor family. The other devices under the thyristor family are also used for the construction of various phase-controlled rectifiers.
(c) Power devices are available in a single unit or a module. The power converter usually requires two, three, four, or six power devices based on the topology of the circuit. Various power modules such as dual (half-bridge configuration), quad (in full bridge configuration), and six (in 3-Φ) are available. The power modules offer advantages such as lower on-state loss, high voltage and current switching, and high speed than conventional devices.

(d) To construct a converter along with the power module and other accessories like gate drive circuit, isolation device, protection and diagnostic circuits for excess current, open load, overheating, etc. All such devices form one module called intelligent module or smart power.

References and suggested readings

1. Ramamoorty M., An Introduction to Thyristors and their Applications, East-West Press Pvt. Ltd., New Delhi, ISBN: 8185336679.
2. Rashid, Muhammad, Power Electronics Circuits Devices and Applications, Pearson Education India, Noida, ISBN: 978-0133125900.
3. Rashid, M. H. (2014). Power Electronics: Devices, Circuits, and Applications: International Edition. United Kingdom: Pearson Education.
4. Singh, M. D. and Khanchandani, K.B., Power Electronics, Tata McGraw Hill Publishing Co. Ltd, New Delhi, 2008 ISBN: 9780070583894.
5. Mohan Ned, Undeland Tore.M., Robins William P., Power Electronics – Converters, Applications and Design, John Wiley & Sons. Inc.
6. Bimbhra, P.S., Power Electronics, Third Edition (2005), Khanna Publishers, ISBN No.81-7409-056-8.
7. Bhattacharya, S.K., Fundamentals of Power Electronics, Vikas Publishing House Pvt. Ltd. Noida. ISBN: 978-8125918530.
8. Sugandhi, Rajendra Kumar and Sugandhi, Krishna Kumar, Thyristors: Theory and Applications, New Age International (P) Ltd. Publishers, New Delhi, ISBN: 978-0-85226- 852-0.
9. Jain & Alok, Power Electronics and its Applications, Penram International Publishing (India) Pvt. Ltd, Mumbai, ISBN: 978-8187972228.
10. Fewson, D. (1998). Introduction to Power Electronics. United Kingdom: Elsevier Science.
11. Williams, B. W. (1992). Power electronics: devices, drivers, applications, and passive components. United Kingdom: McGraw-Hill.
12. Robbins, W. P., Mohan, N., Undeland, T. M. (2003). Power Electronics: Converters, Applications, and Design. India: John Wiley & Sons.

5 Industrial Control Circuits

UNIT SPECIFICS

This unit covers the following aspects:

- *Introduction to power electronics application in industrial control circuits.*
- *Battery charger circuits using SCR.*
- *Emergency lighting system using an uncontrolled full-wave rectifier and SCR.*
- *Temperature controller circuit using SCR.*
- *Illumination control using TRIAC.*
- *Introduction to switched-mode power supply (SMPS) and its classification.*
- *Flyback SMPS.*
- *Push-pull SMPS.*
- *Half Bridge Converter (SMPS).*
- *Full Bridge Converter (SMPS).*
- *Introduction to the uninterrupted power supply (UPS) and its various types.*
- *Short-break UPS or off-line (line preferred) UPS.*
- *No-break UPS or online (inverter preferred) UPS.*
- *Line-interactive UPS.*
- *UPS Battery and its ratings.*
- *Static AC and DC circuit breakers and its various types.*
- *Burglar's alarm system.*
- *Fan speed control using TRIAC.*

The application of power semiconductor devices for some industrial control circuits is presented and the working of these circuits for generating further curiosity and creativity as well as improving problem-solving capacity with some numerical problems.

Besides giving a few multiple-choice questions as well as questions of short and long answer types marked in two categories following the lower and higher order of Bloom's taxonomy, assignments through a few numerical problems, a list of references, and suggested readings are given in the unit so that one can go through them for practice.

After the detailed description of the selected topics, based on the content, there is a "Know More" section appended. This section has been designed to supplement additional information and higher learning skills on power electronics applications.

RATIONALE

This fundamental unit on industrial control circuits using various semiconductor devices helps students to get a primary idea about the application of power electronic devices in various industrial control circuits.

Some selected industrial control circuits with different power semiconductor devices are presented in this unit.

The physics behind various circuits for industrial applications using power semiconductor devices are discussed at length to develop the basic idea about these circuits.

Some related problems are pointed out after each section with their solutions which can help further for getting a clear idea of the concerned topics. The mathematics behind a few applications will certainly help students with numerical problem-solving.

As a student in the field of electrical engineering, this unit on industrial control circuits helps students to grasp the basic knowledge of applications of power electronic devices in industrial applications.

PRE-REQUISITES

ESC101: Basic Electrical Engineering

UNIT OUTCOMES

After completion of Unit-3 students will be able to:

U5-O1: *Identify various power electronics circuits for industrial application.*

U5-O2: *Explain the working of various industrial control circuits like battery chargers, switched-mode power supplies (SMPS), illumination controllers, uninterrupted power supply (UPS), temperature control circuits, emergency lighting circuits, Fan speed control circuits, and Burglar's alarm system.*

U5-O3: *Understand the working of various SMPS, UPS, static circuit breakers, and battery chargers.*

Unit-5 Outcomes	EXPECTED MAPPING WITH COURSE OUTCOMES (1- Weak Correlation; 2- Medium correlation; 3- Strong Correlation)				
	CO-1	CO-2	CO-3	CO-4	CO-5
U5-O1	3	2	...	...	3
U5-O2	...	...	3	2	3
U5-O3	1	2	2	2	3

5.1. INTRODUCTION

Over the past 40 years, power electronics and motor drives have experienced swift technological advancements, resulting in a vast and diverse field that requires interdisciplinary expertise. As technology advances and equipment costs decrease, their usage is growing in various areas such as industrial, commercial, residential, military, aerospace, and utilities due to increased reliability. The advancements are attributed to various innovations, including power semiconductor devices, converter topologies, analytical and simulation techniques, electrical motors, and control and estimation techniques. However, the incorporation of artificial intelligence techniques like fuzzy logic and artificial neural networks has raised the bar and made the work of power electronic engineers more challenging. In the 21st century, power electronics will play a significant role in global industrial automation, energy generation, and conservation efforts. As power-electronics technology advances and becomes more affordable, its practical applications are expanding. These applications include switching mode power supplies (SMPS), uninterrupted power supply (UPS), electrochemical processes, heating and lighting, static var compensation, active filtering, high voltage DC systems, photovoltaic systems, and variable frequency motor drives. One of the most significant benefits of power electronics is the increased energy efficiency of electrical equipment, which helps to save electricity consumption. With the world's energy consumption rising, particularly in industrialized nations, there is a pressing need to improve the energy efficacy of electrical equipment to enhance human living standards. Currently, a large proportion of energy is generated by burning fossil fuels like coal, natural gas, and oil, which contributes to environmental pollution and global warning. By utilizing power electronics to increase energy efficiency, we can not only reduce electricity consumption but also indirectly reduce environmental pollution by decreasing the need for power generation.

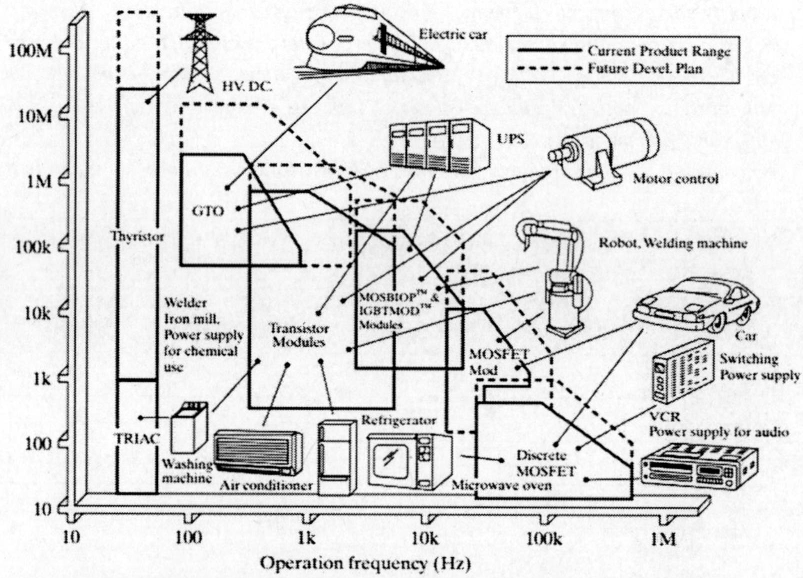

Figure 5.1 Uses for power equipment. (Courtesy of Powerex, Inc.)

In Figure.5.1, the frequency spectrum and uses of power devices are depicted. It is recommended to explore the available power devices, as their capabilities are continually improving. A high-performance power device would ideally possess four features: (1) zero on-state voltage, (2) infinite off-state voltage withstand capability, (3) ability to handle infinite current, and (4) instantaneous on-off switching speed, leading to infinite switching speed. The introduction of SiC-based power devices is expected to bring about significant improvements, such as greatly reduced on-state resistance and switching time, as well as an almost tenfold increase in voltage rating. Consequently, it is expected that there will be a shift in the applications of power devices shown in Figure.5.1. In this Unit some important industrial control circuits will be presented.

5.2. BATTERY CHARGER USING SCR

Portable devices and communication devices need a low-power battery. These batteries need to be recharged. Hence low-power battery chargers are required to recharge the batteries. The application of power electronic devices like SCR makes it possible to construct automatic battery chargers. Figure 5.2 illustrates an automatic battery charging circuit employing SCR.

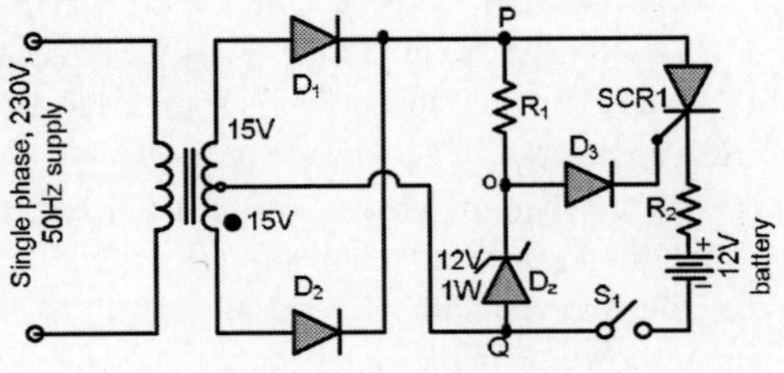

Figure.5.2 Battery Charger using SCR

The various components and their functions of this circuit are as follows. A centre-tapped step-down transformer voltage rating of 230V/ (15-0-15) V is used to step down the available AC voltage. The power diodes D_1 and D_2 are used to form an uncontrolled full-wave rectifier. The pulsating DC voltage appears across terminals *P* and *Q*. The positive terminal of the 12V discharged battery is connected to the cathode of SCR1 in series with resistance R_2. The anode of SCR1 is connected to terminal *P*. The Switch S_1 is connected between the negative terminal of the 12V battery and terminal *Q*. The circuit also requires a Zener diode (D_z), diode D_3, and resistance R_1, and their connections are shown in Figure.5.2.

When SCR1 is in the OFF state, its cathode is maintained at the same potential as the discharged battery. When switch S_1 is ON, the SCR circuit is completed. During the positive half cycle, when the voltage at the midpoint *O* reaches a sufficient level, the diode D_3 and gate-cathode junction forward biased, and the SCR1 is triggered. As a result, the charging current passed through the battery during the entire positive half cycle and the battery is charging. The D_z maintained maximum voltage (12V) at point *O*. When the battery is fully charged (12V), the cathode is maintained at 12V, the D_3, gate-cathode junction of SCR1 is reverse biased and SCR1 is not triggered. Thus, after full charging the battery is stopped.

5.2.1. Battery Charger using SCR with trickle charging facility

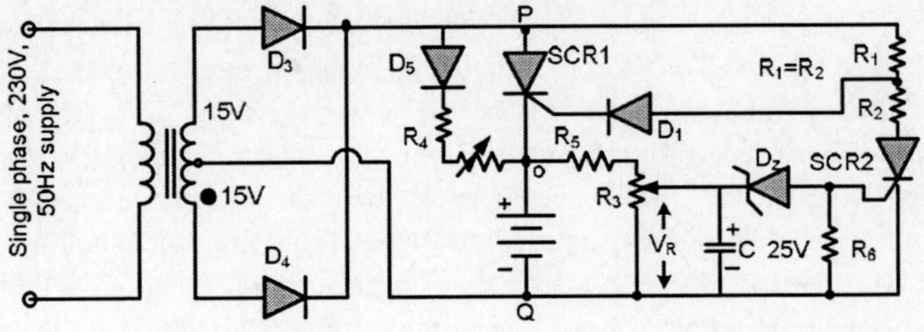

Figure.5.3 Battery charger using SCR with trickle charging facility

A battery charger circuit with a trickle charging facility is shown in Figure.5.3. The same center-tapped transformer as used in Figure.5.2 is used here to step down the available voltage level to 15-0-15V. The discharged battery (12V) is connected between mod point O and point Q. The diode D_3 and D_4 form the full wave uncontrolled rectifier. This rectifier generates DC voltage. This DC voltage is applied to the battery charger circuit from terminals P and Q. A Zener diode D_z having a breakdown voltage less than 12V (Say 11V) is connected and the variable point of resistance is R_3. When the battery voltage is less than the breakdown voltage of D_z, the SCR2 cannot be triggered since no gate current flows. Hence, SCR2 is off-state. During the positive half cycle, SCR1 is triggered since it received the gate current via the forward-biased diode and resistance R_1 and the battery is charging during the entire positive half cycle.

When the battery voltage exceeds 11V, reaching a point where the pick-off voltage V_{R3} equals the breakdown voltage V (11V in this case) of the Zener diode D_Z plus the gate voltage, the breakdown of Zener diode D_Z occurs. As a result, the gate current is supplied to SCR2 and it is turned on. Initially, SCR2 triggers at a 90° phase angle coinciding with the peak supply voltage, peak charging current, and maximum battery voltage. As the charging process proceeds through SCR1, the battery voltage rises, and the triggering angle of SCR2 advances in each half-cycle, triggering before the main SCR1. At this point, with SCR2 ON, the voltage divider arrangement of resistors R_1 and R_2 ensures that the voltage at the anode of diode D_1 remains lower than the voltage at the cathode of SCR1. Thus, diode D_1 becomes reverse-biased, preventing gate current from being supplied to SCR1 and turned off. For trickle charging, a branch can be added that includes diode D_5 and resistor R_4. Thus, when SCR1 is in the OFF state, a low charging current is maintained through diode D_5 and resistor R_4, enabling slow charging of the battery. Heavy charging can only resume after the battery voltage drops to a level where V falls below V_Z, causing the triggering of SCR2 to cease in each cycle.

5.3. EMERGENCY LIGHTING SYSTEM USING UNCONTROLLED FULL-WAVE RECTIFIER AND SCR

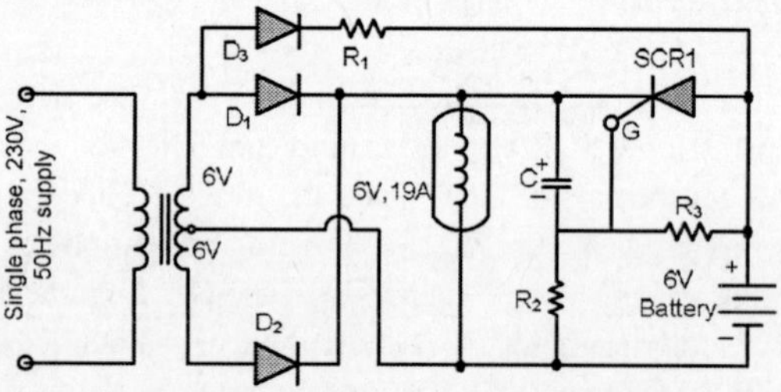

Figure.5.4 Emergency Lighting System

An emergency lighting system for household application is shown in Figure.5.4. The system utilizes the available single-phase supply (say 230V, 50Hz) for operation. A centre-tapped step-down transformer (230/ (6-0-6) V) is used at the input to step down the voltage level. The diodes D_1 and D_2 form the uncontrolled full-wave rectifier. It rectifies the input AC into DC voltage. The output DC (6V) is supplied to a 6V lamp from terminals P and O. The lamp receives 6V DC when AC is present. At the same time, a pulsating current passed through the diode D_3 and resistance R_1. This facilitates trickle battery charging. The capacitor C is connected across the gate and cathode of SCR1. The capacitor is charged with its upper plate positive charge. The voltage of the capacitor is below 6V (Output voltage of the rectifier). This voltage reverse-biased the gate to the cathode junction. With the battery voltage at the anode and the rectifier output voltage slightly higher at the cathode, SCR1 remains reverse-biased and cannot conduct. As a result, the lamp illuminates due to the rectifier's DC output voltage. The operation of the Emergency Lighting System can be summarized as follows:

(a) The lamp remains illuminated even if the AC supply is disconnected in the emergency lighting system.
(b) Initially, the input AC supply of 230 V is reduced to 6-0-6 V using a step-down transformer.
(c) The full-wave rectifier, consisting of D_1 and D_2, converts the 6V AC supply into a 6V DC supply, ensuring that the lamp remains in the ON state. Another current flows through D_3 and Resistor R_3, which charges the battery.
(d) When the AC power supply is turned OFF, the voltage at the cathode of the SCR decreases. This decrease, combined with the battery potential and a positive gate pulse, causes the SCR to turn ON. Consequently, when this battery potential is applied across the lamp, it turns ON.

5.4. TEMPERATURE CONTROLLER USING SCR

A Temperature Controller functions by comparing a sensor signal with a set point and carrying out calculations based on the difference between these values. It is employed to regulate a heater or other equipment. Controllers capable of managing sensor signals for factors like humidity, pressure, and flow rate are referred to as Controllers. Electronic controllers are specifically termed Digital Controllers. The Temperature controller finds application in various scenarios, including:
(a) Monitoring and regulating the temperature of devices, rooms, and electronic components.
(b) Implementing heat reduction measures in car engines, and controlling the cooling pads in computers and laptops.
(c) Playing a crucial role in the process control of chemical reactions, as the reactions are highly dependent on temperature.

The mercury-in-glass thermostat is a highly sensitive measuring device that can detect temperature changes effectively. A circuit for temperature control using SCR is shown in Figure.5.5.

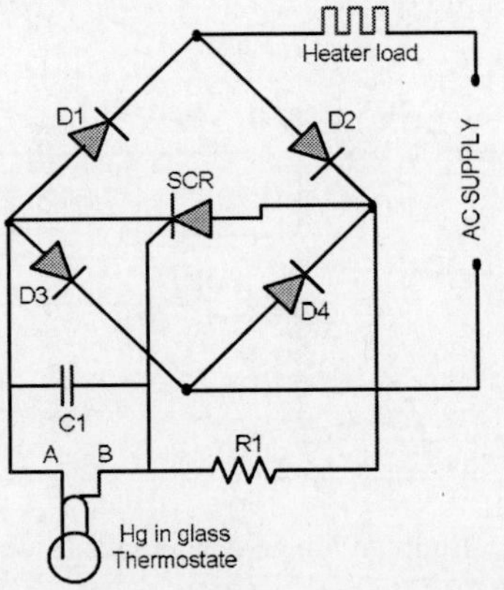

Figure 5.5 Temperature Controller using SCR

In this setup, a diode bridge rectifier is connected to the AC power source via a heater. This leads to a fully rectified voltage across the thyristor T_1. When the thermostat is open, the voltage across capacitor C_1 triggers thyristor T_1 during each half cycle of the input supply, allowing current to pass through the heater. The charging time constant is determined by the product of R_1 and C_1 networks. As the temperature rises, the conductive thermostat creates a short circuit across the capacitor, preventing it from charging and triggering the thyristor. Consequently, no current flows through the heater. There are two modes of operation. They are given below.

Mode 1: When the temperature is less than the desired value: In this operating mode, when the temperature is below the desired value, the mercury in the glass thermostat cannot create a short circuit between electrodes *A* and *B*. Consequently, the gate current (I_g) for the SCR is provided during both half cycles. As a result, when the thermostat opens, the SCR will trigger during each half-cycle and supply power to the heater load.

Mode 2: **When the desired temperature is reached:** In this mode, when the desired temperature is achieved, electrodes *A* and *B* are connected through the mercury, creating a short circuit. This short circuit effectively connects the gate and cathode terminals of the SCR. As a result, the SCR is turned OFF, causing the heater to also turn off.

5.5. ILLUMINATION CONTROL USING TRIAC

Light sources, such as tube lights or bulbs, flicker by their specific wattage rating. If you require brighter light, you must use a high-wattage bulb, while a low-wattage bulb is suitable for lower light requirements. To control the voltage supplied to the light, a light dimmer circuit or illumination control is employed. For many decades, variable transformers and resistors have been used to achieve this circuit. These methods have been utilized in various settings such as theatres, stage lighting, homes, conference halls, and restaurants. However, traditional light control methods are bulky, less efficient, and challenging to manage in remote areas. The advent of power electronics between the 1960s and 1970s introduced TRIACs and thyristors to the market, enabling the design of small, cost-effective light dimmer circuits with high efficiency.

A light dimmer operates by selectively cutting portions of the AC voltage waveform. This permits only specific segments of the waveform to reach the lamp, thereby controlling its brightness. The power transferred to the lamp determines its brightness, so the more the waveform is chopped, the dimmer the light becomes. By cutting the waveform at the zero-crossing point, smooth dimming can be achieved without causing the lamp to flicker. The circuit for illumination control using TRIAC is shown in Figure.5.6.

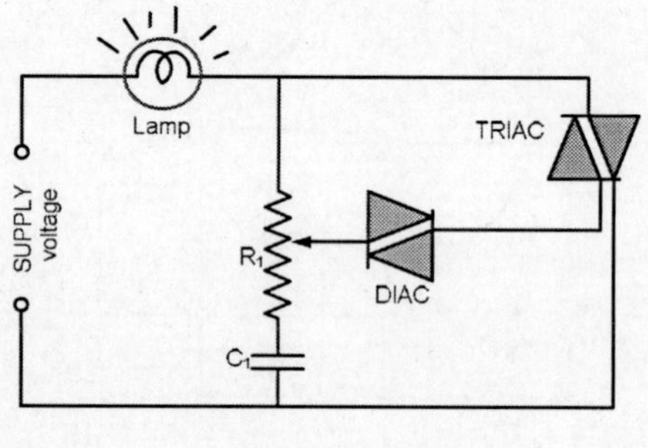

Figure 5.6 Illumination control circuit using TRIAC

The operation of the Illumination Control using TRIAC can be summarized as follows:

(a) The light is a load that operates on AC voltage. To control the brightness of the light, it is necessary to apply controlled AC voltage to the TRIAC. Phase control using DIAC-TRIAC is employed to achieve illumination control.

(b) In this circuit, the DIAC generates trigger pulses for the TRIAC.

(c) During the positive half cycle of the AC waveform, the TRIAC requires a positive gate pulse to turn it ON. This pulse is provided by a capacitor when its voltage exceeds the breakdown voltage of the DIAC. The capacitor charges through the path consisting of a resistor (*R*) and a capacitor (*C*).

(d) By adjusting the value of resistor *R*, the firing angle (α) of the TRIAC can be changed. As the firing angle (α) increases, less voltage is applied to the light, resulting in dimmer illumination.

Conversely, as the firing angle (α) decreases, more voltage is applied to the light, resulting in brighter illumination. This is how the Light dimmer circuit using DIAC-TRIAC operates.

5.6. SWITCHED MODE POWER SUPPLY (SMPS)

With the advancement of electronics, the need for a DC power supply has increased. Mostly the DC power supply is required for various integrated circuits (ICs) that are used in various equipment like computers. A lightweight compact device called switched mode power supply (SMPS) was developed by the National Aeronautics and Space Administration (NASA) in the 1960s for use in its space vehicle. The SMPS is a lightweight compact power supply. In the latter stages, the SMPS became popular. Presently nearly 80% of the total production power supply is the SMPS. Though, the phase-controlled rectifier can be used for DC supply, due to the filter circuit requirement to obtain ripple-free DC, the phase-controlled rectifier base DC supply is inefficient, bulky, and weighty. The SMPS works like a chopper. By operating the ON/OFF switch very rapidly the rise in AC ripple can be eliminated easily by L and C filter circuits. These components are of small size and light in weight. In SMPS, the DC output voltage is controlled by controlling the duty cycle of the chopper by PWM of FM techniques. The chopper circuit is not included in the syllabus of this book. Since SMPS operates on the principle of the chopper, a brief introduction of the same is provided here.

The chopper is a semiconductor device that converts fixed DC to adjustable DC. This device is used in two different ways. Namely AC link chopper and DC chopper. In the case of the AC link chopper, the conversion of fixed DC to variable DC involved two stages. The available DC supply is first converted to AC using an inverter, and the output of the inverter is stepped down using a transformer which is then converted to DC by a diode rectifier. Thus, the system becomes costly, bulky, and less efficient. In the case of the DC chopper, there is only one stage. The fixed DC is converter to variable DC using a semiconductor device. Figure.5.7(a) and Figure.5.7(b) show the block diagram of the AC Link chopper and DC chopper respectively.

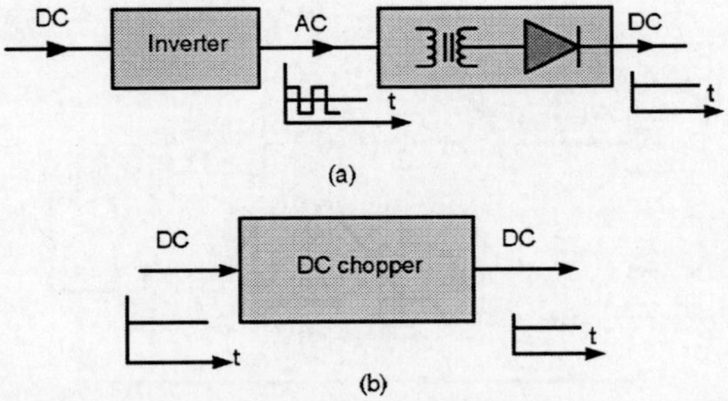

Figure.5.7 Block diagram representation of chopper (a) AC link chopper, (b) DC chopper

A chopper is a high-speed ON/OFF semiconductor switch. It may be a power BJT, power MOSFET, GTO, or forced commutated thyristor. The chopper connects the source to the load and disconnects the load from the source at a very fast speed. Thus, a chopped load voltage is obtained. An elementary chopper circuit is shown in Figure.5.8(a). The output voltage and current waveform are shown in Figure.7.8(b). Here, V_s and V_o are the input and output DC voltage respectively. I_o is the output DC current. T_{on} and T_{off} are the on-and-off periods of the switch.

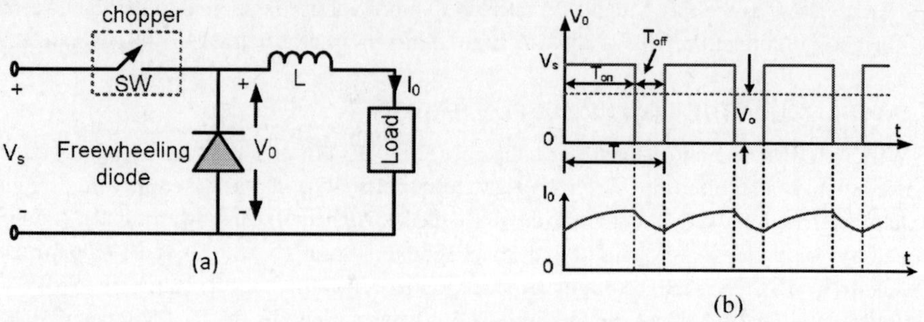

Figire.5.8 Circuit and waveform of an elementary chopper (a) Circuit, (b) voltage and current waveform

Here, the shopper is represented by switch *SW*. Which may be on or off as desired. During T_{on}, *SW* is on (close), and during T_{off}, *SW* is off (open). The total time period or chopping period *T* is the sum of T_{on} and T_{off}. The average output voltage is given by,

$$V_0 = \frac{T_{on}}{T_{on} + T_{off}}.V_s = \frac{T_{on}}{T}V_s = DV_s \tag{5.1}$$

where $D = T_{on}/T$ is the duty cycle of the chopper. Thus, by varying the *D* we can get variable output DC voltage. The value of *D* is between 0 and 1. The circuit of Figure.5.8(a) is called a step-down chopper since it steps down the DC voltage.

There is another chopper that steps up the voltage. This is called a step-up chopper. An elementary circuit diagram is shown in Figure.5.9. The output voltage equation of the step-up chopper is given by

$$V_0 = V_s.\frac{T}{T - T_{on}} = V_s.\frac{1}{1-D} \tag{5.2}$$

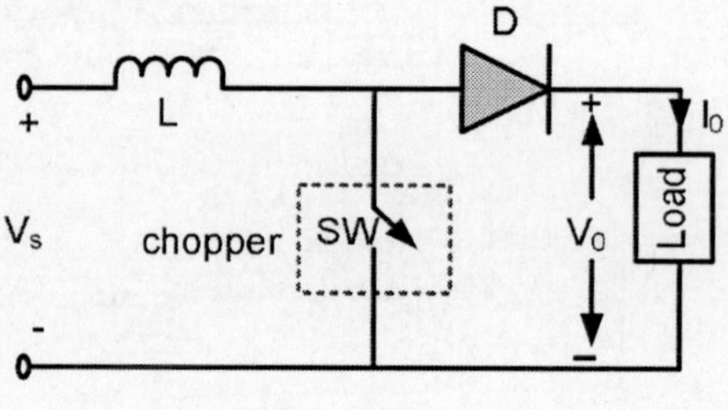

Figure.5.9 An elementary circuit of a step-up chopper

5.6.1 Classification of SMPS

There are four different types of SMPS. They are

(a) Flyback SMPS
(b) Push-pull SMPS
(c) Half bridge SMPS
(d) Full Bridge SMPS

All are discussed in brief in the following subsections.

5.6.2 Flyback SMPS

The circuit diagram of Flyback SMPS is shown in Figure.5.10(a). The components of this circuit are MOSFET (*M1*), an isolation transformer (*T1*), a diode *D*, capacitor *C*, and the load. The uncontrolled rectifier converts available AC into DC. It is assumed transformer is not demagnetized during chopping period *T*. When the MOSFET switch *M1* is on, supply voltage V_s is applied to the transformer primary (i.e $v_1 = V_s$), and a corresponding output voltage v_2 is found at the secondary with polarity as shown. The secondary voltage v_2 is given by $v_2 = (N_2/N_1)$. Vs. The v_2 reverse biases the diode *D*. The equivalent circuit for this on condition of M1 is shown in Figure.5.10(b). The capacitor *C* is a filter capacitor and it is large enough so that the voltage across it is equal to load voltage (V_O). When *M1* is off, a voltage of opposite polarity is induced in the primary and secondary as shown in Figure.5.10(c). The voltage across the secondary is now $v_2 = -V_O = -(N_2/N_1).V_s$. Diode D is now forward-biased and start conducting. The energy stored in the transformer is partly delivered to the load and capacitor *C*. Various waveforms for voltage and currents such as primary and secondary voltages (v_1, v_2), magnetizing current of the transformer (i_m), and diode current (i_D) are depicted in Figure.5.10(d). At the start ($t = 0$) of T_{on}, the i_m is not zero but it rises linearly from the initial value i_{m0} to i_{m1} at $t = T_{on}$. The magnetic energy stored in the transformer core during T_{on}.

The equation for transformer magnetizing current at time *t* is given by equation (5.3), in which *L* is the transformer magnetizing inductance.

$$i_m(t) = I_{mo} + \frac{V_s}{L}t, 0 < t < T_{on} \tag{5.3}$$

The equation for transformer magnetizing current at $t = T_{on}$ is obtained by putting $t = T_{on}$ in equation (5.3). The resultant equation is

$$i_m(\mathrm{T}_{on}) = I_{mo} + \frac{V_s}{L}\mathrm{T}_{on} = I_{m1} \tag{5.4}$$

The fall in transformer magnetizing current during the time T_{off} can be expressed as

$$i_m(t) = I_{m1} - \frac{V_0.N_1}{N_2}.\frac{1}{L}(t - T_{on}),.....T_{on} < t < T \tag{5.5}$$

The transformer magnetizing current at $t = T$ is obtained by substituting $t = T$ in equation (5.5) and given by equation (5.6).

$$i_m(\mathrm{T}) = I_{m1} - \frac{V_0.N_1}{N_2}.\frac{1}{L}(T - T_{on}) = I_{mo} + \frac{V_s}{L}\mathrm{T}_{on} - \frac{V_0.N_1}{N_2}.\frac{1}{L}(T - T_{on}) \tag{5.6}$$

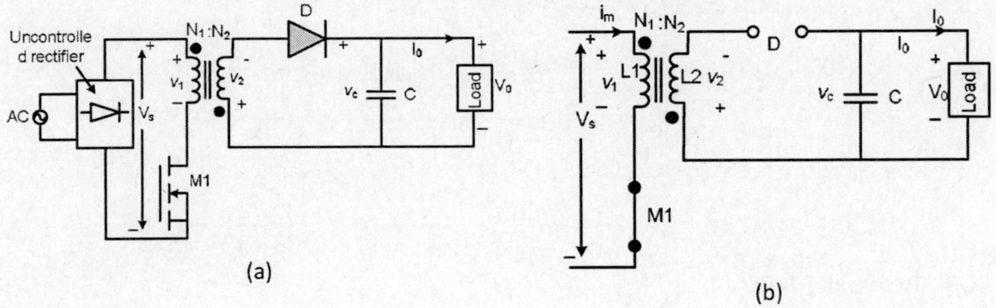

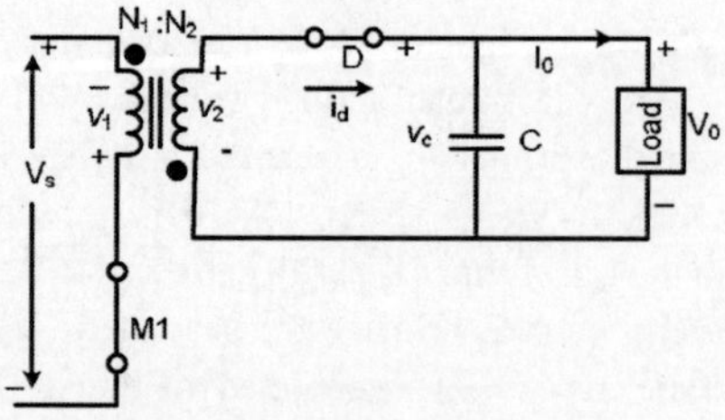

(c)

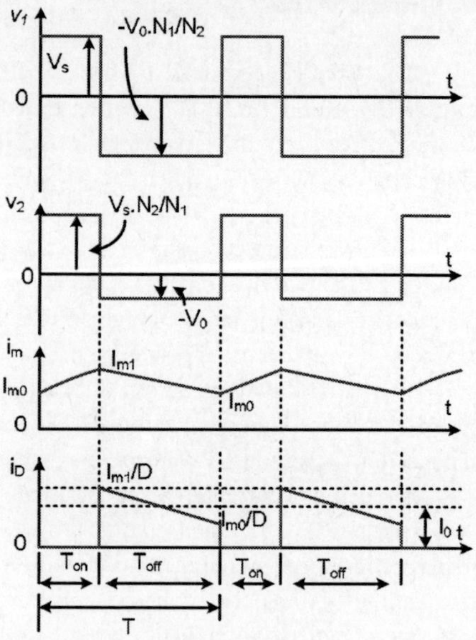

(d)

Figure.5.10 Circuit diagram and voltage and current waveform of Flyback SMPS (a) Circuit, (b) equivalent for T_{on}, (c) equivalent circuit for T_{off}, (d) Voltage and current waveform

The energy stored in one chopping period is zero, hence i_m at $t = 0$, and at $t = T$ are equal. Hence, from equation (5.3) (at $t = 0$) and equation (5.6) (at $t = T$), we will get

$$I_{mo} = I_{mo} + \frac{V_s}{L}\mathrm{T}_{on} - \frac{V_0.N_1}{N_2}.\frac{1}{L}(T - T_{on})$$

$$\Rightarrow \frac{V_s}{L}\mathrm{T}_{on} = \frac{V_0.N_1}{N_2}.\frac{1}{L}(T - T_{on}) = V_s.\mathrm{T}_{on} = \frac{V_0}{a}.(T - T_{on})$$

Thus, the load voltage is given by equation (5.7), in which $a = N_2/N_1$, transformer turn ratio, and $D =$ duty cycle of the Flyback SMPS $= T_{on}/T$

$$V_0 = \frac{aV_s.\mathrm{T}_{on}}{(T - T_{on})} = \frac{aV_s.\mathrm{D}}{(1 - D)} \tag{5.7}$$

From Figure.5.10(b), it is also seen the open circuit voltage across *M1* is

$$V_{oc} = v_1 + V_s = \frac{V_0}{a} + V_s = \frac{V_s.D}{1 - D} + V_s = \frac{V_s}{1 - D} \tag{5.8}$$

The diode current (i_D) is derived as follows

$$i_D(t) = i_m(t).\frac{N_1}{N_2} = \frac{N_1}{N_2}\left[I_{m1} - \frac{V_0.N_1}{N_2}.\frac{1}{L}(t - T_{on})\right] = \frac{I_{m1}}{a} - \frac{V_0}{a^2.L}.(t - T_{on}) \tag{5.9}$$

5.6.3 Push-pull SMPS

The circuit diagram of a push-pull type SMPS is shown in Figure.5.11. It has two power MOSFETs *M1* and *M2*, a transformer having midpoint tapings at both primary and secondary, two diodes D_1 and D_2, a load, and the control circuit. The uncontrolled diode rectifier is used at the input to convert the available AC supply into DC. The DC output of the uncontrolled diode rectifier is the input voltage V_s to the push-pull SMPS. By switching on the MOSFET M1, the V_s is applied to the lower half of the primary winding. It is denoted by v_1. The corresponding output voltage at the secondary (both upper and lower half), v_2 is (N_2/N_1). V_s. Now, the D_1 is forward-biased. The output voltage is given by

$$V_0 = \frac{N_2}{N_1} V_s = aV_s \tag{5.10}$$

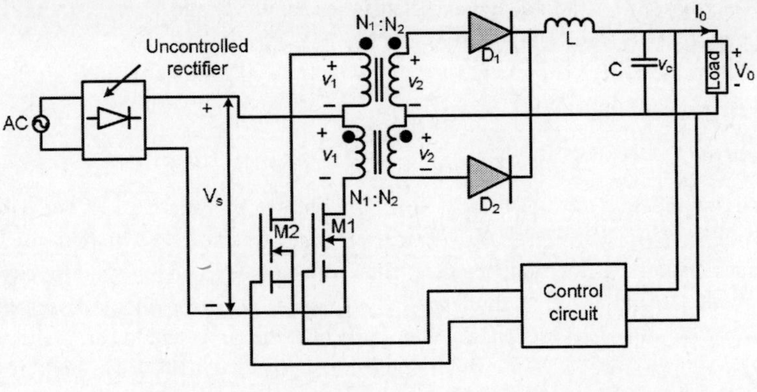

Figure.5.11 Circuit diagram of Push-pull type SMPS

By switching on the MOSFET $M2$, the voltage $v_1 = -Vs$ is applied to the upper primary winding, and the corresponding secondary voltage at both the upper and lower secondary winding will be $v_2 = -(N_2/N_1)$. V_s. The diode D_2 is forward-biased and the output voltage is given by

$$V_0 = \frac{N_2}{N_1} V_s = aV_s \tag{5.11}$$

Thus, it is seen that the primary voltage V_s swings from $+V_s$ to $-V_s$ when $M1$ and $M2$ switch on respectively. The power MOSFETs $M1$ and $M2$ operate at $D = 0.5$. when $M1$ is off, the voltage across the $M1$ terminal is $2V_s$.

5.6.4 Half Bridge Converter (SMPS)

The circuit diagram of half-bridge SMPS is shown in Figure.5.12. The circuit has one uncontrolled rectifier, two MOSFETs $M1$ and $M2$, capacitors $C1$ and $C2$ of equal values, a transformer with mid-point taping in the secondary, diodes D_1 and D_2, filter inductance L, filter capacitance C, and the control circuit. Since, the value of capacitance of $C1$ and $C2$, hence voltage across them is $V_s/2$.

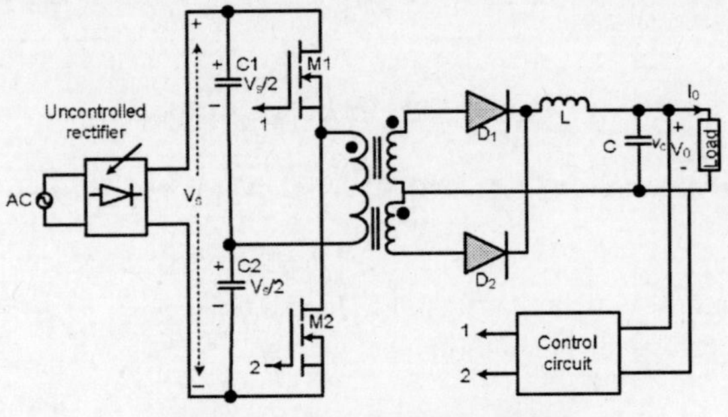

Figure.5.12 Half bridge SMPS

When $M1$ is on, the voltage across the primary is, $v_1 = V_s/2$, and across the secondary, $v_2 = (V_s/2).(N_2/2N_1)$ and diode D_1 is forward biased. When $M2$ is on, the voltage across the primary reverse voltage is $v_1 = -V_s/2$, and across the secondary, $v_2 = -(V_s/2).(N_2/2N_1)$ and diode D_2 is forward biased. It implies that the transformer's secondary voltage swings from $+V_s/2$ to $-V_s/2$. The average output voltage is given by

$$V_0 = \frac{V_s.N_2}{2N_1} = 0.5aV_s \qquad (5.12)$$

During the off period of *M1*, the voltage across it is the input voltage *Vs*. (i.e $V_{OC} = V_s$). Similarly, during the off period of *M2*, the voltage across it is V_s.

The half-bridge SMPS is preferred for high-voltage applications whereas push-pull SMPS is preferred for low-voltage applications.

5.6.5 Full Bridge Converter (SMPS)

The circuit diagram of full-bridge SMPS is shown in Figure.5.13. The circuit has one uncontrolled rectifier, four MOSFETs *M1, M2, M3,* and *M4*, a transformer with mid-point taping in the secondary, diodes D_1 and D_2, filter inductance *L*, filter capacitance *C*, and the control circuit.

When both *M1* and *M2* are simultaneously switched on, input voltage V_s appears across the transformer primary ($v_1 = Vs$), and secondary voltage on both the upper and lower half is $v_2 = (N_2/N_1).V_s = aV_s$. Diode D_1 is forward-biased and the output voltage is $V_o = aV_s$. When both *M3* and *M4* are simultaneously switched on, input voltage V_s appears across the transformer primary is reversed ($v_1 = -Vs$) and secondary voltage on both the upper and lower half is $v_2 = -(N_2/N_1).V_s = -aV_s$. Diode D_2 is forward-biased and the output voltage is $V_o = aV_s$. The open circuit voltage that appears across the MOSFETs when they are switched off is equal to the input voltage V_s.

The full bridge SMPS operates at the minimum voltage and current as compared with the other three SMPS discussed above. Normally full bridge SMPS is preferred for high-voltage applications.

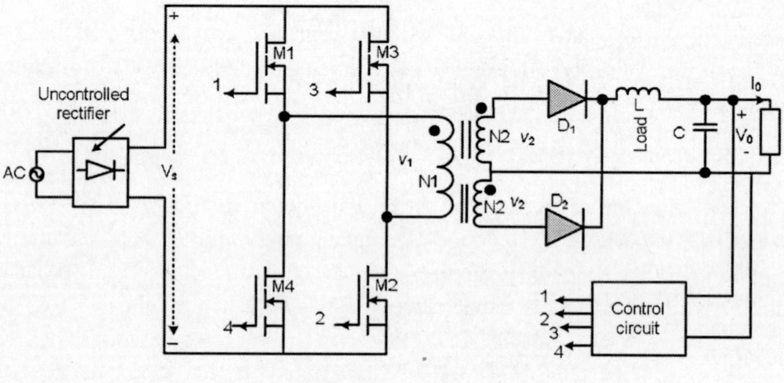

Figure.5.13 Full bridge SMPS

5.6.6 Advantages and disadvantages of SMPS over conventional power supply

Advantages

(a) Smaller in size, light in weight, and high efficiency.

(b) Less sensitive to variations in input voltage.

Disadvantages

(a) Produce higher output ripple and poor regulation.

(b) It is the source of electromagnetic and radio interference.

(c) Require filters for radiofrequency noise.

5.7. UNINTERRUPTED POWER SUPPLY (UPS)

It is very much necessary to provide uninterrupted power supply to some critical loads like hospital intensive care units (ICUs), communication systems, safety monitoring, etc. The uninterrupted power supply (UPS) is a device that supplies continuous power supply to the customer. Nowadays, static-type UPS is used.

5.7.1 UPS configuration

There are mainly two types of static UPS. They are short break type and no break type. In the case of short break type load gets disconnected for a short duration but in the case of no break type UPS, continuous power is supplied by the same. Additionally, there is one more configuration called line interactive UPS.

(a) Short-break UPS or off-line (line preferred) UPS

Figure.5.14 (a) shows the configuration of a short-break UPS. This configuration is also called off-line (line preferred) UPS. It has a rectifier, a battery bank, an inverter, a filter, and normally on and normally off switches. When AC supply is available, the AC is supplied to the loads via the normally ON switch, and at the same time the AC is rectified by the rectifier and the DC output of the rectifier charges the batteries of the battery bank. Also, during the available AC supply, the normally OFF switches are open. The output of the rectifier also supplies DC to the inverter. The inverter converts DC into AC. The AC output is filtered and ready to supply the loads. When the AC supply is interrupted, the normally OFF switch operates and switched ON and supplies power to the load. A momentary interruption in the supply (4 to 5ms) to the load is observed in the case of lamps and fluorescent tubes are part of the loads. When the normally ON switch is open and the normally OFF switch is closed, the lamps will have a transient dip in the illumination, and fluorescent tubes momentarily off and then turn on. When the AC supply appears, the critical loads are connected through the normally ON switch to the main supply. Here also, a momentary interruption in illumination is noticed.

(b) No-break UPS or online (inverter preferred) UPS

In case, no break in supply to the load is required, the arrangement shown in Figure.5.14(b) is used. It has a rectifier, a battery bank, an inverter, a filter, and normally ON and normally OFF transfer switches. The available AC supply is rectified by the rectifier, the output DC of the rectifier charges the batteries and supplies DC to the inverter continuously, and the output of the inverter supplies power to the loads via a filter to load without any break. The AC main supply is connected via a normally ON static transfer switch. The rectifier charges the battery bank only. When there is no supply, the battery supplies the load through a static transfer switch. The switch is normally OFF. When there is no power this transfer switch is ON and supplies power to the load.

(c) Line-interactive UPS

Figure.5.15 shows the block diagram of the interactive UPS configuration. The battery charger and inverter are the heart of this type. In this configuration, the static transfer switch is directly connected to the AC mains and inductance L. When the main supply is available, the static switch is closed and the load is directly supplied from the mains. At the same time, the battery bank charges through inverter/charger block. When the supply fails, the static switch is turned off, the battery bank supplies power to the load via inverter.

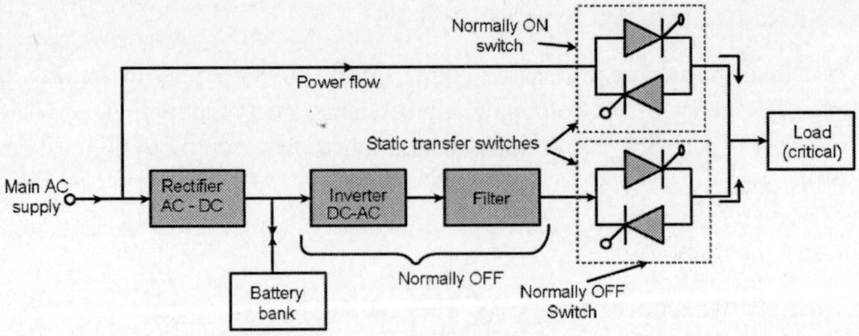

Figure.5.14 (a) Short-break UPS

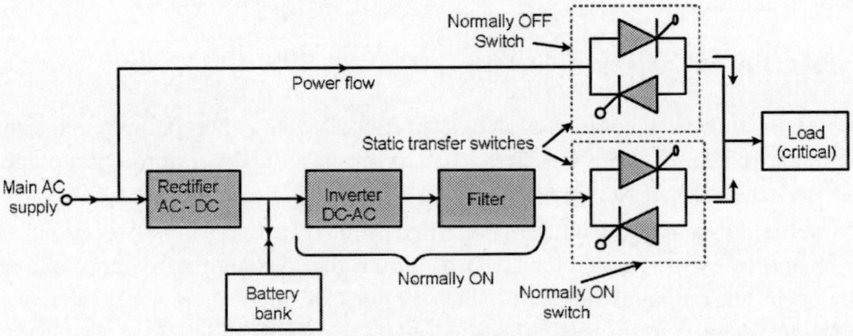

Figure.5.14(b) No-break UPS

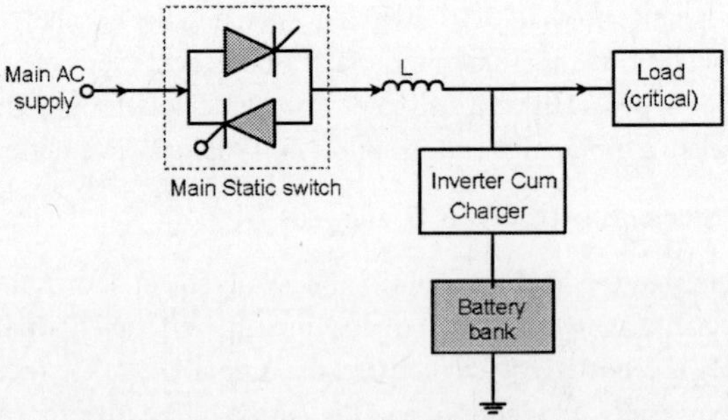

Figure.5.15 Line interactive UPS configuration

5.7.2 UPS battery

The battery is the most important component of the UPS system. The reliability of the UPS system is dependent on the battery. The life of the system is also affected by the selection of the battery. There are various types of batteries such as lead acid, nickel-cadmium, etc. The lead-acid battery is commonly used. Capacity and efficiency are the important parameters of a battery.

(a) Battery capacity

Generally, the capacity of the battery is expressed in ampere-hour (AH or A-H). AH tells us the amount of current that a battery can supply in one hour. The capacity of the battery depends on some factors like rate of discharge, temperature, and density of the electrolyte. The AH rating decreases with an increase in the rate of discharge. The AH rating increases with an increase in temperature. The capacity also increases with the density of the electrolytes. The capacity of the battery is calculated using the following.

$$\text{Capacity in Kw} = \text{Capacity in Kw} = \frac{\text{Load kVA x power factor}}{\text{Inverter efficiency}} \tag{5.13}$$

$$\text{Numbers of cells} = \frac{\text{Minimum allowable battery voltage}}{\text{Final voltage/cell}} \tag{5.14}$$

$$\text{Cell capcity in kW/cell} = \frac{\text{Battery capacity in kW}}{\text{Numbers of cells}} \tag{5.16}$$

(b) Battery efficiency

The battery efficiency can be expressed in two ways. They are given below.

$$\text{Ampere hour (AH) efficiency} = \frac{\text{AH during discharge}}{\text{AH input while charging}} \tag{5.17}$$

Watt-hour (WH) (Energy)

$$= \text{AH efficiency x } \frac{\text{Average value of cell voltage while discharging}}{\text{Average value of cell voltage while charging}} \tag{5.18}$$

The typical value of AH efficiency is 90-95%. Similarly, WH efficiency is about 70-80%.

5.8. STATIC AC AND DC CIRCUIT BREAKER

The static circuit breakers are semiconductor-based circuit breakers (CBs). SCR is utilized for the construction of static CBs. These CBs provide fast and reliable interruption to the continuous current. There are mainly two types of static CBs namely static AC circuit breakers (SACCB) and static DC circuit breakers (SDCCB).

(a) Static AC circuit breakers

A simple configuration of a static AC circuit breaker (SACCB) is depicted in Figure.5.16(a). the SCR T1 and SCR T2 are used in this configuration. They are turned on when the load current is passed through the Zero. To break the circuit, the triggering pulse is removed to bring the SCRs to the off state. When the SCRs are in an off state, no current can pass to the load and the circuit is interrupted. When turning off the signal received by the control unit due to some faults, the triggering signal is withdrawn from SCR T1 or SCR T2 and the circuit is interrupted. The related waveform for output voltage and current along with the triggering signal is shown in Figure.5.16(b). Example: say the triggering signal i_{g1} is not applied at $t = 4\pi+\Phi$ i.e. withdrawn. However, if the turn-off command is received before $t = 4\pi+\Phi$, say at $t = 3\pi+\Phi$, the current continues to flow till $4\pi+\Phi$. Thus, the maximum time delay for breaking the circuit is π/ω second after the turn-off signal is received by the control unit due to some exigencies in the system.

(b) Static DC circuit breakers

A simple circuit for a Static DC circuit breaker is depicted in Figure.5.17. The circuit is similar to that of a class - C commutation circuit. When the input voltage is DC, the commutation of the SCRs must be forced commutation. When the SCR *T1* is turned on, the load voltage is equal to the source voltage (V_s in this case), and the capacitor starts charging through the path V_s-R_2-C-SCR *T1*. To break the circuit auxiliary SCR *T2* is turned on. The capacitor C immediately applies the reverse voltage V_s across SCR *T1* and it is turned off. After this capacitor *C* charges from $+V_s$ to $-V_s$, through the path V_s-load-C-SCR *T1*. At a full charge of capacitor *C* with negative V_s, i.e. $-V_s$ (left plate +Ve and right plate

-Ve, the current through the load is zero. The current through R_2 is also less than the holding current of SCR *T2* and hence SCR *T2* is turned off.

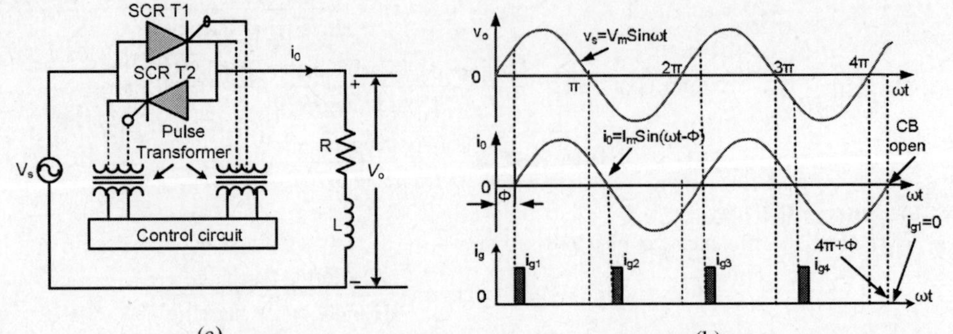

(a) (b)

Figure.5.16 (a) SACCB, (b) waveform for output voltage and current, and gate current

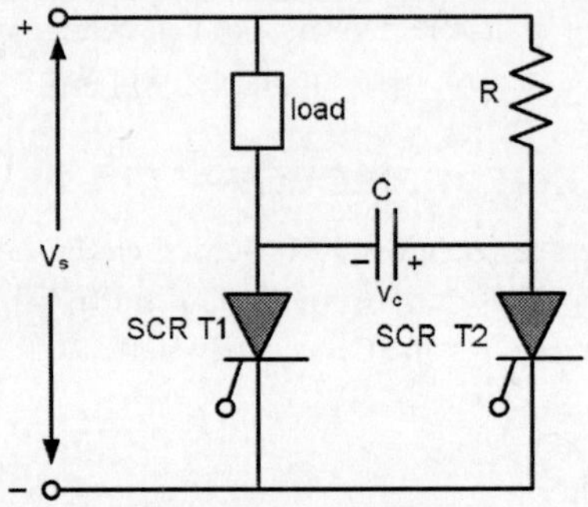

Figure.5.17 SDCCB

5.9. BURGLAR'S ALARM SYSTEM

A Burglar's alarm system is a security system that detects intrusion. This is a security alarm used for protection against burglary, property damage, and personnel protection against intruders. There are various types of such alarm systems and most of them use one or more sensors to detect the intruders, alerting devices for indicating the intruders. Typical security burglar alarm system includes a premises control unit, a control panel, sensors, security devices, and alerting devices, etc. In some cases, security devices are coupled with monitoring devices. If some events occur, the premises control unit sends a signal to central monitoring units. The operator present on the premises will take necessary action like informing the property owner, police, etc.

There are many types of Burglar alarm systems. They are wired, wireless, and smart alarm. The wired one is the most traditional. It requires physical wires to connect various pieces of equipment. Wireless type uses radio frequencies to communicate among the components. A few factors need to be considered while choosing a burglar's alarm system. Firstly, a decision needs to be taken on how many entry points want to protect. This will decide how many sensors are required. Secondly, it is required to decide on the type of monitoring system. A circuit diagram of Burger's alarm system is shown in Figure.5.18.

The various components of the systems are four resistances *R1, R2, R3,* and *R4,* the diodes *D1,* and *D2*, transistors *T1* and *T2*, a capacitor *C1*, a speaker, and a power supply. Two switches *S1* and *S2* are placed in different places. When the switch *S1* closed *D1* via the resistances *R1, R2* switched the transistor *T1* and *T2*. To form an oscillation circuit the transistor *T1* and *T2* gets positive feedback provided by capacitor *C1*. When *S1* is pressed, any intruder is detected by low-frequency tone. Due to some other conditions when *S2* is pressed, *D2* conducts and supplies power to *T1* and *T2* and are in working and result in sounds from the speaker. This sound that is observed by the speaker can only be stopped by switching off the power supply.

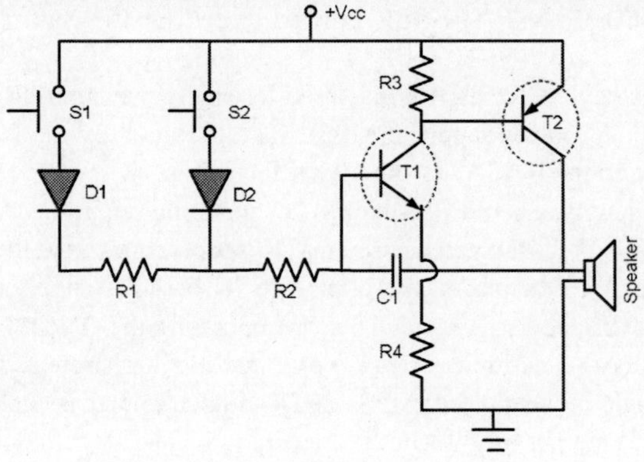

Figure.5.18 Burglar's Alarm System

5.10. FAN SPEED CONTROL USING TRIAC

The fan speed is regulated using a fan regulator. Nowadays most of the fan regulators are electronic regulators. The electronic fan regulator is an electronic device consisting of DIACs, TRICs, as well a potentiometer. The same can be used to control the speed of drives like a mixer grinder, solution mixer, etc. However current rating is different from that fan regulator. The TRIAC is operated similarly to two reverse parallel thyristors having a common gate terminal. The TRIC can trigger conduction in both positive and negative polarity voltages. It acts like a full-wave thyristor. Either positive or negative polarity voltage can be applied to the gate. The triggering of TRICs can be done by DIAC circuits. An isolation transformer is used as an isolation transformer to protect against electric shock for safety. The unit operates at 230V, 50 Hz. The circuit diagram of this electronic regulator is shown in Figure.5.19.

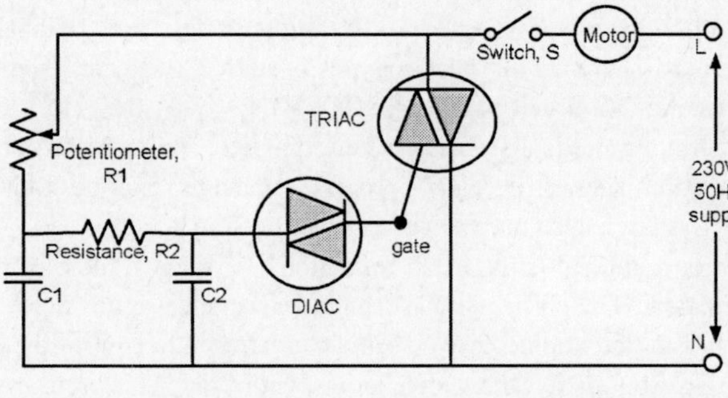

Figure.5.19 Electronic fan regulator using DIAC and TRIAC

Unit Summary

This Unit explores some important industrial control circuits as mentioned in the syllabus. Following is the summary of this unit.

1. The low-power automatic battery chargers can be constructed using power electronics devices like SCR. The circuits of the SCR-based battery charger are presented in this unit.
2. A Temperature Controller functions by comparing a sensor signal with a set point and carrying out calculations based on the difference between these values. It is employed to regulate a heater or other equipment. Controllers capable of managing sensor signals for factors like humidity, pressure, and flow rate are referred to as Controllers. Electronic controllers are specifically termed Digital Controllers. The SCR is widely used for the control circuits in temperature control circuits.
3. Power electronic device like SCR is also used for emergency lighting systems.
4. The switched mode power supply (SMPS) was developed by the National Aeronautics and Space Administration (NASA) in the 1960s for use in its space vehicle. It is a lightweight compact power supply. The SMPS supplies DC to various equipment like computers. Though, the phase-controlled rectifier can be used for DC supply, due to the filter circuit requirement to obtain ripple-free DC, the phase-controlled rectifier base DC supply is inefficient, bulky, and weighty. The SMPS works like a chopper. By operating the ON/OFF switch very rapidly the rise in AC ripple can be eliminated easily by L and C filter circuits. These components are of small size and light in weight. In SMPS, the DC output voltage is controlled by controlling the duty cycle of the chopper by PWM of FM techniques. The MOSFET is commonly used as the switching component in SMPS.
5. There are four different types of SMPS. They are
   (a) Flyback SMPS
   (b) Push-pull SMPS
   (c) Half bridge SMPS
   (d) Full Bridge SMPS
6. It is very much necessary to provide uninterrupted power supply to some critical loads like hospital intensive care units (ICUs), communication systems, safety monitoring, etc. The uninterrupted power supply (UPS) is a device that supplies continuous power supply to the customer. Nowadays, static-type UPS is used. There are two types of static UPS. They are short break type and no break type. In the case of short break type load gets disconnected for a short duration but in the case of no break type UPS, continuous power is supplied by the same. There are two types of UPS. Shor- break and no break UPS. A power semiconductor device like SCR is used in UPS.
7. The static circuit breakers are semiconductor-based circuit breakers (CBs). SCR is utilized for the construction of static CBs. These CBs provide fast and reliable interruption to the continuous current. There are mainly two types of static CBs namely static AC circuit breakers (SACCB) and static DC circuit breakers (SDCCB).
8. A Burglar's alarm system is a security system that detects intrusion. This is a security alarm used for protection against burglary, property damage, and personnel protection against intruders. This system is also used to power electronic switches.
9. The fan speed is regulated using a fan regulator. Nowadays most of the fan regulators are electronic regulators. The electronic fan regulator is an electronic device consisting of DIACs, TRICs, as well as a potentiometer. The same can be used to control the speed of drives like a mixer grinder, solution mixer, etc. However current rating is different from that fan regulator.
10. The TRIAC can also use for dimmer circuits to control the illumination.

Exercises

Example.5.1

A UPS is used to supply a load of 600W at 0.8 power factor lagging. The efficiency of the inverter is 80%. The voltage rating of the battery is 24V. There is no separate charger for the battery. Find (a) the kVA rating of the inverter, (b) the wattage of the rectifier, and (c) the Rating of the battery in Ampere-hour for 30 minutes back up.

Solution: Given data are battery voltage =24V, battery back-up time = 30minutes, efficiency of the inverter = 80%, Load = 600W at 0.8 pf lagging

(a) kVA rating of the inverter

Total RMS power is given by

$$\text{Total RMS power} = \frac{\text{Active power in watt}}{\text{power factor}} = \frac{600}{0.8} = 750\text{VA}$$

Thus, kVA rating of the inverter is 0.75kVA.

(b) Wattage of the rectifier

The wattage of the rectifier can be obtained as follows

$$\text{Wattage of the rectifier} = \frac{\text{kVA rating of UPS xpower factor}}{\text{Efficiency of the inverter}} = \frac{0.75\text{kVA x } 0.8}{0.8} = 0.75\text{kW}$$

(c) Ampere-hour rating can be obtained as follows

$$\text{Battery voltage rating x current} = k\text{W rating of battery}$$

$$\therefore \text{ Battery current} = \frac{\text{kW of the battery}}{\text{Battery voltage}} = \frac{0.75\text{x}10^3}{24} = 31.25\text{A}$$

If we assume that the efficiency of the battery is 100%, the Ampere-hour is

Ampere-hour of the battery = battery current x back-up time in hours = 31.25x0.5 = 15.625 Ampere-hour

Example.5.2

The rating of a UPS is 20KVA and its backup time is 15 minutes. The efficiency of the inverter is 85%. Load to be supplied by the UPS having 0.8 power factor lagging. The voltage range of the battery is 147-190V. The battery is of the lead-acid type having 6 cell groups per jar. The final voltage on the cell is 1.75V/cell. Select the battery for the UPS system.

Solution:

Battery power in kW is given by

$$\text{Battery wattage (kW)} = \frac{\text{Load in kVA x power factor}}{\text{inverter efficiency}} = \frac{20\text{x}0.8}{0.85} = 18.824\text{kW}$$

The number of cells in the battery and the number of jars are obtained as

$$\text{Numbers of cells} = \frac{\text{Minimum battery voltage}}{\text{Final voltage/cell}} = \frac{147}{1.75} = 84$$

$$\text{Numbers of jars} = \frac{\text{Numbers of cells}}{\text{Numbers of cells/jar}} = \frac{84}{6} = 14$$

The kW rating of the cell is obtained as

$$\text{kW cell size} = \frac{\text{kW of the battery}}{\text{Numbers of cells}} = \frac{18.824}{84} = 0.224\text{kW}$$

A battery discharge table is normally provided. Corresponding to 15 minutes backup time, the battery discharge rate in kW is 0.255 and the corresponding battery is 90 ampere-hours (A-H) is selected. This kW rating is greater than that obtained from the calculation (0.224kW).

Example.5.3

The discharge rate of a battery is 35A for 5 hours at an average emf= 1.95V. The voltage rating of the battery is 2V. The battery is again charged to 45A for 4 hours to restore its original voltage. Assuming a lead acid type battery, Find (i) A-H efficiency, and (ii) WH efficiency.

Solution:

(a) Using the equation, $\text{Ampere hour (AH) efficiency} = \frac{\text{AH during discharge}}{\text{AH input while charging}}$

We will get,

$$\text{Ampere hour (AH) efficiency} = \frac{35\text{x}5}{45\text{x}4} = 97.22\%$$

(b) Using the equation,

$$\text{Watt-hour (WH) (Energy)} = \text{AH efficiency x } \frac{\text{Average value of cell voltage while discharging}}{\text{Average value of cell voltage while charging}}$$

We will get,

$$\text{Watt-hour (WH) (Energy)} = 97.22 \text{ x } \frac{1.95}{2} = 94.79\%$$

Multiple Choice Questions

1. In No break UPS, the static transfer switch connected with the main AC supply is
   (a) Normally on
   (b) Normally off
   (c) Off continuously
   (d) None of the above
2. The correct statement related to SMPS is
   (a) It is less sensitive to input voltage variations
   (b) It is smaller as compared to rectifiers
   (c) It has a low input ripple
   (d) It is a source of radio interference
3. Fill in the blank in the statement, "The .... is used for critical loads"
   (a) SMPS
   (b) UPS
   (c) MPS
   (d) RCCB
4. The machine used in rotating type UPS is
   (a) DC motor
   (b) Self-excited DC generator
   (c) Alternator

(d) Battery bank

5. Which converter is used in static UPS
   (a) Rectifier
   (b) Inverter
   (c) Both the inverter and rectifier
   (d) None of the above
6. No discontinuity is found in
   (a) Short break static UPS configuration.
   (b) Long break static UPS configuration.
   (c) No break static UPS configuration.
   (d) Rotating type UPS configuration.
7. A momentary discontinuity observed in
   (a) Short break static UPS
   (b) Long break static UPS
   (c) No break static UPS
   (d) Rotating UPS
8. Which batteries are suitable for UPS
   (a) NC
   (b) Li-On
   (c) Lead acid
   (d) All of the above
9. The HVDC transmission is better as compared to HVAC transmission, because of
   (a) smaller transformer size
   (b) smaller conductor size
   (c) higher corona loss
   (d) smaller power transfer capabilities
10. The half-bridge SMPS or full-bridge SMPS preferred for
    (a) High voltage application
    (b) Low voltage application
    (c) Medium voltage application
    (d) None of the above
11. The push-pull SMPS is preferred for
    (a) Low voltage application
    (b) High voltage application
    (c) Medium voltage application
    (d) None of the above
12. Select the advantage of SMPS over conventional DC supply
    (a) Smaller in size, light in weight, high efficiency, and less sensitive to variation of input voltage.
    (b) Larger in size, light in weight, high efficiency, and less sensitive to variation of input voltage.
    (c) Smaller in size, Heavy, high efficiency, and less sensitive to variation of input voltage.
    (d) Smaller in size, Heavy, high efficiency, and more sensitive to variation of input voltage.
13. Select the disadvantages of SMPS over conventional DC supply
    (a) Produce higher output ripple, poor regulation, source of electromagnetic and radio interference, and require filters

(b) Produce smaller output ripple, poor regulation, source of electromagnetic and radio interference, and require filters.
(c) Produce higher output ripple, good regulation, source of electromagnetic and radio interference, and require filters.
(d) Produce higher output ripple, poor regulation, source of electromagnetic and radio interference, and do not require filters.

14. In short break UPS, the inverter and filter are
    (a) Normally on
    (b) Normally off
    (c) Off continuously
    (d) None of the above
15. In no break UPS, the inverter and filter are
    (a) Normally on
    (b) Normally off
    (c) Off continuously
    (d) None of the above
16. In short break UPS, the static transfer switch in series with the inverter and filter is
    (a) Normally on
    (b) Normally off
    (c) Off continuously
    (d) None of the above
17. The switched mode power supply (SMPS) is used to
    (a) Obtain a controlled ac power supply
    (b) Obtain a controlled dc power supply
    (c) Store dc power
    (d) Switch from one to another source
18. In No break UPS, the static transfer switch in series with the inverter and filter is
    (a) Normally on
    (b) Normally off
    (c) Off continuously
    (d) None of the above
19. In short break UPS, the static transfer switch connected with the main AC supply is
    (a) Normally on
    (b) Normally off
    (c) Off continuously
    (d) None of the above
20. The principle used by SMPS is
    (a) Phase control
    (b) Chopper
    (c) Integral control
    (d) MOSFET
21. A Burglar's alarm system is a security system that uses.......... to detect intrusion
    (a) Alarm
    (b) Clock
    (c) Light
    (d) None of the above
22. Which power electronic device is mainly used to trigger SCR and TRIAC
    (a) DIAC

(b) MOSFET
(c) BJT
(d) Diode

23. Static circuit breaker uses
(a) Controlled device
(b) Uncontrolled device
(c) Both controlled and uncontrolled device
(d) None of the above

24. The speed induction motor can be controlled by
(a) A cyclo converter
(b) A chopper
(c) An inverter
(d) All the above
(e) None of the above

25. Which one of the following operates in an emergency light
(a) An inverter
(b) A rectifier
(c) A cyclo converter
(d) Both inverter and rectifier

Answers to multiple-Choice Questions

1.	**(b)**	**6.**	**(c)**	**11.**	**(a)**	**16.**	**(b)**	**21.**	**(a)**
2.	**(c)**	**7.**	**(a)**	**12.**	**(a)**	**17.**	**(b)**	**22.**	**(a)**
3.	**(b)**	**8.**	**(c)**	**13.**	**(a)**	**18.**	**(a)**	**23.**	**(a)**
4.	**(c)**	**9.**	**(b)**	**14.**	**(b)**	**19.**	**(a)**	**24.**	**(a)**
5.	**(c)**	**10.**	**(a)**	**15.**	**(a)**	**20.**	**(b)**	**25.**	**(b)**

Short and Long Answer Type Questions

1. With the help of the necessary diagram, describe the operation of static UPS.
2. Explain the operation of an online UPS. Prove the necessary diagram of the online UPS.
3. Why online UPS is costlier?
4. Why offline UPS is cheaper?
5. Draw the schematic diagram of an offline UPS and explain the operation of the same.
6. List the merits and demerits of online and offline UPS.
7. Explain the basic principle of SMPS with the necessary block diagram.
8. Explain the operation of flyback SMPS with the necessary diagrams.
9. Explain the operation of push-pull SMPS with the necessary diagrams.
10. Explain the operation of half-bridge SMPS with the necessary diagrams.
11. Explain the operation of full bridge SMPS with the necessary diagrams.
12. With the help of the necessary circuit and waveforms, describe the operation flyback SMPS.
13. With the help of necessary circuits and waveforms, describe the operation of push-pull SMPS.
14. With the help of necessary circuits and waveforms, describe the operation of half-bridge SMPS.
15. With the help of necessary circuits and waveforms, describe the operation of full bridge SMPS.
16. Describe the operation of a battery charger with the necessary diagrams.
17. Draw the circuit diagram of different battery chargers using SCR. Explain the operation of the circuit.
18. Draw the circuit diagram of the emergency lighting system using a full wave diode rectifier and SCR. Explain the operation of the circuit.

19. With the circuit diagram, explain the operation of the temperature controller using SCR.
20. With the circuit diagram, explain the illumination control using TRIAC.
21. Explain the control of the speed of a fan using TRIAC.
22. Classify SMPS.
23. What is the function of the Burglar alarm system?
24. Draw the circuit diagram of a burglar alarm system and explain.
25. What are the types of static circuit breakers?
26. Explain the operation of static AC and static DC circuit breakers.

Numerical problems

1. A UPS is used to supply a load of 500W at 0.85 power factor lagging. The efficiency of the inverter is 85%. The voltage rating of the battery is 14V. There is no separate charger for the battery. Find (a) the kVA rating of the inverter, (b) the wattage of the rectifier, and (c) the rating of the battery in Ampere-hour for 30 minutes back up.
2. A UPS is used to supply a load of 600W at 0.85 power factor lagging. The efficiency of the inverter is 90%. The voltage rating of the battery is 14V. There is no separate charger for the battery. Find (a) the kVA rating of the inverter, (b) the wattage of the rectifier, and (c) the rating of the battery in Ampere-hour for 30 minutes back up. Assume battery efficiency is 90%.
3. The rating of a UPS is 20KVA and its back-up time is 15 minutes. The efficiency of the inverter is 80%. Load to be supplied by the UPS having 0.8 power factor lagging. The voltage range of the battery is 150-190V. The battery is of the lead-acid type having 6 cell groups per jar. The final voltage on the cell is 1.70V/cell. Select the battery for the UPS system.
4. The discharge rate of a battery is 30A for 4 hours at an average emf = 1.92V. The voltage rating of the battery is 2V. The battery is again charged to 40A for 3 hours to restore its original voltage. Assuming a lead acid type battery, Find (i) A-H efficiency, and (ii) WH efficiency.

Know More

Here some additional applications of power electronics are presented. Students are advised to follow the reference books listed at the end.

1. The AC system is normally used for the generation, transmission, and distribution of electrical power. In recent years, the high voltage DC (HVDC) system is also used for long-distance transmission is also used because it is economical. There are many such transmission systems presently available. Two AC systems are linked by such an HVDC transmission system. The AC system input voltage is first rectified by the rectifier to a Very high voltage DC and the power is transmitted via an HVDC transmission system. At the receiving end, the HVDC is converted to AC by using inverters. Both rectifier and inverter are power electronics based.
2. There are various types of HVDC transmission systems namely monopolar, bipolar, homopolar, and back-to-back tie station. In a monopolar HVDC transmission system, there is one pole (one conductor) and the earth. One line conductor with earth or sea as return conductor whole power is transmitted. In the bipolar HVDC system there are two poles, one is positive and the other is negative. In homopolar there are two poles of the same polarity and the earth is the return conductor. In the back-to-back HVDC tie system, there are no poles. Two back-to-back converters are used at the same location. Using this HVDC tie station, two adjacent AC networks are tied together.
3. Power electronics devices are used for static var compensator (SVC). The SVCs are used to prevent voltage flickers caused by industrial loads that cause frequent changes in reactive power.

4. In radio frequency (RF) heating high frequency AC is used. High-frequency AC is obtained from a power electronics converter. There are two types of RF heating. They are Induction heating and dielectric heating.
5. Nowadays, Electronic ballasts are used instead of inductive ballast in fluorescent lamps. These lamps improve energy efficiency. The operating frequencies are more than 25KHz. This is possible by using an uncontrolled AC-DC converter along with a filter followed by a high-frequency DC-AC inverter.
6. The time delay circuit uses SCR and UJT. The SCR and TRIC are suited for flasher circuits.
7. The power electronics are the key components for AC and DC drives. Till a couple of decades back, variable drives were not used because of the factors like poor efficiency, large size, and low speeds. With the advent of power electronics, presently the drives are variable speed drives, highly efficient, and smaller in size.
8. Power electronics devices are used to construct AC voltage controllers. AC voltage controllers are thyristor-based devices that convert fixed AC voltage into variable AC voltage without changing the frequency. Some of such applications of AC voltage controllers are industrial and domestic heating, transformer tap changing, lighting control, speed control of single-phase and three-phase AC drives, etc.

References and suggested readings

1. Ramamoorty M., An Introduction to Thyristors and their Applications, East-West Press Pvt. Ltd., New Delhi, ISBN: 8185336679.
2. Rashid, Muhammad, Power Electronics Circuits Devices and Applications, Pearson Education India, Noida, ISBN: 978-0133125900.
3. Rashid, M. H. (2014). Power Electronics: Devices, Circuits, and Applications: International Edition. United Kingdom: Pearson Education.
4. Singh, M. D. and Khanchandani, K.B., Power Electronics, Tata McGraw Hill Publishing Co. Ltd, New Delhi, 2008 ISBN: 9780070583894.
5. Mohan Ned, Undeland Tore.M., Robins William P., Power Electronics – Converters, Applications and Design, John Wiley & Sons. Inc.
6. Bimbhra, P.S., Power Electronics, Third Edition (2005), Khanna Publishers, ISBN No.81-7409-056-8.
7. Bhattacharya, S.K., Fundamentals of Power Electronics, Vikas Publishing House Pvt. Ltd. Noida. ISBN: 978-8125918530.
8. Sugandhi, Rajendra Kumar and Sugandhi, Krishna Kumar, Thyristors: Theory and Applications, New Age International (P) Ltd. Publishers, New Delhi, ISBN: 978-0-85226- 852-0.
9. Jain & Alok, Power Electronics and its Applications, Penram International Publishing (India) Pvt. Ltd, Mumbai, ISBN: 978-8187972228.
10. Fewson, D. (1998). Introduction to Power Electronics. United Kingdom: Elsevier Science.
11. Williams, B. W. (1992). Power electronics: devices, drivers, applications, and passive components. United Kingdom: McGraw-Hill.
12. Robbins, W. P., Mohan, N., Undeland, T. M. (2003). Power Electronics: Converters, Applications, and Design. India: John Wiley & Sons.
13. https://www.electronicshub.org/security-alarm-circuit/
14. Md. Barkat Ullah et al, "Design and Development of Low-Cost Security Alarm Using Optoelectronic Device, *Journal of Electronics and Communication Engineering,* Vol 10, No. 4, Ver. I (Jul - Aug .2015), pp 19-23, DOI:10.9790/2834-10411923

APPENDICES

APPENDIX-A

Some Important Mathematical Formulas

A1: Trigonometric formulas

$sin(A \pm B) = \sin A \cos B \pm \cos A \sin B$

$\cos(A \pm B) = \cos A \cos B \mp \sin A \sin B$

$$\tan(A+B) = \frac{\tan A + \tan B}{1 - \tan A.\tan B}$$

$$\tan(A-B) = \frac{\tan A - \tan B}{1 + \tan A.\tan B}$$

$$\cot(A+B) = \frac{\cot A.\cot B - 1}{\cot B + \cot A}$$

$$\cot(A-B) = \frac{\cot A.\cot B + 1}{\cot B - \cot A}$$

$\sin 2A = 2\sin A \cos B$

$\cos 2A = 1 - 2\sin^2 A = 2\cos^2 A - 1$

$$\sin A + \sin B = 2\sin\frac{A+B}{2}.\cos\frac{A-B}{2}$$

$$\sin A - \sin B = 2\cos\frac{A+B}{2}.\sin\frac{A-B}{2}$$

$$\cos A + \cos B = 2\cos\frac{A+B}{2}.\cos\frac{A-B}{2}$$

$$\cos A - \cos B = 2\sin\frac{A+B}{2}.\sin\frac{B-A}{2}$$

$$\cos A.\cos B = \frac{1}{2}[\cos(A-B) + \cos(A+B)]$$

$$\sin A.\sin B = \frac{1}{2}[\cos(A-B) - \cos(A+B)]$$

$\sin(-\theta) = -\sin\theta$; $\cos(-\theta) = \cos\theta$;

$$\sin A.\text{co } B = \frac{1}{2}[\sin(A-B) + \sin(A+B)]$$

$\tan(-\theta) = -\tan\theta$

$\sin 3A = 3\sin A - 4\cos^3 A$

$\cos 3A = 4\cos^3 A - 3\cos A$

$$\tan 3A = \frac{3\tan A - \tan^3 A}{1 - \tan^2 A}$$

A2: Differentiating formulas

$$\frac{d}{dx}(x^n) = n.x^{n-1}$$

$$\frac{d}{dx}(ax+b)^n = na(ax+b)^{n-1}$$

$$\frac{d}{dx}(\sin x) = \cos x$$

$$\frac{d}{dx}(\cos x) = -\sin x$$

$$\frac{d}{dx}(\tan x) = \sec^2 x$$

$$\frac{d}{dx}(\cot x) = -\text{cosec}^2 x$$

$$\frac{d}{dx}(\sec x) = \sec x.\tan x$$

$$\frac{d}{dx}(\text{cosec}\, x) = -\text{cosec}x.\cot x$$

$$\frac{d}{dx}(a^x) = a^x \log_e a$$

$$\frac{d}{dx}(\log_e x) = \frac{1}{x}$$

$$\frac{d}{dx}(e^x) = e^x$$

$$\frac{d}{dx}(Cu) = C\frac{du}{dx};\ \frac{d}{dx}(C) = 0$$

$$\frac{d}{dx}(u+v+w) = \frac{du}{dx} + \frac{dv}{dx} + \frac{dw}{dx}$$

$$\frac{d}{dx}(u.v) = u.\frac{dv}{dx} + v.\frac{du}{dx}$$

A3: Integration formulas

$$\int x^n dx = \frac{x^{n+1}}{n+1}$$

$$\int a^x dx = \frac{a^x}{\log_e a}$$

$$\int e^x dx = e^x$$

$$\int \cos x dx = \sin x$$

$$\int \sin x dx = -\cos x$$

$$\int \sin mx dx = -\frac{\cos mx}{m}$$

$$\int \cos mx dx = \frac{\sin mx}{m}$$

$$\int \sin mx dx = -\frac{\cos mx}{m}$$

$$\int \sin mx . \sin nx dx = \frac{\cos(m-n)x}{2(m-n)} - \frac{\sin(m+n)x}{2(m+n)} \; for \; m \neq n;$$

$$\int \tan x dx = \log(\sec x)$$

$$\int \cos mx . \cos nx dx = \frac{\sin(m-n)x}{2(m-n)} + \frac{\sin(m+n)x}{2(m+n)} \; for \; m \neq n;$$

$$\int \cot x dx = \log(\sin x)$$

$$\int (u+v+w)dx = \int udx + \int vdx + \int wdx$$

$$\int C dx = Cx$$

APPENDIX-B

Laplace Transform

1. If f(t) is impulse, δ(t), the Laplace transform, F(s) =1
2. If f(t) is unit step, u(t), the Laplace transform, F(s) $=\frac{1}{s}$
3. If f(t) = t, then the Laplace transform, F(s) = $\frac{1}{s^2}$
4. If f(t) = e^{-at}, then the Laplace transform, F(s) = $\frac{1}{s+a}$
5. If f(t) = $\sin \omega t$, then the Laplace transform, F(s) = $\frac{\omega}{s^2+\omega^2}$
6. If f(t) = $\cos \omega t$, then the Laplace transform, F(s) = $\frac{s}{s^2+\omega^2}$
7. If $f(t) = t.e^{-at}$, then the Laplace transform, $F(s) = \frac{1}{(s+a)^2}$
8. If $f(t) = e^{-at} \sin \omega t$, then the Laplace transform, $F(s) = \frac{\omega}{(s+a)^2+\omega^2}$
9. If $f(t) = e^{-at} \cos \omega t$, then the Laplace transform, $F(s) = \frac{s+a}{(s+a)^2+\omega^2}$

APPENDIX-C

Fourier Analysis

The output of power converters is a periodic function of time and is normally expressed by (C.1), where T is the time period of the periodic function.

$$v_o(t) = v_o(t+T) \tag{C.1}$$

The angular frequency, ω is given by ω is given by (C.2), where f is the frequency of the voltage in Hz.

$$\omega = \frac{2\pi}{T} = 2\pi f \tag{C.2}$$

Thus, in terms of angular frequency, the output voltage equation becomes

$$v_o(\omega t) = v_o(\omega t + 2\pi) \tag{C.3}$$

The Fourier theorem states that a periodic function of time can be expressed as a sum of a constant term and an infinite series comprised of sine and cosine of frequency n, where n is an integer. Accordingly, the output voltage of a converter given by equation (C.1) becomes

$$v_o(t) = \frac{a_0}{2} + \sum_{n=1,2,...}^{\alpha} (a_n \cos n\omega t + b_n \sin n\omega t) \tag{C.4}$$

Here, the first term ($a_0/2$) is the average value of output voltage $v_o(t)$. The expression for constant terms a_0, a_n, and b_n are given by equation (C.5), (C.6) and (C.7) respectively.

$$a_0 = \frac{2}{T}\int_0^T v_0(t)dt = \frac{1}{\pi}\int_0^{2\pi} v_0(\omega t)d(\omega t) \tag{C.5}$$

$$a_n = \frac{2}{T}\int_0^T v_0(t)\cos n\omega t\, dt = \frac{1}{\pi}\int_0^{2\pi} v_0(\omega t)\cos n\omega t\, d(\omega t) \tag{C.6}$$

$$b_n = \frac{2}{T}\int_0^T v_0(t)\sin n\omega t\, dt = \frac{1}{\pi}\int_0^{2\pi} v_0(\omega t)\sin n\omega t\, d(\omega t) \tag{C.7}$$

The output voltage may be of analytical function. If so, the above constants can be found by a single integration. Generally, the output voltage of the converters is discontinuous and, in this case, few integrations need to be performed to find the constants. The value of the terms ($a_n \cos n\omega t + b_n \sin n\omega t$) can be expressed by (C.8).

$$a_n \cos n\omega t + b_n \sin n\omega t = \sqrt{(a_n^2 + b_n^2)}\left[\frac{a_n}{\sqrt{(a_n^2 + b_n^2)}}\cos n\omega t + \frac{b_n}{\sqrt{(a_n^2 + b_n^2)}}\sin n\omega t\right] \tag{C.8}$$

Or

$$a_n \cos n\omega t + b_n \sin n\omega t = \sqrt{(a_n^2 + b_n^2)}\left[\sin\phi_n \cos n\omega t + \cos\phi_n \sin n\omega t\right] \tag{C.9}$$

Or

$$a_n \cos n\omega t + b_n \sin n\omega t = \sqrt{(a_n^2 + b_n^2)}\sin(n\omega t + \phi_n) \tag{C.10}$$

where,

$$\phi_n = \tan^{-1}\frac{a_n}{b_n} \tag{C.11}$$

Thus the equation for voltage, $v_o(t)$ can be expressed in series by (C.12),

$$v_o(t) = \frac{a_0}{2} + \sum_{n=1,2,3...}^{\alpha} C_n \sin(n\omega t + \phi_n) \tag{C.12}$$

where

$$C_n = (a_n + b_n)^{1/2} \tag{C.13}$$

Here, C_n and Φ_n are the peak value and delay angle of the nth harmonic component of output voltage respectively.

REFERENCES FOR FURTHER LEARNING

1. Ramamoorty M., An Introduction to Thyristors and their applications, East-West Press Pvt. Ltd., New Delhi, ISBN: 8185336679.
2. Sugandhi, Rajendra Kumar and Sugandhi, Krishna Kumar, Thyristors: Theory and Applications, New Age International (P) ltd. Publishers, New Delhi, ISBN: 978-0-85226- 852-0.
3. Bhattacharya, S.K., Fundamentals of Power Electronics, Vikas Publishing House Pvt. Ltd. Noida. ISBN: 978-8125918530.
4. Jain & Alok, Power Electronics and its Applications, Penram International Publishing (India) Pvt. Ltd, Mumbai, ISBN: 978-8187972228.
5. Rashid, Muhammad, Power Electronics Circuits Devices and Applications, Pearson Education India, Noida, ISBN: 978-0133125900.
6. Singh, M. D. and Khanchandani, K.B., Power Electronics, Tata McGraw Hill Publishing Co. Ltd, New Delhi, 2008 ISBN: 9780070583894.
7. Zbar, Paul B., Industrial Electronics: A Text –Lab Manual, McGraw Hill Publishing Co. Ltd., New Delhi, ISBN: 978-0070728226.
8. Grafham D.R., SCR Manual, General Electric Co., ISBN: 978-0137967711.
9. Mohan Ned, Undeland Tore.M., Robins William P., Power Electronics – Converters, Applications and Design, John Wiley & Sons. Inc,
10. https://www.google.com/search?q=Instruments+for+testing+power+transistor.
11. https://www.researchgate.net/publication/307475653_Single_Electron_Transistor_SET_Operation_and_Application_Perspectives/stats
12. Monika Gupta, "A Study of Single Electron Transistor (SET)", International Journal of Science and Research (IJSR) ISSN (Online),
13. https://www.google.com/search?q=picture+of+heat+sink+for+power+device.
14. Rashid, M. H. (2014). Power Electronics: Devices, Circuits, and Applications: International Edition. United Kingdom: Pearson Education.
15. Undeland, T. M., Mohan, N., Robbins, W. P. (2003). Power Electronics: Converters, Applications, and Design. India: John Wiley & Sons.
16. Fewson, D. (1998). Introduction to Power Electronics. United Kingdom: Elsevier Science.
17. Williams, B. W. (1992). Power electronics: devices, drivers, applications, and passive components. United Kingdom: McGraw-Hill.
18. Robbins, W. P., Mohan, N., Undeland, T. M. (2003). Power Electronics: Converters, Applications, and Design. India: John Wiley & Sons.
19. https://www.youtube.com/watch?v=d4sbVc-r7I4
20. https://www.youtube.com/watch?v=ekFmK3xtbK8
21. https://www.youtube.com/watch?v=9i61RREAWjI
22. Bimbhra, P.S., Power Electronics, Third Edition (2005), Khanna Publishers, ISBN No.81-7409-056-8.
23. https://www.electronicshub.org/security-alarm-circuit/
24. Md. Barkat Ullah et al, "Design and Development of Low-Cost Security Alarm Using Optoelectronic Device, *Journal of Electronics and Communication Engineering,* Vol 10, No. 4, Ver. I (Jul - Aug .2015), pp 19-23, DOI:10.9790/2834-10411923

Dynamic QR Code for Further Reading

Book, P.S. Bimbhra

PE handbook by Rashid

Single Electron transistors

Power transistor

Thyristor

Book by Mohan

CO AND PO ATTAINMENT TABLE

Course outcomes (COs) for this course can be mapped with the programme outcomes (POs) after the completion of the course and a correlation can be made for the attainment of POs to analyze the gap. After proper analysis of the gap in the attainment of POs necessary measures can be taken to overcome the gaps.

Table for CO and PO attainment

Course Outcomes	Attainment of Programme Outcomes *(1- Weak Correlation; 2- Medium correlation; 3- Strong Correlation)*											
	PO-1	PO-2	PO-3	PO-4	PO-5	PO-6	PO-7	PO-8	PO-9	PO-10	PO-11	PO-12
CO-1	3	2	2	2	1	-	-	-	-	-	-	-
CO-2	3	1	2	1	-	-	-	-	-	-	-	-
CO-3	3	3	2	2	2	-	-	-	-	-	-	-
CO-4	3	3	3	2	1	-	-	-	-	-	-	-
CO-5	3	3	3	1	2	-	-	-	-	-	-	-

The data filled in the above table can be used for gap analysis.

INDEX